Beginning R 4

From Beginner to Pro

Matt Wiley
Joshua F. Wiley

Apress®

Beginning R 4: From Beginner to Pro

Matt Wiley
Victoria College,
Victoria, TX, USA

Joshua F. Wiley
Turner Institute for Brain and Mental Health,
Monash University, Melbourne, Australia

ISBN-13 (pbk): 978-1-4842-6052-4
ISBN-13 (electronic): 978-1-4842-6053-1
https://doi.org/10.1007/978-1-4842-6053-1

Managing Director, Apress Media LLC: Welmoed Spahr
Acquisitions Editor: Steve Anglin
Development Editor: Matthew Moodie
Coordinating Editor: Mark Powers

Cover designed by eStudioCalamar

Cover image by Shutterstock

Distributed to the book trade worldwide by Apress Media, LLC, 1 New York Plaza, New York, NY 10004, U.S.A. Phone 1-800-SPRINGER, fax (201) 348-4505, e-mail orders-ny@springer-sbm.com, or visit www.springeronline.com. Apress Media, LLC is a California LLC and the sole member (owner) is Springer Science + Business Media Finance Inc (SSBM Finance Inc). SSBM Finance Inc is a **Delaware** corporation.

For information on translations, please e-mail booktranslations@springernature.com; for reprint, paperback, or audio rights, please e-mail bookpermissions@springernature.com.

Apress titles may be purchased in bulk for academic, corporate, or promotional use. eBook versions and licenses are also available for most titles. For more information, reference our Print and eBook Bulk Sales web page at http://www.apress.com/bulk-sales.

Any source code or other supplementary material referenced by the author in this book is available to readers on GitHub via the book's product page, located at www.apress.com/9781484260524. For more detailed information, please visit http://www.apress.com/source-code.

Printed on acid-free paper

Table of Contents

About the Authors

Matt Wiley leads institutional effectiveness, research, and assessment at Victoria College, facilitating strategic and unit planning, data-informed decision making, and state/regional/federal accountability. As a tenured, associate professor of mathematics, he won awards in both mathematics education (California) and student engagement (Texas). Matt holds degrees in computer science, business, and pure mathematics from the University of California and Texas A&M systems.

Outside academia, he has co-authored three books about the popular R programming language and was managing partner of a statistical consultancy for almost a decade. His programming experience is with `R`, `SQL`, `C++`, `Ruby`, `Fortran`, and `JavaScript`.

A programmer, a published author, a mathematician, and a transformational leader, Matt has always melded his passion for writing with his joy of logical problem solving and data science. From the boardroom to the classroom, he enjoys finding dynamic ways to partner with interdisciplinary and diverse teams to make complex ideas and projects understandable and solvable. Matt enjoys being found online via Twitter (`@matt_math`) or `http://mattwiley.org/`.

Joshua F. Wiley is a lecturer in the Turner Institute for Brain and Mental Health and School of Psychological Sciences at Monash University. He earned his PhD from the University of California, Los Angeles, and completed his postdoctoral training in primary care and prevention. His research uses advanced quantitative methods to understand the dynamics between psychosocial factors, sleep, and other health behaviors in relation to psychological and physical health. He develops or co-develops a number of R packages including `varian`, a package to conduct Bayesian scale-location structural equation models; `MplusAutomation`, a popular package that links R to the commercial

Mplus software; `extraoperators` for faster logical operations; `multilevelTools` for diagnostics, effect sizes, and easy display of multilevel/mixed effects model results; and miscellaneous functions to explore data or speed up analysis in `JWileymisc`. Joshua enjoys being found online via Twitter (`@WileyResearch`) or `http://joshuawiley.com/`.

About the Technical Reviewer

Rachel Winkenwerder is an associate professor of mathematics at Victoria College and serves as assistant director of institutional effectiveness, research, and assessment. With extensive mathematical teaching experience in China and Texas for both secondary and tertiary education, Rachel understands the practical development of a contextualized curriculum. Her most recent work in higher education includes co-chairing her institution's curriculum and instructional council, reviewing regional accreditation narratives, and leading academic assessment. Her proven classroom track record teaching statistical methods courses combines with her current work in the R programming language and her earned degrees in computer science, mathematics, and education.

Acknowledgments

We would like to sincerely thank all our students through and across the years, who collectively taught us A Great Many Important Things.

Foreword

This book was, in part, written to be rigorously suitable for a sole mathematics course for undergraduates for what may be called "core." In that respect, it may be suited to fit into a statistical methods course (usually with either an elementary or introductory modifier at the front). Additionally, depending on how far through the last chapters one goes, it may also be suitable for an upper-division undergraduate course for the social sciences (e.g., psychology or sociology).

Beyond these objectives, however, this book is meant to empower learners to physically experience statistical thinking through the hands-on use of the R programming language. While theory will not be ignored, techniques and theory will be first motivated through models, visuals, and an intuitive approach. The true objective is to share a language for communicating complex numerical facts in understandable terms. This practical application – sometimes called empirical and quantitative skills – is designed to enable students to critically think and explore increasingly complex data sets, successfully describe and summarize large quantities of information, and accurately analyze, model, and communicate results to both technical and lay audiences.

To make this happen, this text has two distinct parts. Part 1 is designed to efficiently walk the reader step by step through installing and understanding the essential minimums of the R programming language's computer environment. We will do our best to avoid "techno speak," maintain an everyday language that jump-starts the reader as fast as possible through the initial stages, and move as smartly as possible to studying actual statistics. Part 2 is a methods approach to introductory statistics. Populations, samples, descriptive statistics, probability, distributions, correlation, regression, confidence intervals, hypothesis testing, and *an*alysis *o*f *va*riance (ANOVA) are all treated in turn. While not shying away from technical mathematical theory, all statistical ideas are first introduced conceptually. From visualizations to hands-on activities to help "get your hands dirty," the goal is to build a solid, real-world, contextual intuition that makes the theory more relatable. Your authors' goal is to help learners "live a statistical life" – and completely avoid writing a book used to pass one course and then quickly forgotten.

Lastly, while we have said this text is suitable for an undergraduate course, it will without a doubt increase a learner's ability to use a highly popular programming language and lay the foundation for using R for research, data science, machine learning, dynamic reporting, and bespoke visualizations. As such, it is also suitable for practiced data analysts looking to make the transition to R, for graduate students looking to both gain a powerful skill and refresh their knowledge of statistics, and for anyone who enjoys learning about data and statistics.

Thank you for spending your time and attention on our book. Please be sure to download source code and engage with this learning hands-on, and do not hesitate to reach out to us should anything not be working.

CHAPTER 1

Installing R

You are here to learn statistics, and the first step is *installing* some *program*. It can almost feel like a magician's trick. You need to know some applied statistics, and somehow you need to learn how to program instead. While this can look scary, never fear! This chapter will walk you through step by step what you need to do to have a pain-free installation. Regardless of whether you are learning on your personal computer (be that a PC or Mac) or on your company/institution's networked machine, this guide will let you know in everyday language what to do and yet stay complete enough to give to your IT department if needed (for corporate/institutional learners).

It is worth mentioning that by learning how to think statistically using the R [16] language, you not only learn on *free*, open source software, you also start *using* statistics from day 1 the way it is used in "real" life. In addition to statistical knowledge, you gain some quite useful applied skills.

This chapter helps you learn to be able to

- Download the latest version of R from CRAN [2] and install on either a Windows or Mac computer.
- Download and install RStudio on either a Windows or Mac computer.
- Understand the RStudio project environment.
- Evaluate if your installations and understanding work through applying some basic R code principles.

If you already know something about R and are eager to get into learning statistics, we start by listing the software required for this book to work in Table 1-1. On the other hand, if you wish to journey through the walk with us, please proceed through the chapter.

M. Wiley and J. F. Wiley, *Beginning R 4*, https://doi.org/10.1007/978-1-4842-6053-1_1

Table 1-1. *Beginning R Tech Stack*

Software	URL
R 4.0.2	`https://cran.r-project.org/`
RStudio 1.2.x	`https://rstudio.com/products/rstudio/download/`
Windows 10	`www.microsoft.com/`
MacOS	`www.apple.com/macos`

1.1 Your Tech Stack

Because the software in Table 1-1 tend to all require each other in a certain order and are technologies, the list of software required for a project to work/run is sometimes referred to as a *Tech Stack*. If you find yourself having trouble installing `R`, one of the first things you would want to share with anyone helping you is your Tech Stack. In particular, your computer's operating system will make a difference as may the version of `R` you are running. While this book is written for `R.Version()[["version.string"]]`, most likely future versions will not change so very much that this book will not work.

1.2 Updating Your Operating System

While it is not an absolute requirement, your computer's operating system (e.g., Windows or macOS) should be up to date. An up-to-date operating system helps for a few reasons. Firstly, we are going to walk you through installing the latest version of `R` and RStudio. Those are for sure tested on the current operating system versions. Secondly, most "tech" types tend to keep their software updated; thus, it helps to have a system similar to what they are currently familiar with should you run into trouble.

Windows

To ensure Windows 10 is up to date is not difficult at all. In the operating system search bar (a magnifying glass most often to the right of the Windows logo), type `check for updates` and select the option titled *Check for updates System settings* with your mouse cursor. This should open Windows Update, and there will be a box titled `Check for updates` to click. If successful, the *Last checked: Today* text will include the current time.

Before you move on to actually installing R, now is a good time to verify your computer is a 64-bit machine. It is possible, although not as likely these days, that you may have a 32-bit computer. In that same *Type here to search* Windows search box, go ahead and type `About your PC`. This should bring up an option titled *About your PC System settings*. Selecting that will open up a screen with some information about your system. In particular, under Device specifications, you're looking to see if your *System type* is either 64-bit or 32-bit. In either case, keep that in mind later on when you install R.

MacOS

Getting R running on *macOS* requires ensuring you have the latest version of macOS. This is important as the installation process differs across versions. At the time of writing, the latest version is *macOS Catalina*, with *macOS Big Sur* or later likely to be released by the time you are reading this. You can get help with upgrading from Apple: `www.apple.com/au/macos/how-to-upgrade/`.

1.3 Downloading and Installing R from CRAN

With your operating system up to date, it is time to download the latest version of R. As of this writing, it is `R version 4.0.2 (2020-06-22)` and can be found at `https://cran.r-project.org/`. Depending on your operating system, you will install a slightly different version of R. Thus, we ask you choose your own adventure.

Windows

To install R on Windows 10, use your favorite web browser (e.g., Chrome, Firefox, or Edge) to visit `https://cran.r-project.org/`. There should be a box on that web page titled *Download and Install R*, and you will want to select the *Download R for Windows* link `https://cran.r-project.org/bin/windows/`. From there, your goal is to install `base R`, so go ahead and select the *base* link `https://cran.r-project.org/bin/windows/base/`. The first link at the top should be *Download R 4.0.2 for Windows (XX megabytes, 32/64 bit)*.

It is both likely and possible that by the time you read this book, the version will be a larger number than 4.0.2. The megabytes may also likely be slightly different. Go ahead and download that latest version. Most likely, as long as the major version is still 4.x.x, everything in this book should stay fully or almost fully true.

After downloading, go ahead and navigate to your Downloads folder where the file `R-4.0.2-win.exe` is saved. Most often, your web browser will have a pop-up reminding you where the file is. Otherwise, you may press and hold the windows key on your keyboard followed by the E key which is sometimes called the *`Win + e`*. Not capital E, just the lowercase e key while still holding down the windows key button. This will open Windows Explorer, and most often there is a folder on the left titled *Downloads*.

Once you have found your `R-4.0.2-win.exe` file, you will want to double-click it to start the install process. For now, go ahead and accept all the default options, clicking Next and OK as needed.

MacOS

To get R running most flexibly on macOS, several additional tools are suggested.

Before you can install R, you need to install Xcode from App Store. After installation, **open it** to accept terms; otherwise, it may not work.

Once you have Xcode installed, you also need to install command-line tools. Open `Terminal` (if you cannot find it, try spotlight search) and type in `xcode-select --install` and then press *Enter* to run. If you run into any access issues, you may need to enable root. You can do that by typing in Terminal the following: `dsenableroot`, and pressing *Enter/Return* to enable root user.

Note At the terminal, if you are asked to enter a password, type the password you use to log in to your Mac and press Enter/Return. When typing your password in the terminal, no characters will appear, but it is still being entered.

Install `XQuartz/X11`. Visit `www.xquartz.org/`, download and run the file, and follow any on-screen instructions.

Go to `https://mac.r-project.org/tools/` and follow their instructions to get the mandatory tools and libraries. An example of the instructions includes, install `gfortran` by going to `https://github.com/fxcoudert/gfortran-for-macOS/releases` and download gfortran 8.2. Download and install following any instructions.

Although not needed for R itself, for many add-on packages that extend R's functionality, you will want some additional tools.

Install `homebrew` for macOS at `https://brew.sh/` and follow the "Install Homebrew" steps. If you run into any access issues, you may need to enable root. You can do that by typing in Terminal the following: `dsenableroot`, and pressing *Enter/Return* to enable root user.

Note At the terminal, if you are asked to enter a password, type the password you use to log in to your Mac and press Enter/Return. When typing your password in the terminal, no characters will appear, but it is still being entered.

Install `openssl` which allows `R` to securely download files and packages from the Internet. Do this by opening the terminal (you can search for "terminal" or look in the launchpad) and type this code once the terminal opens and press *Enter*: `brew install openssl`

Install `libgit2` which is needed for one of our graphing packages. Do this by opening the terminal (you can search for "terminal" or look in the launchpad) and type this code once the terminal opens and press *Enter*: `brew install libgit2`

Finally, download `R` by visiting `https://cloud.r-project.org/` and clicking *Download R for (Mac) OS X*. Then download version 4.0.2. Once you have downloaded `R`, please be sure to install it to your apps.

1.4 Downloading and Installing RStudio

Congratulations! You have now installed `R`. However, there is still one more thing to install. While `R` by itself is powerful, it is not a convenient working environment. In programming, one often works in something called an IDE (*Integrated Development Environment*). In our case, we want to install RStudio Desktop, which is going to add a lot of easy-to-use visuals to help us "see" `R` better. RStudio will make it easier to focus on learning statistics.

Please visit `https://rstudio.com/products/rstudio/download/` and select the *RStudio Desktop Free* option. As before, please catch up with us in the section that matches your operating system.

Windows

Having visited `https://rstudio.com/products/rstudio/download/` and selected the *RStudio Desktop Free* option, you are now ready to install RStudio Desktop for Windows. As of this writing, the most up-to-date version is `1.3.1056.exe`. Again as before, you will want to simply download the most recent version which should be step 2 on this website `https://rstudio.com/products/rstudio/download/#download`.

Again, simply click the download button, and note where your browser is saving the file. To install, double-click the texttttRStudio-1.3.1056.exe file. Also, as with `R` itself, you will want to go with all the default options as they are selected, clicking *Next* and *OK* as needed.

Please note we are supposing you have a `64-bit` operating system. If that is not the case, you will need one of the older copies of RStudio which may be found at this link: `https://rstudio.com/products/rstudio/older-versions/`.

MacOS

Having visited `https://rstudio.com/products/rstudio/download/#download` and selected the *RStudio Desktop Free* option, you are now ready to install RStudio Desktop for macOS. As of this writing, the most up-to-date version is `1.3.1056.dmg`. Make sure `R.app` and `RStudio.app` are able to access disk resources needed.

Follow this guide `www.r-bloggers.com/escaping-the-macos-10-14-mojave-filesystem-sandbox-with-r-rstudio` to give them the necessary permissions.

1.5 Using RStudio

Now that you have `R` and RStudio installed, it is time to open and run RStudio for the first time. Go ahead and click the RStudio icon to open RStudio. What you should see is a small strip of icons along the top along with three large tiles or boxes that make up the bulk of the program screen. One large window pane on the left should be called *Console*, and inside that pane, there should be some text that reads *R version 4.0.2* (or whatever the version of `R` you just downloaded and installed is).

On the right side, there should be two smaller window panes. On top is the *Environment* which should be empty. On the lower-right side should be the *Files/Plots/Packages/Help/Viewer* pane. It should be showing you a file directory.

Your first chore (and you will only do this once) is to set some default options to make sure we share the same settings.

On the very top menu ribbon, find and click the *Tools* link, and then select *Global Options* from the drop-down menu. The Options menu should open, and you should be already in the *General* tab as in Figure 1-1. You will want to make sure the following options are **not** selected in that *General* tab ➤ *Basic*:

- **Restore most recently opened project at startup: Not checked**
- **Restore previously open source documents at startup: Not checked**
- **Restore .RData into workspace at startup: Not checked**
- **Save workspace to .RData on exit: Never**

Go ahead and click *Apply*.

Before you leave this Options menu, in order to have a little fun while you learn, please click *Appearance* as in Figure 1-2. Now, select the following options:

- **RStudio theme: Modern**.
- **Zoom: Set however you like**.
- **Editor font: Choose your favorite font**.
- **Editor font size: Big enough to read, small enough to optimize monitor space**.
- **Editor theme: Vibrant Ink** (we disagree on this – truly choose your favorite).

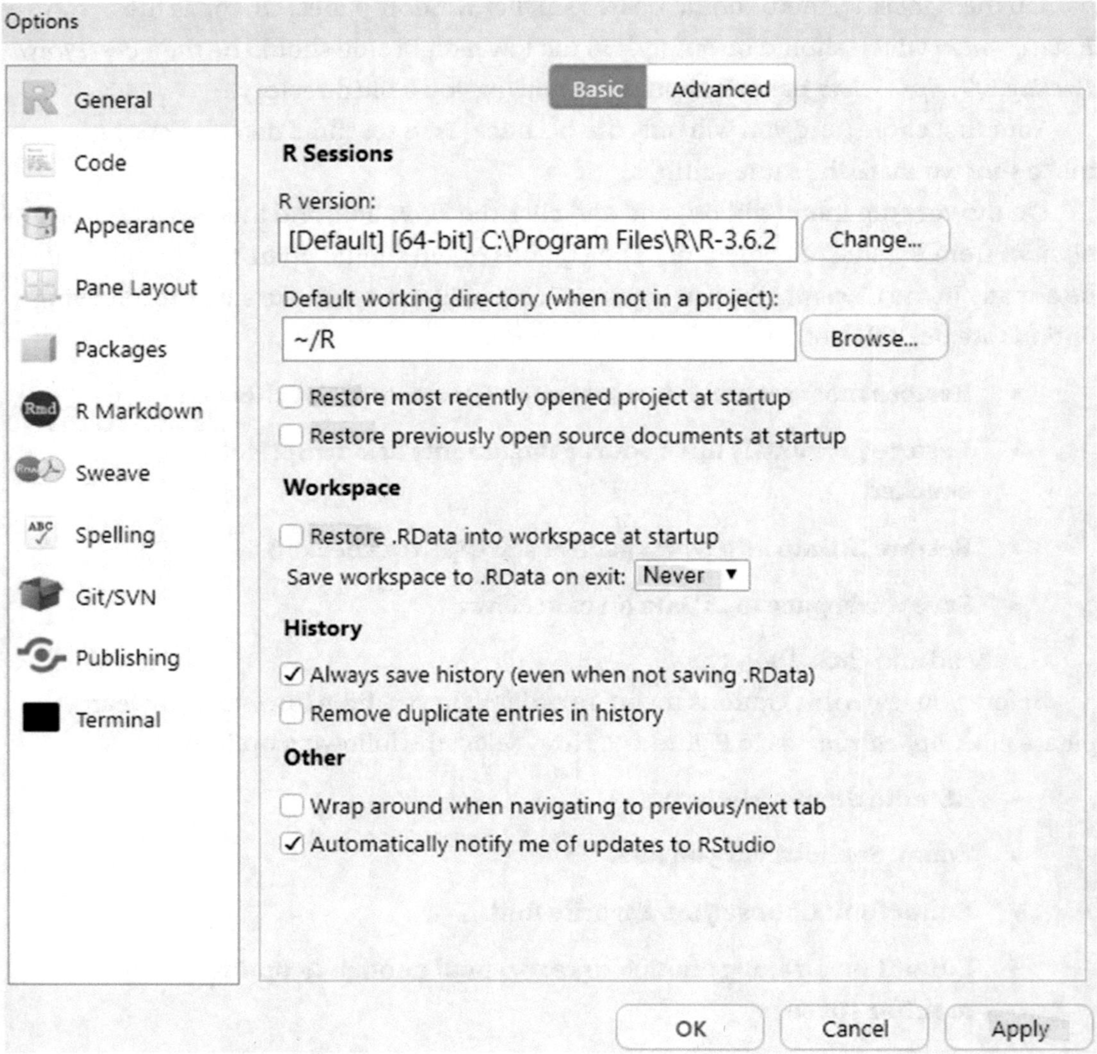

Options

General
Code
Appearance
Pane Layout
Packages
R Markdown
Sweave
Spelling
Git/SVN
Publishing
Terminal

Basic | Advanced

R Sessions

R version:
[Default] [64-bit] C:\Program Files\R\R-3.6.2 | Change...

Default working directory (when not in a project):
~/R | Browse...

☐ Restore most recently opened project at startup
☐ Restore previously open source documents at startup

Workspace

☐ Restore .RData into workspace at startup
Save workspace to .RData on exit: Never ▾

History

☑ Always save history (even when not saving .RData)
☐ Remove duplicate entries in history

Other

☐ Wrap around when navigating to previous/next tab
☑ Automatically notify me of updates to RStudio

OK | Cancel | Apply

Figure 1-1. *RStudio general options*

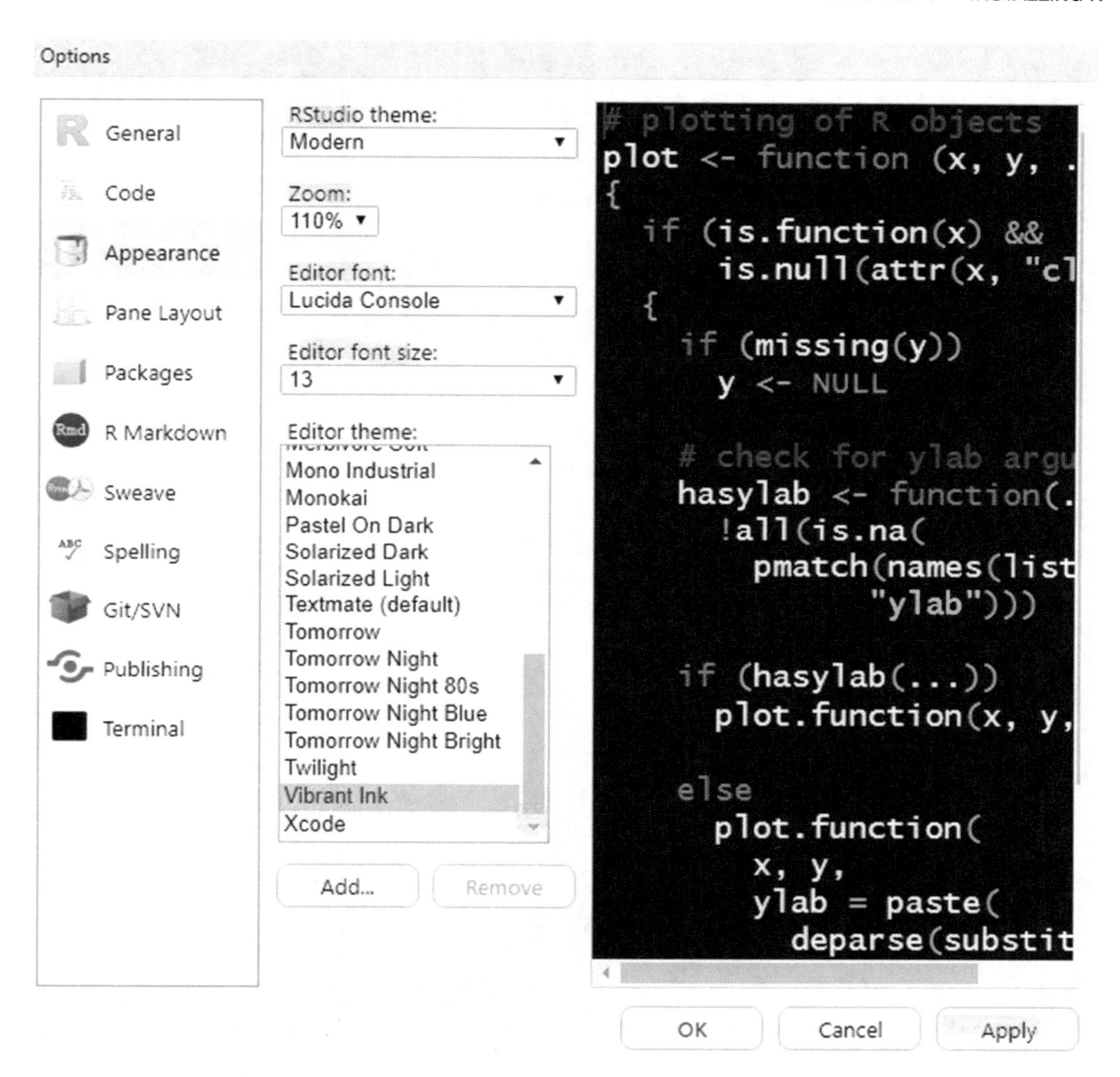

***Figure 1-2.** RStudio Appearance*

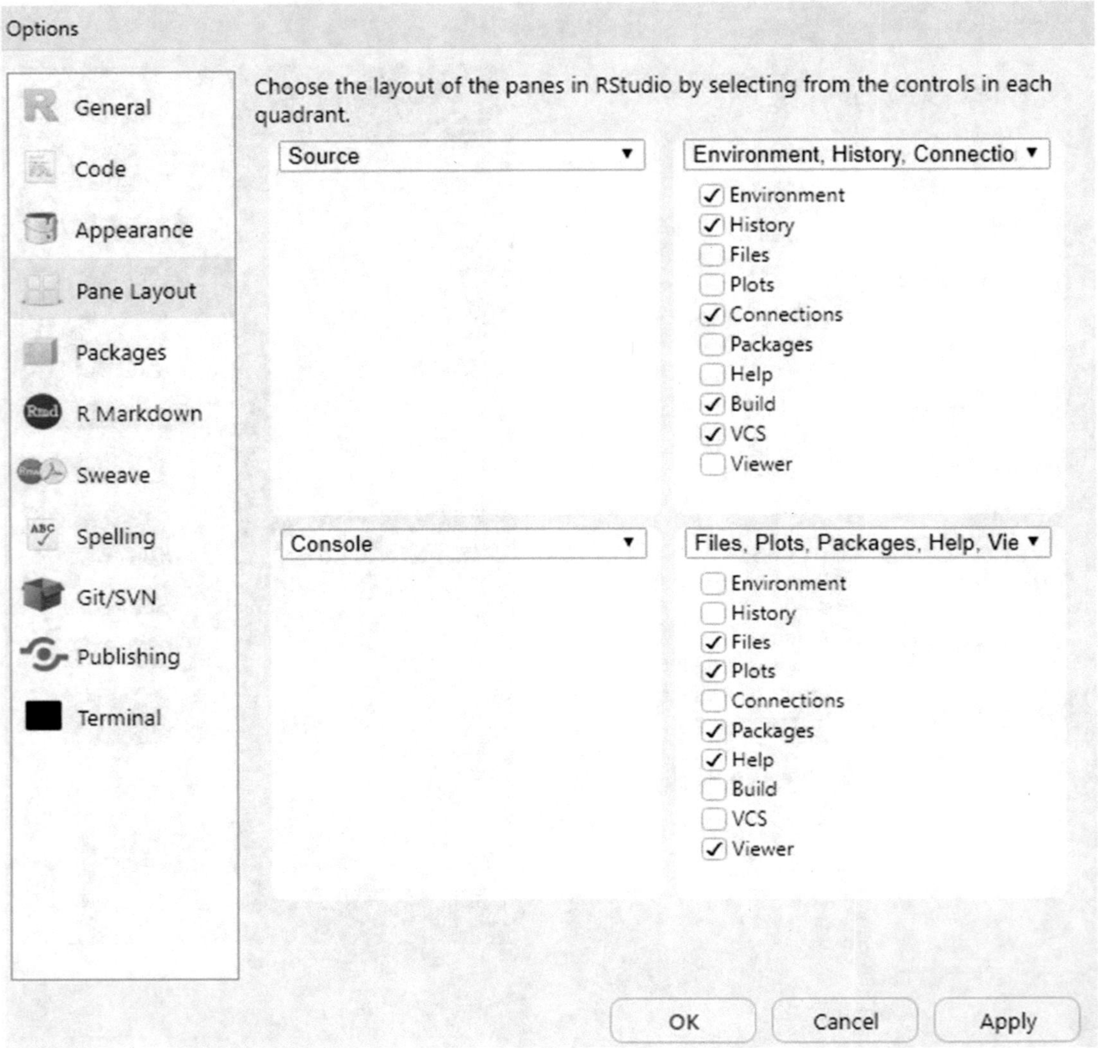

Figure 1-3. *RStudio Pane Layout*

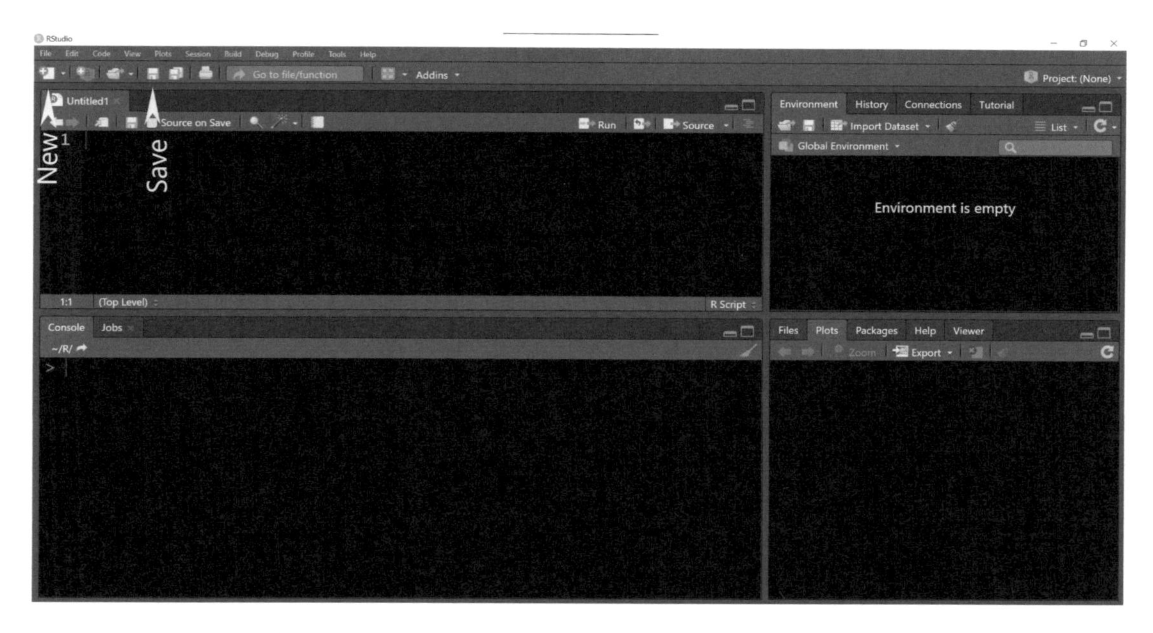

Figure 1-4. *RStudio New and Save example*

Lastly, please click *Pane Layout* and ensure your options match ours as in Figure 1-3.

Now, at last, you are done with setup. Please click *OK*, and welcome to a world where we spend a great deal of time.

Lastly, if you want to learn some more about using RStudio as an interface, the following two-page "cheat sheet" is a good place to go. It diagrams all the many parts of RStudio and how they can help you do your work: `https://github.com/rstudio/cheatsheets/raw/master/rstudio-ide.pdf`.

New Projects

Now that RStudio has some good defaults, it is time to start your first **project**. Creating a project builds a folder that will hold your work, and it has some files unique to RStudio which will allow you to easily work on a related set of ideas. There is no right or wrong answer for how many R files should live in a project. If you are taking this book as part of a class, it may make sense to build one project for each chapter. In that case, a good project name would be the title of the chapter. On the other hand, this entire book was written in a single project titled "BeginningR_2020."

For now, let us go ahead and build one project for each chapter. At the very least, you will get comfortable starting, closing, and opening projects.

To start a new project, on the upper-left menu ribbon, select *File* and click *New Project*. You will want to then use the New Project wizard to choose *New Directory* ➤ *New Project* and then pick a *Directory name*. For now, we recommend `01Installing`. Once that is typed into the *Directory name* field, select *Create Project*.

You are now in your first project!

Your first task is to create a new, blank R file. To do that, on the top ribbon, right under the word *File*, there is a small icon with a plus sign on top of a blank bit of paper. Click that icon, and then, from the menu of new files, pick the first one titled *R Script*. This blank file is now visible because you should have a tab in that top-left pane titled *Untitled1*. Go ahead and click the floppy disk–shaped *save* icon. Name this file `MyFirstRScript.R` and click *Save*. A screenshot showing what the new script and save look like is in Figure 1-4.

If all has gone well, you now have four panes as we described earlier in the setup. Your top-left pane is titled *MyFirstRScript.R* and is a script or code pane. This is the area where we will do most of our work. Moving clockwise, your top-right pane is the Environment pane; it is still empty. The lower-right pane is your *Files/Plots/Packages/Help/Viewer* pane. There should now be three files in your *project directory* which is in particular the *Files* tab. Those three files should be *.Rhistory*, *01Installing.Rproj*, and *MyFirstRScript.R*. Lastly, on the lower left is your *Console* pane, and right now it should still be telling you about your current R version. That is all about to change. Let us briefly discuss the purpose of each of these panes in turn.

The script pane, which currently holds your *MyFirstRScript.R*, is where you will type R code. Think of this just like a Word file or a PowerPoint slide. It will store and save any code you write there, and until you activate or *run* your code, nothing happens. If you like, this script area is your program. At present, your program is blank. We will change that very soon. You will write your first program in R, and become a programmer.

The *Environment* pane shows the data or *objects* or *variables* your program has stored in memory. Right now, there is nothing there. Most commonly in this book, you will find some data in that spot.

The *Files* tab shows your project's *working directory*. Any files you want to use for this project need to live in this directory. This is important! R needs to know where to look for any Excel files or data files you use to get information for statistical analysis. Those files all need to live in the directory folder. However, this pane does more than just show you files. Notice there are tabs for *Plots*, *Packages*, *Help*, and *Viewer*. Go ahead and click through each of those in turn. For now, *Plots* and *Files* and *Help* will be your most common tabs here. Go ahead and get back to *Files*.

The last pane is the *Console*. This is where your code will run. There are two places code lives in RStudio. One is in the script area – that is where code is saved. However, to run code, it will always run in the console.

With that, you have met the four areas that you will use to learn statistics. Before we are ready to wrap up this chapter, we want to make sure everything is working. It is time to write your first program.

1.6 My First R Script

For a first program, we want to make sure R is working and hopefully see some interesting results. In this book, we are going to show you each bit of code, along with the output. When you follow along in your script, you will need to type out the code into your script area, use your mouse to click and drag to select the text (or hold *Shift* and use arrow keys on your keyboard to highlight the code) you want to run, and then press *Ctrl+Enter* to run the code on Windows or *Cmd+Enter* on macOS.

Please note when we use the convention *Ctrl+Enter,* what we mean is to press and hold Ctrl and then press Enter and then let go of both keys. On macOS, it means press and hold the Cmd key and then press Enter/Return and then let go of both keys. This is the easiest way to run code from the script area. In the long term, it will save you a lot of time compared to using your mouse to click Run each time.

In statistics, we are often in the business of trying to find a connection or a relationship between two parts of the same type of object. For example, we might imagine that as a vehicle's *weight* increases, the *miles per gallon* (a measure of fuel efficiency) decreases. To see how we can use R to explore such a physical concept, let us take a brief look at our first data set.

The data set `mtcars` is a car magazine dataset from quite some time ago. Go ahead and type `mtcars` into your script area, highlight that word, and press *Ctrl+Enter*. You should see the following 32 rows and 11 columns of data on your screen that are showing on our screens and showing here in this text.

```
mtcars
```

```
##                    mpg cyl disp  hp drat  wt qsec vs am gear carb
## Mazda RX4           21   6  160 110  3.9 2.6   16  0  1    4    4
## Mazda RX4 Wag       21   6  160 110  3.9 2.9   17  0  1    4    4
## Datsun 710          23   4  108  93  3.8 2.3   19  1  1    4    1
```

```
## Hornet 4 Drive       21  6  258 110  3.1 3.2   19  1  0    3    1
## Hornet Sportabout    19  8  360 175  3.1 3.4   17  0  0    3    2
## Valiant              18  6  225 105  2.8 3.5   20  1  0    3    1
## Duster 360           14  8  360 245  3.2 3.6   16  0  0    3    4
## Merc 240D            24  4  147  62  3.7 3.2   20  1  0    4    2
## Merc 230             23  4  141  95  3.9 3.1   23  1  0    4    2
## Merc 280             19  6  168 123  3.9 3.4   18  1  0    4    4
## Merc 280C            18  6  168 123  3.9 3.4   19  1  0    4    4
## Merc 450SE           16  8  276 180  3.1 4.1   17  0  0    3    3
## Merc 450SL           17  8  276 180  3.1 3.7   18  0  0    3    3
## Merc 450SLC          15  8  276 180  3.1 3.8   18  0  0    3    3
## Cadillac Fleetwood   10  8  472 205  2.9 5.2   18  0  0    3    4
## Lincoln Continental  10  8  460 215  3.0 5.4   18  0  0    3    4
## Chrysler Imperial    15  8  440 230  3.2 5.3   17  0  0    3    4
## Fiat 128             32  4   79  66  4.1 2.2   19  1  1    4    1
## Honda Civic          30  4   76  52  4.9 1.6   19  1  1    4    2
## Toyota Corolla       34  4   71  65  4.2 1.8   20  1  1    4    1
## Toyota Corona        22  4  120  97  3.7 2.5   20  1  0    3    1
## Dodge Challenger     16  8  318 150  2.8 3.5   17  0  0    3    2
## AMC Javelin          15  8  304 150  3.1 3.4   17  0  0    3    2
## Camaro Z28           13  8  350 245  3.7 3.8   15  0  0    3    4
## Pontiac Firebird     19  8  400 175  3.1 3.8   17  0  0    3    2
## Fiat X1-9            27  4   79  66  4.1 1.9   19  1  1    4    1
## Porsche 914-2        26  4  120  91  4.4 2.1   17  0  1    5    2
## Lotus Europa         30  4   95 113  3.8 1.5   17  1  1    5    2
## Ford Pantera L       16  8  351 264  4.2 3.2   14  0  1    5    4
## Ferrari Dino         20  6  145 175  3.6 2.8   16  0  1    5    6
## Maserati Bora        15  8  301 335  3.5 3.6   15  0  1    5    8
## Volvo 142E           21  4  121 109  4.1 2.8   19  1  1    4    2
```

This is a good example of what is sometimes called *raw* data. There is a lot of information about each car, and each bit of information is in a column. This is a great time to take a moment and appreciate the power of your computer. While we have asked you to spend most this chapter – and indeed the entire first part of this text – practicing R instead of learning statistics right away, the payoff is right here. Can you imagine typing all these numbers into a calculator?!

We do not want all these columns. In fact, we only want to talk about the relationship between weight (wt) and miles per gallon (mpg). To only look at one set of numbers at a time, in R, we can go into our data's columns by using the $ operator. Go ahead and type mtcars$wt on one line, press Enter to move to the next line, type mtcars$mpg, highlight both lines, and then run them by pressing *Ctrl+Enter* (*Cmd+Enter* for macOS) again. Your results should be matching our results printed here:

```
mtcars$wt
```

```
##  [1] 2.6 2.9 2.3 3.2 3.4 3.5 3.6 3.2 3.1 3.4 3.4 4.1 3.7 3.8 5.2 5.4
## [17] 5.3 2.2 1.6 1.8 2.5 3.5 3.4 3.8 3.8 1.9 2.1 1.5 3.2 2.8 3.6 2.8
```

```
mtcars$mpg
```

```
##  [1] 21 21 23 21 19 18 14 24 23 19 18 16 17 15 10 10 15 32 30 34 22 16
## [23] 15 13 19 27 26 30 16 20 15 21
```

From here on out, we will not mention typing into the script, highlighting, or pressing *Ctrl+Enter* (*Cmd+Enter* for macOS). Instead, we will simply discuss ideas, discuss how to turn those ideas into code, and then show you both the code and the output. If you get stuck and feel it is running code in R that is the likely holdup, there are some useful videos online. If you are a student in a class, having taught several thousand of our own students over the years, we encourage you to reach out to your professor or instructor. Especially if you are so unlucky as to have one of us as your instructor, for sure reach out! On the other hand, if that is not possible, go ahead and reach out to us. We do not mind at all, will do our best to answer, enjoy hearing from readers, and can be reached readily enough. Go ahead and take a look at our bios in the front of this book and do not be shy if you need help.

Moving on, having seen both weight and miles per gallon, we are ready to take a look at these two data points visually. If this works, you will have written your first program to plot() and visualize these two data sets. You will have also confirmed that all systems are go with RStudio and R.

To tell R we want to plot something, we use the function called plot(x, y). Functions take one or more *inputs* (in this case, weight and miles per gallon) and give us an output (in this case, a plot of those two number pairs). If you are familiar with x-axis and y-axis from algebra, you will know to expect a graph like the one you see in Figure 1-5. Go ahead and copy the code precisely as you see here in your script pane, highlight it, and run the code! Your graph will show up in your *Plots* pane, which up until now was your *Files* pane. Remember you can click back to Files anytime you want. Notice you can *export* your plot if you like. Now is a great moment to share your first plot with the world. In programming, these types of first steps are sometimes called "hello worlds" (because, back when we first learned to program, graphics and visuals were rare, so instead our first program literally printed "hello world" to the console – tough times!). More than just an old 1970s era dataset, you now can visually see any data set you can get into R. Pretty powerful stuff.

```
plot(x = mtcars$wt, y = mtcars$mpg)
```

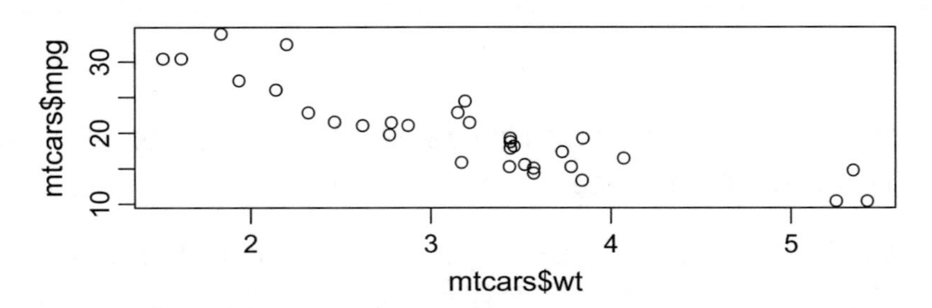

Figure 1-5. *Plot of mtcars weight and miles per gallon*

As you look at your graph, notice some features about this graph. As the x-axis of `mtcars$wt` increases, the y-axis of `mtcars$mpg` decreases. That is what we might expect to see happen; a car's weight should drag down the fuel efficiency. If you run this code on your own system, you may notice that the figure looks a little different. For example, Figure 1-5 is wider than it is tall, but we can change the aspect ratio (how tall vs. how wide the figure is); and although it is the same data, it looks a bit different. To show you what we mean by that, here we make the same figure but now roughly equal height and width in Figure 1-6.

```
plot(x = mtcars$wt, y = mtcars$mpg)
```

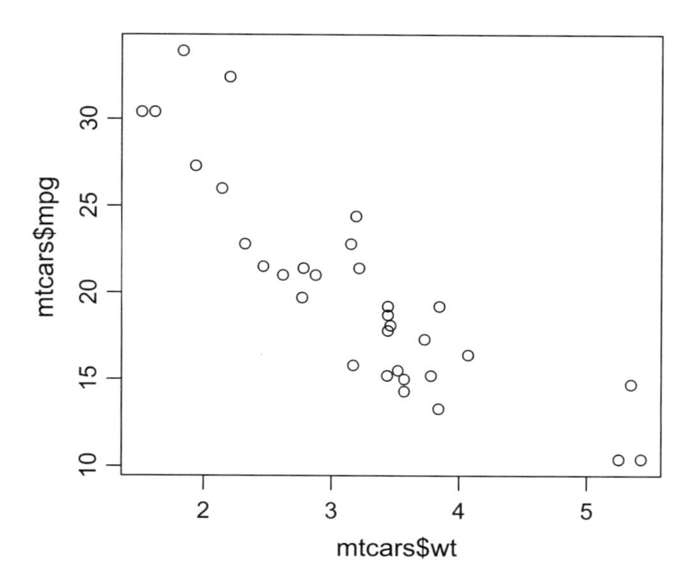

Figure 1-6. *Plot of mtcars weight and miles per gallon with a different aspect ratio that is roughly equal height and width rather than much wider than tall*

In RStudio, you can move your mouse cursor to the edge between panels to where you see multidirectional arrows and then click and draw to resize and reshape the panels of RStudio. The figure will automatically update to the shape you make for it.

As we get into Chapter 2, we will explore some fancier graphics. All the same, this is a good first start. Go ahead and click the save icon once more. From there, select *File* and then *Close Project* which is almost at the end of the file menu. Having closed out your project, you are ready to close RStudio by clicking the *X* on the top right of the screen.

The next time you open RStudio, if you need to get back into your Chapter 1 project, you may select *File*, and either *Recent Projects* will have a link to your *01Installing* project or you may use *Open Project...* to navigate to your project file.

1.7 Summary

In this book, we will wrap up each chapter with a summary such as in Table 1-2 detailing in order of introduction specific facts which may be useful to quickly reference. In particular, these facts may serve as a useful guide for the tools which will help you solve the exercises.

Table 1-2. *Chapter Summary*

Idea	What It Means
R	A statistical programming language.
RStudio	An Integrated Development Environment or IDE for R.
CRAN	Comprehensive R Archive Network.
mtcars	A data set about cars we will use often.
$	One R command to access columns by name.
`plot(x = , y = )`	The function name to make an x-axis and y-axis plot.

1.8 Practice for Mastery

Check your progress and grow through practice by working through some exercises. Comprehension checks ask critical thinking questions that may be best answered with a written or verbal response. Part of the art of statistics is successfully communicating results to your stakeholders or audience. Sometimes that audience is highly technical and other times very much not technical. Exercises are more direct applications of the concepts explored in the chapter.

Comprehension Checks

1. We used the data set `mtcars` in the script pane of RStudio. What happens if you were to go to the *Console* pane of RStudio and, at the prompt ¿, type `mtcars` and press *Enter*? Compare/contrast this behavior to highlighting and running `mtcars` in your saved *MyFirstRScript.R*.

2. In *MyFirstRScript.R*, you ran `plot(x = mtcars`*wt, y = mtcars*`mpg)` to get the graph in Figure 1-6. What happens if you run `plot(mtcars`*wt, mtcars*`mpg)`? Can you run `plot(mtcars`*wt, mtcars*`mpg)` in the Console pane instead? What does this suggest?

Exercises

1. You met the `mtcars` data and plotted weight and miles per gallon. Using the same `plot()` function, change the input to create a graph with miles per gallon on the horizontal x-axis and weight on the vertical y-axis.

2. There is another data set native to `R` named `iris`. Using the methods described in this chapter with `mtcars`, explore `iris`, view the data in the `Sepal.Length` and `Petal.Length` columns, and create a graph with `Sepal.Length` on the horizontal x-axis and `Petal.Length` on the y-axis.

CHAPTER 2

Installing Packages and Using Libraries

You installed R and RStudio and are now familiar with them. Now it is time to discover another feature of R that will help you learn statistics.

In Chapter 1, your first R script used the function `plot(x = mtcars$wt, y = mtcars$mpg)`. Then, just like magic, a graph appeared in your lower-right plot pane – except it was not magic. Behind the scenes, you were secretly using the graphics [15] package.

What is a package? When a programmer writes code meant to be used more than one time, they tend to give that code a name, and we call those names *functions*. An R *package* is usually one or more functions collected together to solve a common or similar task. By collecting those functions together in a package, the package can be shared with many people. Keep in mind R is free and used by many of the best researchers around the world. Each one has skills solving certain types of problems. By sharing their functions in packages, we all benefit together.

This is exactly what happened when you used the `plot()` function. Some programmers (in this case, the R core programming team) got together and wrote in a script some functions to create graphs (including the plot function). They put all that code together in the `graphics` package.

Using `plot()` did not take any extra work on your part, because some packages are so popular they are included in R by default. However, most packages are not included by default. Most packages are not needed by everyone; including them all would make downloading R take far too long. Instead, packages are kept on CRAN [2], ready to download at need. There is more to packages than what we said here. If you would like to learn more about packages (and perhaps write your own), please see *Advanced R: Data Programming and the Cloud* [23] for details.

M. Wiley and J. F. Wiley, *Beginning R 4*, https://doi.org/10.1007/978-1-4842-6053-1_2

By the end of this chapter, you should be able to

- Know packages are a way to add extra features to R.
- Install add-on packages for R with the `install.packages()` function.
- Activate packages for your R scripts by using the `library()` function.
- Understand when to install a package from CRAN and when to use a local library.

If you already know about packages and are eager to get into learning statistics, go ahead and install these packages in Table 2-1 to move ahead to the next chapter. If you would like to follow along with us as we meet these packages you will use throughout this book, go ahead and continue through this chapter.

Table 2-1. *Package List*

Package	Use
`haven` [19]	Input and output common statistical file types from SPSS, Stata, and SAS.
`readxl` [18]	Input Microsoft .xlsx files to R.
`writexl` [14]	Output Microsoft .xlsx files from R.
`data.table` [9]	Efficiently and easily manage data/rows/columns.
`extraoperators` [20]	An R package with operators to help speed up everyday tasks.
`JWileymisc` [21]	An R package with `aces_daily` simulated data.
`ggplot` [17]	Customizable visualizations, graphics, plots for exploratory data analysis (EDA).
`visreg` [7]	Visualization of regression models.
`emmeans` [13]	Summarize statistical models.
`ez` [12]	Easy Analysis and Visualization of Factorial Experiments (we will use this for ANOVA).
`palmerpenguins` [11]	A fresh, fun data set about penguins (currently on GitHub).

2.1 Installing Packages

A package only needs to be installed **once per computer**. In other words, it should not be installed each time a new project is started. It can be difficult, when learning so many new things, to keep track of what to do when. So, provided you are learning R with us using a single computer, you should only need to install these packages once each. It also does not matter if you install a package while you are inside a project or whether you just install from the Console pane. Either way, it will be installed on your computer, and those features/functions will be available to you from any project you had, have, or will have, on that computer. That being said, of course if you go onto a new computer and want to run your script over there on that new machine, you will need to install the packages you use again on that new machine.

Another thing to know about packages is they often *depend* on other packages. Thus, as you install each package, you may find that despite a plan to install just one package, in the Console pane you may see R is in fact installing several. This is all part of the natural process of package installation and nothing about which to worry. R is clever enough to know all the dependencies needed and will only install missing packages (thus, the order packages are installed can make a difference).

To install a new package from CRAN, we use the `install.packages(pkgs = "")` function. R is a case-sensitive language, and CRAN enforces the rule that no package can have precisely the same name as another. However, one must be careful to spell a package exactly correctly; otherwise, the search feature of the install function will not work. Additionally, since R is case sensitive, one must match the upper- or lowercase letters perfectly as well. Thus, each of the following subsections will not follow standard capitalization rules. Rather, we choose to only use the package name in its fully correct form.

As we go through the installation of each package, we also mention briefly just a little bit about how it is used and what additional features it adds to base R.

Note that when you install a package, you may get a lot of output on the console. This can seem overwhelming, but it's not anything you need to worry about. Generally if there is a problem, you will see something about an "error" or sometimes "non-zero exit status" which is a common way in technical worlds of meaning there was at least one, but maybe more, error or problem. Although it may seem confusing, it comes from the idea that "zero exit status" would mean zero issue by the time it exited the installation process. From that view, "non-zero exit status" means there are some errors or problems. If you do not have errors or "non-zero exit status," there is a good chance that all the information and text shown as packages install is nothing to be concerned about.

Finally, although we will make use of several different add-on packages, you do not need to know everything about these packages. That is, you do not need to study each package. We use sometimes just a few functions from a package to get what we need done as easily and efficiently as possible. It is perfectly okay to install a package to use one function and not learn about any of the other (possibly many) functions in that package.

haven

In the world of statistical computing, R is the increasingly popular newcomer. Because of R's newcomer status, there are often times when importing data comes from some of the classical programs such as SPSS, Stata, or SAS. The haven package [19] provides functions to read data from these sources into R.

Depending on your computer, and if your system has never installed R or any packages before, you may find that more than just the haven package is installed. This is because haven itself depends on some other packages. All the same, if, after you run `install.packages(pkgs = "haven")`, you see the last part of your output matches ours, then your first external package installation will be a success. We save practical applications of haven functions for Chapter 3 on data input and output.

```
install.packages(pkgs = "haven")
```

```
## Installing package into 'C:/Users/.../R/win-library/4.0'
## (as 'lib' is unspecified)

## package 'haven' successfully unpacked and MD5 sums checked
##
## The downloaded binary packages are in
##      C:\...\downloaded_packages
```

readxl

Due to the popularity of the Microsoft Office software suite, Excel-style files are also common sources of data. Due to their popularity, it can be useful to input data from `.xlsx` files. The `readxl` package [18] allows this to happen. As before, you may have a handful of additional packages that R will automatically install for you as dependencies of `readxl`. Notice too that R functions can be somewhat clever about their input. In this case, we are using the fact that the `install.packages()` function is expecting the package name as the **first** input. Thus, we do not need to explicitly state `pkgs = "readxl"`. Going forward, we will use a function's formal input name for the first time we introduce a function. After the first use, we will only use the formal input name when required.

```
install.packages("readxl")
```

```
## Installing package into 'C:/Users/.../R/win-library/4.0'
## (as 'lib' is unspecified)
```

```
## package 'readxl' successfully unpacked and MD5 sums checked
##
## The downloaded binary packages are in
##      C:\...\downloaded_packages
```

writexl

In addition to reading Excel files, when communicating our results to others, it can be helpful to write Excel files too. The `writexl` package [14] allows us to write `.xlsx` files.

```
install.packages("writexl")
```

```
## Installing package into 'C:/Users/.../R/win-library/4.0'
## (as 'lib' is unspecified)
```

```
## package 'writexl' successfully unpacked and MD5 sums checked
##
## The downloaded binary packages are in
##      C:\...\downloaded_packages
```

data.table

Once data are input to R, often we must work with the data for a bit prior to being able to perform statistical analytics. This can range from calculating summary tables to only grabbing out a subset of the total data available. In order to allow us to handle even large amounts of data with ease (not every data set is as short as mtcars), we use the data.table package [9] to allow us to manipulate or "munge" data.

```
install.packages("data.table")

## Installing package into 'C:/Users/.../R/win-library/4.0'
## (as 'lib' is unspecified)

## package 'data.table' successfully unpacked and MD5 sums checked
##
## The downloaded binary packages are in
##      C:\...\downloaded_packages
```

extraoperators

When working with larger data sets, it can be useful to study smaller subsets of information. While base R has the ability to readily chain together multiple criteria for such subsetting, it is not always clear to read such code. Additionally, such code can be long. The extraoperators package [20] saves keystrokes and makes various logical comparisons less convoluted.

```
install.packages("extraoperators")

## Installing package into 'C:/Users/.../R/win-library/4.0'
## (as 'lib' is unspecified)

## package 'extraoperators' successfully unpacked and MD5 sums checked
##
## The downloaded binary packages are in
##      C:\...\downloaded_packages
```

JWileymisc

The JWileymisc package [21] has a *simulated* dataset (aces_daily) based on a 12-day study of research participants' self-reported measures including stress and affect. The participants reported in three times per day, although, as usual for real-world research (or simulations thereof), there may be missing data. This data set helps us understand statistical methods beyond the "classroom-perfect" examples more commonly seen.

```
install.packages("JWileymisc")
```

```
## Installing package into 'C:/Users/.../R/win-library/4.0'
## (as 'lib' is unspecified)

## package 'JWileymisc' successfully unpacked and MD5  sums checked
##
## The downloaded binary packages are in
##      C:\...\downloaded_packages
```

ggplot2

Your first program was a plot using the base graphics package. We often have need of more custom or more advanced graphics. Much like R is a statistical programming *language*, the ggplot2 package [17] uses what is called the *grammar of graphics*.

```
install.packages("ggplot2")
```

```
## Installing package into 'C:/Users/.../R/win-library/4.0'
## (as 'lib' is unspecified)

## package 'ggplot2' successfully unpacked and MD5 sums checked
##
## The downloaded binary packages are in
##      C:\...\downloaded_packages
```

visreg

A common statistical method is taking data and finding some mathematical relationships between the inputs or predictors of a model and the output or the target. This allows researchers to understand the relationship between independent variables that can be controlled or adjusted and dependent variables. We saw a simple example of this with a predictor of car weight to an output or dependent variable of miles per gallon. The `visreg` package [7] makes it very easy to visualize parts of those models. It can be easier to understand statistical ideas with the right picture, and this package will both help you learn how a model works and how to communicate your results effectively.

```
install.packages("visreg")
```

```
## Installing package into 'C:/Users/.../R/win-library/4.0'
## (as 'lib' is unspecified)

## package 'visreg' successfully unpacked and MD5 sums checked
##
## The downloaded binary packages are in
##      C:\...\downloaded_packages
```

emmeans

The emmeans package [13] helps us summarize and get follow-up tests on statistical models to aid understanding and displaying results in a way that is easy to interpret. We will install this package now.

```
install.packages("emmeans")
```

```
## Installing package into 'C:/Users/.../R/win-library/4.0'
## (as 'lib' is unspecified)

## package 'emmeans' successfully unpacked and MD5 sums checked
##
## The downloaded binary packages are in
##      C:\...\downloaded_packages
```

ez

Our penultimate package is the ez package [12] which helps us perform a statistical method called *an*alysis *o*f *va*riance (ANOVA). We discuss ANOVA near the end of this book. All the same, we will install this package now.

```
install.packages("ez")
```

```
## Installing package into 'C:/Users/.../R/win-library/4.0'
## (as 'lib' is unspecified)
```

```
## package 'ez' successfully unpacked and MD5 sums checked
##
## The downloaded binary packages are in
##      C:\...\downloaded_packages
```

palmerpenguins

Our last package gives us data rather than any statistical computations. The palmerpenguins package [11] includes a popular data set about penguins!

```
install.packages("palmerpenguins")
```

```
## Installing package into 'C:/Users/.../R/win-library/4.0'
## (as 'lib' is unspecified)
```

```
## package 'palmerpenguins' successfully unpacked and MD5  sums checked
##
## The downloaded binary packages are in
##      C:\...\downloaded_packages
```

2.2 Using Packages

Now that these packages have all been installed, it is time to make sure the installation worked. You may recall that packages only need to be installed once per computer. However, every time you are wanting to use a function from a package, you will have to let R know which packages to search. This is because when you use a function (remember a function is just a name for some specific code that does a specific task),

R needs to match the function name you type to the code that does the work. Rather than waste time searching all the packages you may have installed (we have a couple hundred on our computers), we tell R which packages to search for a particular script by using the `library()` call. This needs to be run each time you start up one of your scripts in the script pane. Thus, convention says to put that call near the top of your code. This gives other users of your script a warning that, if they do not have one of those packages installed, they must install it before using your code.

It is enough to run the library call **once per session**. Every time you start RStudio and open a project, you need to run any library calls. It is easy enough to run the calls, and, going forward, we start each chapter with any library calls needed for that chapter's code. To show you what a successful call looks like, we go ahead and run the `readxl` package:

```
library(readxl)
```

As you can see, it is not especially exciting. If the package is already installed, a good library call usually does not display any text. However, sometimes, a successful library call may give you just a bit of information, as the `data.table` package does:

```
library(data.table)

## data.table 1.13.0 using 6 threads (see ?getDTthreads). Latest news:
r-datatable.com
```

In contrast, trying to use a package that is not installed will result in an error:

```
library(fake)

## Error in library(fake): there is no package called 'fake'
```

You have now completed the installation of all the background software and packages needed for the rest of this book. Congratulations! From here on out, we will first learn some code together and then onward into full statistics.

2.3 Summary

In this book, each chapter concludes with a summary such as in Table 2-2 detailing in order of introduction specific facts which may be useful to quickly reference.

Table 2-2. *Chapter Summary*

Idea	What It Means
`install.packages(pkgs = "")`	Installs packages named; need only be run once per computer.
`library()`	Loads a package into memory for use in a script; need only be run once per session.

2.4 Practice for Mastery

Check your progress and grow through practice by working through some exercises. Comprehension checks ask critical thinking questions that may be best answered with a written or verbal response. Part of the art of statistics is successfully communicating results to your stakeholders or audience. Sometimes that audience is highly technical and other times very much not technical. Exercises are more direct applications of the concepts explored in the chapter.

Comprehension Checks

1. Suppose you share some `R` code you wrote with a friend. Would that friend need to install any packages your code used on their computer?

Exercises

1. Visit `https://rdatatable.gitlab.io/data.table/` and download the `data.table` cheat sheet from the package author.
2. Using the search engine of your choice, search this phrase – `R ggplot2` – and see if you can find a cheat sheet.

CHAPTER 3

Data Input and Output

In general, when performing a statistical analysis, the majority of time spent ends up being dealing with data. This involves successfully inputting data as well as organizing or *cleaning* data. For now, our goal is to recognize various types of data files and use those types to import and export data to and from `R`.

By the end of this chapter, you should be able to

- Recognize common data file types by suffix.
- Apply R packages and functions to *import* data into `R`.
- Apply R packages and functions to *export* data out of `R`.

3.1 Setup

In order to practice inputting and outputting the various types of files you may use, we need to do some setup. In order to continue practicing creating and using projects, we will start a new project for this chapter. Inside that project, we also will create some folders and download some data files from this book's `GitHub` site.

If necessary, review the steps in Chapter 1 to create a new project. After starting RStudio, on the upper-left menu ribbon, select *File* and click *New Project*. Choose *New Directory* ➤ *New Project*, with *Directory name* `03DataIO`, and select *Create Project*. To create an `R` script file, on the top ribbon, right under the word *File*, click the small icon with a plus sign on top of a blank bit of paper, and select the *R Script* menu option. Click the floppy disk–shaped *save* icon, name this file `DataIO.R`, and click *Save*.

In the lower-right pane, you should see your project's two files, and right under the *Files* tab, click the button titled *New Folder*. In the *New Folder* pop-up, type `data` and select *OK*. In the lower-right pane, click your new *data* folder. Repeat the folder creation process, making a new folder titled `ch03`.

M. Wiley and J. F. Wiley, *Beginning R 4*, https://doi.org/10.1007/978-1-4842-6053-1_3

Next, you need to download some files from this book's site. They should live on *GitHub* at `http://github.com/apress/beginning-r-4`. However, the permanent link can also be found on the website `http://mattwiley.org/publications.html` by clicking *Code* for *Beginning R*. Navigate to this chapter's folder and download the Texas counties data [3] as well as the American Community Survey [4] data into the ch03 folder you just created. In all, you should have six files in your project's subfolders.

With that set up, you are ready to practice importing and exporting data.

If you need any refresher or quick tips on where to find things in RStudio as you go through the chapter, look at this short "cheat sheet" which has visuals and labels for most of RStudio's features: `https://github.com/rstudio/cheatsheets/raw/master/rstudio-ide.pdf`.

3.2 Input

When inputting files to `R`, the usual way data arrive is in some sort of table format where there are named columns or *variables* and numbered rows or *observations*. Each column thus represents a category, and each row represents a specific measurement of those variables.

The usual challenges to inputting these types of "mostly decent" data tend to be successfully reading the **type** of file storing the data and adapting to any unique aspects of different data files (such as the first few rows being of a different format than expected). An example of such a foible might be that a data gatherer/creator may include some notes or comments in the first few lines of an Excel file. It is only after these notes that the actual columns and rows of data start.

Note When we say the **type** of file, we mean things like an Excel spreadsheet or a comma-separated text file or a dataset from another proprietary piece of software. Almost always, the **type** of a file is indicated by the *extension*. The file extension are the last few characters of the file name after the period. For example, an Excel file might be called "data.xlsx". In this case, the file name is "data" and the extension is ".xlsx"; and the extension, the ".xlsx" part, is what tells us that it is an Excel file. Other common extensions are ".csv" which means a text file with cells separated by commas and ".rds" which is R's own format for storing data. All of these can be told by looking at the file extension.

Thus, our first task is to identify the correct function (and package) to read in our data as well as determine if we need to skip some rows. The majority of functions written to input data have at least three variables as options that include the location of the data file, the number of rows to *skip*, and whether the data has a *header* with column names. It is worth noting most data input functions generally default to assuming the first row is a column name header and no rows should be skipped.

Most input functions include many options - and in general most of those do not need to be used. In fact, the easiest way to take care of some messiness in a file is often to open it yourself by hand, fix it manually, and then import into `R`. On the other hand, if you routinely get a file (perhaps some sort of daily update from some company or third party) and that file is predictable in the way it does not work nicely, it may be worth exploring additional function options to code an automatic read.

Manual Entry

While we usually focus on input as something that involves some external file to read in, it is possible to input data by hand in `R`. This can be useful for several scenarios, including setting up some variables all the way to more extensive data input. In `R`, the *assignment* operator is `<-`. These two symbols together, the less than character and the `-`, are both found on the right-hand side of your keyboard, near M and P, respectively. It helps to give our data a name that makes sense; for example, if using the year 2020 as a variable, we may call it `academicYear`.

Note To help code be pronounceable, we often make data variables nouns. For the same reason, functions are often verbs (`e.g., plot()`).

To read the following line of code out loud, we would say "the variable `academicYear` is assigned the number 2020":

```
academicYear <- 2020
```

Notice that after running the preceding line, in your top-right *Environment* pane, there is now a value for `academicYear` that shows the value `2020`. If we need to use the academic year in our code, we could now easily just type `academicYear` and recover our result:

```
academicYear
```

```
## [1] 2020
```

Revisiting a familiar data set, mtcars, let us explore how we might use assignment in another way. Recall the wt variable was the weight values of the cars. Should we wish to use those values, as we saw before, we could specifically call out that single column by assigning it to a variable name. Again, after assignment, we are allowed to use our variable directly:

```
mtcarsWeight <- mtcars$wt
mtcarsWeight
```

```
##  [1]  2.620 2.875 2.320 3.215 3.440 3.460 3.570 3.190 3.150 3.440 3.440
## [12]  4.070 3.730 3.780 5.250 5.424 5.345 2.200 1.615 1.835 2.465 3.520
## [23]  3.435 3.840 3.845 1.935 2.140 1.513 3.170 2.770 3.570 2.780
```

While manual assignment is simple, it would be very tedious to type in all the data we would ever use directly into R. We move on to using some the packages we installed in Chapter 2.

CSV: .csv

Perhaps the most common data type is a table of information stored in the *comma-separated values* format which has the file extension **.csv**. While R can read such files natively, the data.table package has a function named fread() that reads in CSV files very fast indeed. This function has many inputs or arguments that are possible, although we only discuss three now. This function's first input or argument is file = "". Because an RStudio project defaults to the project as the *working directory* (which we showed how to set in Chapter 1), we need only give the folder file path data/ch03/ and the file name Counties_in_Texas.csv to this argument. The second argument, header, defaults to an automatic feature which usually does a good job of guessing if there is a header. Should you know your file does or does not have a header, then you can manually set this argument to either TRUE or FALSE. The third argument we discuss is skip. It takes an integer (i.e., whole number, without any decimal places, like 0, 1, 2, 3) input that is the number of rows you wish to skip. The default is zero; therefore, like header, you can often safely ignore this argument. You cannot skip part of a row, so something like skip = 1.25 or skip = 3.5 is not valid.

While packages are normally placed at the beginning of a set of code, in this case, we include the `library()` call to the `data.table` package here, to help remind us that without that library call, `fread()` will not work. We meet the assignment operator again, as we assign our data to a variable named `csvData`:

```
library(data.table)

## data.table 1.13.0 using 6 threads (see ?getDTthreads). Latest news:
r-datatable.com

csvData <- fread(file = "data/ch03/Counties_in_Texas.csv",
                 header = TRUE,
                 skip = 0)
```

Notice that after running the preceding line of code, while there is no output to the *Console*, our *Environment* pane definitely has 254 observations of 10 variables. You may click the arrow icon just left of `csvData`, to see the types of data just read. For example, `CountyName` is of type `chr` which stands for *character* data, while `CountyNumber` is of type `int` for *integer* or whole number. Look at `https://github.com/rstudio/cheatsheets/raw/master/rstudio-ide.pdf` for more ways to interact with data through RStudio rather than just through code.

With 254 observations or rows, our data might be a bit long to visually inspect all rows. A newer way to explore data can be seen in your top-right Environment pane. There are both a dot with an arrow in it on the left side of each data set and a square table grid on the right side. Clicking the left-side dot and arrow will allow one to see what types of data are in each data set. In addition to showing the *column names*, it will also show the types of data (e.g., **int**eger or **Date** or **num**eric). Another option is to click the square table grid on the far right. This will open the data set for viewing in a new tab. Closing that tab should get you back to your code area.

Often, it is enough to view just the first few rows. The `head()` function allows us to inspect the first six rows by default:

```
head(csvData)

##     CountyName FIPSNumber CountyNumber PublicHealthRegion
## 1:    Anderson          1            1                  4
## 2:     Andrews          3            2                  9
## 3:    Angelina          5            3                  5
## 4:     Aransas          7            4                 11
```

```
## 5:      Archer          9           5                   2
## 6:   Armstrong         11           6                   1
##    HealthServiceRegion MetropolitanStatisticalArea_MSA
## 1:                4/5N                              --
## 2:                9/10                              --
## 3:                4/5N                              --
## 4:                  11                  Corpus Christi
## 5:                 2/3                   Wichita Falls
## 6:                   1                        Amarillo
##    MetropolitanDivisions MetroArea NCHSUrbanRuralClassification_2006
## 1:                    -- Non-Metro                      Micropolitan
## 2:                    -- Non-Metro                      Micropolitan
## 3:                    -- Non-Metro                      Micropolitan
## 4:                    --     Metro                      Medium Metro
## 5:                    --     Metro                       Small Metro
## 6:                    --     Metro                       Small Metro
##    NCHSUrbanRuralClassification_2013
## 1:                      Micropolitan
## 2:                      Micropolitan
## 3:                      Micropolitan
## 4:                      Medium Metro
## 5:                       Small Metro
## 6:                      Medium Metro
```

While it may not quite seem like it yet, you are now also doing statistics. Exploratory data analysis (EDA) is the first step in most analytics. It may not be fancy, but `head()` is an example of such exploration.

Excel: .xlsx or .xls

Microsoft's Excel format is another common data format. The `read_excel()` function from the `readxl` package inputs such files and again has some useful default options. In addition to the usual three features of file path, column names, and skip, we discuss a fourth argument of `sheet`. Excel files often have more than one sheet. The default is to the first sheet, although you may enter either a number showing which *order* your sheet is in or quote text *string* which names the sheet (Excel often defaults to "Sheet1" for the first sheet).

For our data set, the variable excelData is assigned the output of the read_excel() function for the .xlsx version of the counties in Texas data. While we set sheet = 1 to capture our sheet by sheet order, we could just as well have used sheet = "Counties_in_Texas. To show this in our code itself, we use a *comment* which is done by using the *pound sign* or *hashtag symbol* character. Comments in our code are for humans to read and are **not** run by R:

```
library(readxl)
excelData <- read_excel(path = "data/ch03/Counties_in_Texas.xlsx",
                        sheet = 1, #or sheet = "Counties_in_Texas"
                        col_names = TRUE,
                        skip = 0
                        )
```

This is the same data set as was read from the CSV file. Or is it? Using the head() function on our new data set, compare the FIPSNumber columns and notice they look slightly different. A closer inspection (such as via the arrow drop-down icon in the Environment pane) shows that FIPSNumber is of type int in csvData while it is of type chr in excelData:

```
head(excelData)
```

```
## # A tibble: 6 x 10
##   CountyName FIPSNumber CountyNumber PublicHealthReg~ HealthServiceRe~
##   <chr>      <chr>             <dbl>            <dbl> <chr>
## 1 Anderson   001                   1                4 4/5N
## 2 Andrews    003                   2                9 9/10
## 3 Angelina   005                   3                5 4/5N
## 4 Aransas    007                   4               11 11
## 5 Archer     009                   5                2 2/3
## 6 Armstrong  011                   6                1 1
## # ... with 5 more variables: MetropolitanStatisticalArea_MSA <chr>,
## #   MetropolitanDivisions <chr>, MetroArea <chr>,
## #   NCHSUrbanRuralClassification_2006 <chr>,
## #   NCHSUrbanRuralClassification_2013 <chr>
```

When using functions written by different package authors, some default settings, and some choices, are often not the same. When reading data into R, it is generally a wise step to inspect your data and understand a little more about just what has been read. Another difference you may notice is that our Excel data mentions the word *tibble*. We save an in-depth discussion of *data structure* choices for our *Advanced R: Data Programming and the Cloud* [23] book. While it is good to notice this difference, we shall ignore it for now.

RDS: .rds

We started off showing how to read in data sets from CSV or Excel files, because those are probably the most common way of storing and sharing data sets. They are popular in part because almost everyone on any computer can open and see them, without needing extra software they do not already have.

R does also have its own data format named *R Data "Storage"* which has the extension *.rds*. The ability to read and write `.rds` files is built into R, so we do not need any extra packages. Additionally, because these files come from R, there is only one argument of interest, the `file =` argument. R automatically knows how to read them in, what indicates variables, the type of variable, what the variable names vs. data are, any notes, and so on:

```
rData <- readRDS(file = "data/ch03/Counties_in_Texas.rds")
```

It would get repetitive to see the same data too many times. Thus, we introduce the `tail()` function which shows the last six rows of our data set:

```
tail(rData)
```

```
##    CountyName FIPSNumber CountyNumber PublicHealthRegion
## 1:       Wise        497          249                  3
## 2:       Wood        499          250                  4
## 3:     Yoakum        501          251                  1
## 4:      Young        503          252                  2
## 5:     Zapata        505          253                 11
## 6:     Zavala        507          254                  8
##    HealthServiceRegion MetropolitanStatisticalArea_MSA
```

```
## 1:                    2/3     Dallas-Fort Worth-Arlington
## 2:                   4/5N                              --
## 3:                      1                              --
## 4:                    2/3                              --
## 5:                     11                              --
## 6:                      8                              --
##      MetropolitanDivisions MetroArea
## 1: Fort Worth-Arlington MD     Metro
## 2:                      -- Non-Metro
## 3:                      -- Non-Metro
## 4:                      -- Non-Metro
## 5:                      -- Non-Metro
## 6:                      -- Non-Metro
##    NCHSUrbanRuralClassification_2006
## 1:                Large Fringe Metro
## 2:                           Noncore
## 3:                           Noncore
## 4:                           Noncore
## 5:                           Noncore
## 6:                           Noncore
##    NCHSUrbanRuralClassification_2013
## 1:                Large Fringe Metro
## 2:                          Non-core
## 3:                          Non-core
## 4:                          Non-core
## 5:                      Micropolitan
## 6:                          Non-core
```

Other Proprietary Formats

There are several other pieces of statistical software that are often used to store and analyze data. If you have a colleague who uses one of these programs, you may get a data file that uses that software's data format. We are going to look at three of these very briefly, so that if you ever come across such data, you know how to get them into R.

SPSS: .sav

The last three formats we discuss all use the haven package. Keep in mind there is no need to run `library(haven)` more than once per R session. Thus, we run it just here. By now, the usual arguments are familiar. The IBM Corporation owns the data software known as *SPSS*. Files from this software can have the extension *.sav*. The function call from haven to read this format is `read_spss()`:

```
library(haven)
spssData <- read_spss(file = "data/ch03/Counties_in_Texas.sav",
                      skip = 0)
```

Stata: .dta

Stata is a statistical data software with a file extension of *.dta*. The haven package also has a `read_stata()` function to read in this type of files:

```
stataData <- read_stata("data/ch03/Counties_in_Texas.dta",
                        skip = 0)
```

SAS: .sas7bdat

Our last statistical software data format is SAS which has a rather long extension of *.sas7bdat*. We use a file from the US Census's American Community Survey [23] for variety. This particular file has over 38 *thousand* observations. Data can get very large these days. All the same, note that R reads this file in fast enough. Data must generally get quite large indeed before there is difficulty with modern memory:

```
sasData <- read_sas("data/ch03/ACS_1yr_Seq_Table_Number_Lookup.sas7bdat",
                    skip = 0)
head(sasData)
```

```
## # A tibble: 6 x 9
##   filed tblid  seq   order position cells total title    subject_area
##   <chr> <chr>  <chr> <dbl>    <dbl> <chr> <dbl> <chr>    <chr>
## 1 ACSSF B00001 0001     NA        7 "1 CE~   NA UNWEIGH~ "Unweighted~
## 2 ACSSF B00001 0001     NA       NA ""       NA Univers~ ""
## 3 ACSSF B00001 0001      1       NA ""       NA Total    ""
```

```
## 4 ACSSF B00002 0001      NA        8 "1 CE~    2 UNWEIGH~ "Unweighted~
## 5 ACSSF B00002 0001      NA       NA ""       NA Univers~ ""
## 6 ACSSF B00002 0001       1       NA ""       NA Total    ""
```

3.3 Output

Getting data into R is usually the first stage of an analysis project. Writing the data out of R to share with stakeholders or collaborators is usually one of the last stages. In an R project, it is often a good idea to separate data we read *in* from the data we write *out*. We already created a folder named data for the files to input. Now, we create a new folder, using the same *New Folder* icon in the lower-right pane, named output.

Any data that is in R is able to be written out. It does not matter where those data come from, only that they are in our global environment. Thus, options for writing out in the following examples include mtcars [1], penguins [11], and csvData, excelData, rData, sasData, spssData, or stataData. These are all part of our current instance of R. Because the mtcars and penguins data sets are built into packages, they do not show up in the Environment tab, but can still be written. In other words, we could write the sasData to a CSV file, and all would be well. While it is not an "official" feature of R, it can be used to convert between file types. To avoid having weird data calls, we will use the R data.

CSV

For universal portability, a CSV file format is most often the answer; most software suites can read in CSV files by default. Keep in mind even R has a base function that reads in CSV files; it is only our choice that leads us to use the data table option. Similarly, most software systems can readily save or write or export CSV files. Again, R has this built-in feature as well, although we choose to use the data.table function fwrite(). This function takes two arguments. The first is set to the data we wish to write out. The second is the file path and name of where our data will be written:

```
fwrite(x = rData,
       file = "output/ch3_txCounties.csv")
```

Excel

Microsoft's Excel format is another widely usable format, and the writexl package provides a function write_xlsx() to achieve this. Two additional features of this function include the option to suppress the column names (the default is as shown TRUE) as well as an option to not center and bold the column names (again the default is as shown TRUE). For office environments where colleagues use Excel, these defaults are usually a very human-friendly way to share data out to decision makers. The default bolding of the column names in particular helps reduce confusion about the meaning of shared data:

```
library(writexl)
write_xlsx(x = rData,
           path = "output/ch3_txCounties.xlsx",
          col_names = TRUE,
           format_headers = TRUE)
```

RDS

Our last export format is R's default format. Sometimes, objects are simply saved for later use. If you intend to reload a data set for later work, the saveRDS() function is a very memory efficient choice. On our system, the RDS format is less than half the size of the CSV file:

```
saveRDS(object = rData,
        file = "output/ch3_txCounties.RDS")
```

3.4 Summary

This chapter discussed ways to read and write data in and out of R. Using the functions and packages summarized in Table 3-1, you will be able to access and share data from many sources of information.

Table 3-1. Chapter Summary

Idea	What It Means
`<-`	Assignment operator.
`fread()`	CSV file read-in function from the `data.table` package.
`head()`	Displays the first six rows of a data set.
`read_excel()`	Excel XLSX or XLS file read-in function from the `readxl` package.
`readRDS()`	RDS (native R format) file read-in function.
`tail()`	Displays the last six rows of a data set.
`read_spss()`	SPSS SAV file read-in function from the haven package.
`read_stata()`	Stata DTA file read-in function from the haven package.
`read_sas()`	SAS SAS7BDAT file read-in function from the haven package.
`fwrite()`	CSV file write-out function from the `data.table` package.
`write_xlsx()`	Excel XLSX file write-out function from the `writexl` package.
`saveRDS()`	RDS (native R format) file write-out function.

3.5 Practice for Mastery

Check your progress and grow through practice by working through some exercises. Comprehension checks ask critical thinking questions that may be best answered with a written or verbal response. Part of the art of statistics is successfully communicating results to your stakeholders or audience. Sometimes that audience is highly technical and other times very much not technical. Exercises are more direct applications of the concepts explored in the chapter.

Comprehension Checks

1. Ask around! Question your classmates, colleagues, or friends about the types of data they use. What file formats are common? What is the column and row organization of their data?

Exercises

1. Using the techniques learned in this chapter, write the `penguins` data set to a `.csv` file.

2. Using the techniques learned in this chapter, write the `mtcars` data set to a `.xlsx` file.

3. There are many datasets available to the public (although not all are free to use in a book). Find a dataset from your company, your local government, or some other group that interests you. Download the file, and use one of the functions you learned in this chapter to read the data into `R`.

CHAPTER 4

Working with Data

This text is designed to give you insight into real-world statistical methods. Unlike many "classroom-ready" examples, real-world data often require advance work before any statistics ever occurs. In this chapter, you will learn how to use the `data.table` [9] and `extraoperators` [20] packages to create the data sets you need for applying statistical methods to analyze data and create data-informed decisions.

That is a tall order, so keep in mind it is normal to find `data.tables` somewhat confusing at first. All the same, keep on exploring, keep on trying, be sure to work through the comprehension checks and exercises at the end of this chapter, and always feel comfortable asking for help!

By the end of this chapter, you should be able to

- Convert `R` data into `data.table` format.
- Understand the three areas of the `data.table` structure.
- Apply logical operators to extract a target sample from a larger data set through *subsetting*.
- Create new columns based on current data.
- Apply `by` group calculations to a data set.

4.1 Setup

In order to practice manipulating `data.table`-style information, we need to do some setup. In order to continue practicing creating and using projects, we will start another new project for this chapter. Inside that project, we also will create some folders and download some data files from this book's `GitHub` site.

If necessary, review the steps in Chapter 1 to create a new project. After starting RStudio, on the upper-left menu ribbon, select *File* and click *New Project*. Choose *New Directory* ➤ *New Project*, with *Directory name* `04WorkingWithData`, and select *Create*

M. Wiley and J. F. Wiley, *Beginning R 4*, https://doi.org/10.1007/978-1-4842-6053-1_4

Project. To create an R script file, on the top ribbon, right under the word *File*, click the small icon with a plus sign on top of a blank bit of paper, and select the *R Script* menu option. Click the floppy disk–shaped *save* icon, name this file `WorkwData.R`, and click *Save*.

In the lower-right pane, you should see your project's two files, and right under the *Files* tab, click the button titled *New Folder*. In the *New Folder* pop-up, type `data` and select *OK*. In the lower-right pane, click your new *data* folder. Repeat the folder creation process, making a new folder titled `ch04`.

Next, you need to download some files from this book's site. They should live on *GitHub* at `http://github.com/apress/beginning-r-4`. However, the permanent link can also be found on the website `http://mattwiley.org/publications.html` by clicking *Code* for *Beginning R*. Navigate to this chapter's folder and download the Texas counties data [3] as well as the American Community Survey [4] data into the ch04 folder you just created. In all, you should have six files in your project's subfolders.

Keep in mind if you already downloaded these files for Chapter 3, you could simply copy the files over from Chapter 3's projects. However, be sure to rename the folders so your data live in this file path: `YourProjectFolder/data/ch04`. If you created a unique project for this chapter, you are already in the `04WorkingWithData` folder and would have two subfolders of `data/ch04`.

For this chapter, we use four of our packages. Remember all these packages were installed on your local computing machine in Chapter 2. There is no need to re-install. However, this is a new project, and we are running this set of code for the first time. Therefore, you need to run these four `library()` calls:

```
library(data.table)

## data.table 1.13.0 using 6 threads (see ?getDTthreads). Latest news:
r-datatable.com

library(extraoperators)
library(JWileymisc)
library(palmerpenguins)
```

Applying what we learned in Chapter 3, we read in the `Counties_in_Texas` data set using the `fread()` function from the `data.table` package. Because this function is already from `data.table`, unlike other data, we do **not** need to convert to the `data.table` format; our data already have the correct structure:

```
texasData <- fread(file = "data/ch04/Counties_in_Texas.csv",
                   header = TRUE,
                   skip = 0)
```

We use two more data sets in this chapter. By having three different types of data, we make sure our early explorations meet enough real-world data "flavors." Comparing and contrasting a specific technique with different data helps you practice and learn how to apply these techniques to *any* data set you meet "in the wild."

Unlike the `texasData` though, these two data sets are already here. We simply need to make sure they are in `data.table` format. For now, do not think too much about the `data()` function. All it is doing is loading information from the `JWileymisc` package into our Environment pane. Next, the `as.data.table()` function takes one input, our new data, and uses the assignment operator `<-` to store that as an object named `acesData`. This variable has over `6,500` observations or rows of data across `19` variables or columns. To make sure we do not confuse objects, we use the `rm()` function to *remove* `aces_daily`:

```
data(aces_daily)
acesData <- as.data.table(aces_daily)
rm(aces_daily)
```

For our final data set, we use the `palmerpenguins` [11] data set. However, we map that data into the `data.table` format again using our `as.data.table()` function:

```
penguinsData <- as.data.table(penguins)
```

With three data objects set up, you are ready to practice *working with data*.

4.2 What Do Our Data Look Like?

Before we are able to practice using our data, it will help to understand what we are seeing. Often, in real-world uses of data, researchers become very familiar with their usual data sets. Our goal is to make these three our usual data sets for the rest of this book. Time spent exploring these data now can pay off later.

As noted in Chapter 3, you can view data by going to your top-right Environment pane in RStudio. There are both a dot with an arrow in it on the left side of each data set and a square table grid on the right side. Clicking the left-side dot and arrow will allow one to see what types of data are in each data set. In addition to showing the *column names*, it will also show the types of data (e.g., **int**eger or **Date** or **num**eric). Another option is to click the square table grid on the far right. This will open the data set for viewing in a new tab. Closing that tab should get you back to your code area.

However, traditionally in R, we use *functions* to get work done. The colnames() function takes data as input and gives the column names as output. The column names are in the order the columns naturally appear in the data set:

```
colnames(acesData)
```

```
##  [1] "UserID"             "SurveyDay"      "SurveyInteger"
##  [4] "SurveyStartTimec11" "Female"         "Age"
##  [7] "BornAUS"            "SES_1"          "EDU"
## [10] "SOLs"               "WASONs"         "STRESS"
## [13] "SUPPORT"            "PosAff"         "NegAff"
## [16] "COPEPrb"            "COPEPrc"        "COPEExp"
## [19] "COPEDis"
```

Once we know the names of our columns, we are able to use them. Notice the column titled EDU and notice there is no information about *specific types of education degrees*. To get a sense of what is stored in the EDU column, we use the unique() function. Think back to our first script and the $ operator to access a specific column of data. The unique() function allows you to see each copy of data *only once*:

```
unique(acesData$EDU)
```

```
## [1]  0 NA  1
```

From our output in the preceding code, what we find is the EDU column is an example of *one-hot* encoding (sometimes called *dummy* coding). If the entry is a 1, then that row of data was entered by a participant who has at least a higher education degree. On the other hand, a 0 is a participant who has not (yet) earned a tertiary education degree. There is one more option, where the participant did not respond either way. This is the null or NA (not applicable) option.

Of course, we saw other columns in our data set. Exploring SurveyDay, we see an example of Date-type data. In R, the usual format for such Date data is **YYYY-MM-DD** for year, month, and day. This date style is almost universal, except in the United States where a different style of date is more common. Always be cautious with date objects, however. It is often essential to understand where the data were first coded and, if they have time data attached, account for time zones:

```
unique(acesData$SurveyDay)

##  [1] "2017-02-24" "2017-02-25" "2017-02-26" "2017-02-27" "2017-02-28"
##  [6] "2017-03-01" "2017-03-02" "2017-03-03" "2017-03-04" "2017-03-05"
## [11] "2017-03-06" "2017-03-07" "2017-02-22" "2017-02-23" "2017-03-08"
## [16] "2017-03-09" "2017-03-10" "2017-03-11" "2017-03-12" "2017-03-13"
## [21] "2017-03-14"
```

While these first two functions allow us to easily see all our column names and explore specific column data, it can help to see the first few rows of a data set. The `head()` function does precisely that and by default shows the first six rows. The `acesData` has 19 columns and thus wraps in the console output. One can see there may be quite a bit of NA data. Often, in-class examples do not show how to cope with such challenges. Thus, you are already exploring how real-life information can be messy:

```
head(acesData)

##     UserID  SurveyDay SurveyInteger SurveyStartTimec11 Female Age
## 1:       1 2017-02-24             2          1.927e-01      0  21
## 2:       1 2017-02-24             3          4.859e-01      0  21
## 3:       1 2017-02-25             1          1.157e-05      0  21
## 4:       1 2017-02-25             2          1.931e-01      0  21
## 5:       1 2017-02-25             3          4.062e-01      0  21
## 6:       1 2017-02-26             1          1.638e-02      0  21
##     BornAUS SES_1 EDU  SOLs WASONs STRESS SUPPORT PosAff NegAff
## 1:        0     5   0    NA     NA      5      NA  1.519  1.669
## 2:        0     5   0 0.000      0      1   7.022  1.505  1.000
## 3:        0     5   0    NA     NA      1      NA  1.564     NA
## 4:        0     5   0    NA     NA      2      NA  1.563  1.357
## 5:        0     5   0 6.925      0      0   6.154  1.127  1.000
## 6:        0     5   0    NA     NA      0      NA  1.338  1.661
##     COPEPrb COPEPrc COPEExp COPEDis
## 1:       NA      NA      NA      NA
## 2:    2.257   2.378   2.414   2.181
## 3:       NA      NA      NA      NA
```

```
## 4:        NA        NA        NA        NA
## 5:        NA        NA     2.034        NA
## 6:        NA        NA        NA        NA
```

Go ahead and try these same three functions on both the penguinsData and the texasData data sets. The examples in the next sections make more sense the more familiar you are with the data being manipulated.

For now, we introduce one last function that is a counterpart to head(). The tail() function shows the *last* six rows of data:

```
tail(penguinsData)
```

```
##      species island bill_length_mm bill_depth_mm flipper_length_mm
## 1: Chinstrap  Dream           45.7          17.0               195
## 2: Chinstrap  Dream           55.8          19.8               207
## 3: Chinstrap  Dream           43.5          18.1               202
## 4: Chinstrap  Dream           49.6          18.2               193
## 5: Chinstrap  Dream           50.8          19.0               210
## 6: Chinstrap  Dream           50.2          18.7               198
##    body_mass_g    sex year
## 1:        3650 female 2009
## 2:        4000   male 2009
## 3:        3400 female 2009
## 4:        3775   male 2009
## 5:        4100   male 2009
## 6:        3775 female 2009
```

4.3 How Does data.table Work?

The data.table structure has three distinct areas. The first area operates on *rows* and the second on *columns*, and the last allows for *groups*. Like a mathematical matrix, these first two areas, being rows and columns, are often called i and j. The last, because it is **by** group, is often called the by. Thus, generic **d**ata.**t**able is often shown as DT[i, j, by]. Notice that instead of the usual function parentheses, we have square brackets []. To understand how to use data in this fashion, we explore together each area of the data table. As we progress through the data, remember it helps to understand what the raw data looked like in order to understand what these new techniques are showing.

How Do Row Operations Work?

The first area in DT[**i**, j, by] is for operations on row elements. This includes ordering rows (think alphabetical or numerical order) and selecting out rows with certain features (think subsetting).

order()

The order() function can be used to organize data in order. This function re-orders each *row* so the rows are in order. However, it must know which column in which to do this. Thus, even though it is a *row* operation, the order() function takes a *column* argument. Recall in the penguinsData, the flipper_length_mm column showed some different values for the last six rows. Based on your visual inspection of penguinsData, you noticed flipper_length_mm ranged from 172 to 231, yet they were not in order. We use the first *row* operation or *i*th space of the DT[**i**, j, by] structure to order() the flipper_length_mm from smallest to largest. We call this *increasing* order:

```
penguinsData[order(flipper_length_mm)]
```

```
##      species    island bill_length_mm bill_depth_mm flipper_length_mm
##   1:  Adelie    Biscoe           37.9          18.6               172
##   2:  Adelie    Biscoe           37.8          18.3               174
##   3:  Adelie Torgersen           40.2          17.0               176
##   4:  Adelie     Dream           39.5          16.7               178
##   5:  Adelie     Dream           37.2          18.1               178
## ---
## 340:  Gentoo    Biscoe           51.5          16.3               230
## 341:  Gentoo    Biscoe           55.1          16.0               230
## 342:  Gentoo    Biscoe           54.3          15.7               231
## 343:  Adelie Torgersen             NA            NA                NA
## 344:  Gentoo    Biscoe             NA            NA                NA
##      body_mass_g    sex year
##   1:        3150 female 2007
##   2:        3400 female 2007
##   3:        3450 female 2009
##   4:        3250 female 2007
##   5:        3900   male 2007
```

```
## ---
## 340:          5500   male 2009
## 341:          5850   male 2009
## 342:          5650   male 2008
## 343:            NA   <NA> 2007
## 344:            NA   <NA> 2009
```

Notice we have not yet used the assignment operator `<-` on our data. The preceding code does **not** *permanently* change the order of `penguinsData`. Notice that `flipper_length_mm` as follows is **not** in order:

```
head(penguinsData)
```

```
##    species    island bill_length_mm bill_depth_mm flipper_length_mm
## 1:  Adelie Torgersen           39.1          18.7               181
## 2:  Adelie Torgersen           39.5          17.4               186
## 3:  Adelie Torgersen           40.3          18.0               195
## 4:  Adelie Torgersen             NA            NA                NA
## 5:  Adelie Torgersen           36.7          19.3               193
## 6:  Adelie Torgersen           39.3          20.6               190
##    body_mass_g    sex year
## 1:        3750   male 2007
## 2:        3800 female 2007
## 3:        3250 female 2007
## 4:          NA   <NA> 2007
## 5:        3450 female 2007
## 6:        3650   male 2007
```

If our work required us to *permanently* change our data, we would need to assign that temporary change permanently. In the prior code (`penguinsData[order(flipper_length_mm)]`), we saw output to our console, and yet there were no permanent changes. In contrast, in the following code, there is no output to the console. All the same, our data are now in increasing order based on flipper length:

```
penguinsData <- penguinsData[order(flipper_length_mm)]
```

We can confirm this change if we need to by just calling penguinsData. As you see, the change is now indeed permanent:

```
penguinsData

##        species    island bill_length_mm bill_depth_mm flipper_length_mm
##    1:  Adelie    Biscoe           37.9          18.6               172
##    2:  Adelie    Biscoe           37.8          18.3               174
##    3:  Adelie Torgersen           40.2          17.0               176
##    4:  Adelie     Dream           39.5          16.7               178
##    5:  Adelie     Dream           37.2          18.1               178
##  ---
##  340:  Gentoo    Biscoe           51.5          16.3               230
##  341:  Gentoo    Biscoe           55.1          16.0               230
##  342:  Gentoo    Biscoe           54.3          15.7               231
##  343:  Adelie Torgersen             NA            NA                NA
##  344:  Gentoo    Biscoe             NA            NA                NA
##       body_mass_g    sex year
##    1:        3150 female 2007
##    2:        3400 female 2007
##    3:        3450 female 2009
##    4:        3250 female 2007
##    5:        3900   male 2007
##  ---
##  340:        5500   male 2009
##  341:        5850   male 2009
##  342:        5650   male 2008
##  343:          NA   <NA> 2007
##  344:          NA   <NA> 2009
```

This ability to control when we keep or do not keep a change is very useful. It allows us to explore our data without changing anything. While penguinsData is very short at only 344 rows of data, our acesData has over 6,500 rows. Far too long to look through one at a time, yet R happily does analysis on however many rows we need.

Common uses of order() include organizing by either numeric or alphabetical order. If we place a minus sign (-) in front of the column name, then we get *decreasing order*. In this example, we use order() to sort penguinsData in decreasing, alphabetical order. Notice we do not assign our data to a variable, so this is an example of a temporary use of our order function. Notice also `flipper_length_mm` is now *out of order*. This is because even though we are organizing alphabetically on `species`, R knows we want to keep the data intact for each observation:

```
penguinsData[order(-species)]

##      species    island bill_length_mm bill_depth_mm flipper_length_mm
##   1:  Gentoo    Biscoe           48.4          14.4               203
##   2:  Gentoo    Biscoe           45.1          14.5               207
##   3:  Gentoo    Biscoe           44.0          13.6               208
##   4:  Gentoo    Biscoe           48.7          15.7               208
##   5:  Gentoo    Biscoe           42.7          13.7               208
## ---
## 340:  Adelie    Biscoe           41.0          20.0               203
## 341:  Adelie     Dream           41.1          18.1               205
## 342:  Adelie     Dream           40.8          18.9               208
## 343:  Adelie Torgersen           44.1          18.0               210
## 344:  Adelie Torgersen             NA            NA                NA
##      body_mass_g    sex year
##   1:        4625 female 2009
##   2:        5050 female 2007
##   3:        4350 female 2008
##   4:        5350   male 2008
##   5:        3950 female 2008
## ---
## 340:        4725   male 2009
## 341:        4300   male 2008
## 342:        4300   male 2008
## 343:        4000   male 2009
## 344:          NA   <NA> 2007
```

More than one column at a time can be put into the `order()` function. In this case, the first column is ordered first, and then the second column is ordered *as much as possible without breaking the first column's order*. While here we only show two columns, in practice, this can be extended to as many columns as needed:

```
penguinsData[order(-species, flipper_length_mm)]
##      species    island bill_length_mm bill_depth_mm flipper_length_mm
##   1:  Gentoo    Biscoe           48.4          14.4               203
##   2:  Gentoo    Biscoe           45.1          14.5               207
##   3:  Gentoo    Biscoe           44.0          13.6               208
##   4:  Gentoo    Biscoe           48.7          15.7               208
##   5:  Gentoo    Biscoe           42.7          13.7               208
## ---
## 340:  Adelie    Biscoe           41.0          20.0               203
## 341:  Adelie     Dream           41.1          18.1               205
## 342:  Adelie     Dream           40.8          18.9               208
## 343:  Adelie Torgersen           44.1          18.0               210
## 344:  Adelie Torgersen             NA            NA                NA
##      body_mass_g    sex year
##   1:        4625 female 2009
##   2:        5050 female 2007
##   3:        4350 female 2008
##   4:        5350   male 2008
##   5:        3950 female 2008
## ---
## 340:        4725   male 2009
## 341:        4300   male 2008
## 342:        4300   male 2008
## 343:        4000   male 2009
## 344:          NA   <NA> 2007
```

Examples where this type of multicolumn order could make sense might be an order on family name, given name, and middle name. In some data used by one of us authors, it is common to order by academic semester, course subject, course number, and then student identifier.

Subsetting

Selecting or *subsetting* certain rows out of the data set is also a useful skill that is a row operation. Thus, we use the first *row* operation or *i*th space of the `DT[i, j, by]` for subsetting. Subsets are achieved by using *logical* operators (sometimes called Boolean operators). The conditions for `TRUE` are written up, and then, the rows that evaluate to

TRUE are selected. To set these conditions, we use operators from both base R and our `extraoperators` package. As a reference, before we give some hands-on examples with code, we show the usual operators and functions in Table 4-1.

Note Interval notation is a compact way to very precisely define the interval of numbers you want. For example, suppose you have a set of numbers: 1, 2, 3, 4, 5. In words, we could write "numbers that are greater than 3 and less than 5," which in our little set would give us the number 4. We could also say "numbers that are at least 3 and less than 5," which in our set of numbers would give us the numbers 3 and 4. The written way of describing these is relatively long, so people often use interval notation. Interval notation uses parentheses, "(" and ")", as well as brackets, "[" and "]". Parentheses indicate greater than, "(", or less than, ")". Brackets indicate greater than or equal to, "[", or less than or equal to, "]". For example, the interval notation for the words "numbers that are greater than 3 and less than 5" is (3, 5). The interval notation for the words "numbers that are greater than or equal to 3 and less than 5" is [3, 5). Many of the logical operators in Table 4-1 are used to select specific intervals.

Table 4-1. *Logical Functions*

Function	What It Does
`==`	Are values/vectors equal?
`!=`	Are values/vectors NOT equal?
`%like%`	Are values like the characters in quotes?
< OR `%l%`	Less than.
<= OR `%le%`	Less than or equal.
> OR `%g%`	Greater than.
>= OR `%ge%`	Greater than or equal.
`%gl%`	Greater than AND less than, (x, y).
`%gel%`	Greater than or equal AND less than, [x, y).

(continued)

Table 4-1. (*continued*)

Function	What It Does
%gle%	Greater than AND less than or equal, (x, y].
%gele%	Greater than or equal AND less than or equal, [x, y].
%in%	In.
%nin%	Not in.
%c%	Chain operations on the RHS together.
%e%	Set operator, to use set notation.
is.na()	A function that selects missing data.

Equality is a fairly common logical operation, and here we attempt this on texasData. Much like order(), even though this is a *row* operation, we must perform it inside a particular *column*. Thus, using the CountyName column, we select *only* those rows that are **exactly matching** Victoria. Victoria happens to be a county in Texas that is home to a college where one of us works. Again, please note that as we have not assigned this result to any variable with the <- operator, we are currently only exploring our data. There is no change to the number of observations in the data set (as may be confirmed in the top-right Environment pane):

```
texasData[CountyName == "Victoria"]
```

```
##    CountyName FIPSNumber CountyNumber PublicHealthRegion
## 1:   Victoria        469          235                  8
##    HealthServiceRegion MetropolitanStatisticalArea_MSA
## 1:                   8                        Victoria
##    MetropolitanDivisions MetroArea NCHSUrbanRuralClassification_2006
## 1:                    --     Metro                       Small Metro
##    NCHSUrbanRuralClassification_2013
## 1:                       Small Metro
```

Sometimes, rather than an exact *match*, we want a **non**-match. This is called "does not equal"; in R, **not** is coded with an exclamation point "!". In this case, back in our penguinsData, we select only data **not** in the Adelie species:

```
penguinsData[species != "Adelie"]
```

```
##         species island bill_length_mm bill_depth_mm flipper_length_mm
##    1: Chinstrap  Dream           46.1          18.2               178
##    2: Chinstrap  Dream           58.0          17.8               181
##    3: Chinstrap  Dream           42.4          17.3               181
##    4: Chinstrap  Dream           47.0          17.3               185
##    5: Chinstrap  Dream           43.2          16.6               187
## ---
## 188:     Gentoo Biscoe           52.1          17.0               230
## 189:     Gentoo Biscoe           51.5          16.3               230
## 190:     Gentoo Biscoe           55.1          16.0               230
## 191:     Gentoo Biscoe           54.3          15.7               231
## 192:     Gentoo Biscoe             NA            NA                NA
##      body_mass_g    sex year
##    1:        3250 female 2007
##    2:        3700 female 2007
##    3:        3600 female 2007
##    4:        3700 female 2007
##    5:        2900 female 2007
## ---
## 188:        5550   male 2009
## 189:        5500   male 2009
## 190:        5850   male 2009
## 191:        5650   male 2008
## 192:          NA   <NA> 2009
```

Before we discuss other row selection operators, there are two ways to *combine* logical values. One way is *and* which is coded with & and the other is *or* which is coded with |. The following code shows two different examples where the same individual logical values (e.g., species == "Gentoo" in addition to flipper_length_mm == 230) are joined first by *and* followed by *or*. The difference in results is instructive. Notice too that character values use quotation marks "" while numerical values do not:

```
penguinsData[species == "Gentoo" &
          flipper_length_mm == 230]

##    species island bill_length_mm bill_depth_mm flipper_length_mm
## 1:  Gentoo Biscoe           50.0          16.3               230
## 2:  Gentoo Biscoe           59.6          17.0               230
## 3:  Gentoo Biscoe           49.8          16.8               230
## 4:  Gentoo Biscoe           48.6          16.0               230
## 5:  Gentoo Biscoe           52.1          17.0               230
## 6:  Gentoo Biscoe           51.5          16.3               230
## 7:  Gentoo Biscoe           55.1          16.0               230
##    body_mass_g  sex year
## 1:        5700 male 2007
## 2:        6050 male 2007
## 3:        5700 male 2008
## 4:        5800 male 2008
## 5:        5550 male 2009
## 6:        5500 male 2009
## 7:        5850 male 2009

penguinsData[species == "Gentoo" |
          flipper_length_mm == 230]

##      species island bill_length_mm bill_depth_mm flipper_length_mm
##   1:  Gentoo Biscoe           48.4          14.4               203
##   2:  Gentoo Biscoe           45.1          14.5               207
##   3:  Gentoo Biscoe           44.0          13.6               208
##   4:  Gentoo Biscoe           48.7          15.7               208
##   5:  Gentoo Biscoe           42.7          13.7               208
## ---
## 120:  Gentoo Biscoe           52.1          17.0               230
## 121:  Gentoo Biscoe           51.5          16.3               230
## 122:  Gentoo Biscoe           55.1          16.0               230
## 123:  Gentoo Biscoe           54.3          15.7               231
## 124:  Gentoo Biscoe             NA            NA                NA
##      body_mass_g    sex year
```

```
##    1:          4625 female 2009
##    2:          5050 female 2007
##    3:          4350 female 2008
##    4:          5350   male 2008
##    5:          3950 female 2008
## ---
## 120:          5550   male 2009
## 121:          5500   male 2009
## 122:          5850   male 2009
## 123:          5650   male 2008
## 124:            NA   <NA> 2009
```

Armed with these data, we explore some more of the operators from Table 4-1. The ability to chain together multiple logical operators is quite useful. In this case, the `%like%` operator selects all rows in which `UserID` has a `19` in any position. This is an example of what is sometimes called a *fuzzy* match. It contrasts with the more exact nature of the other operators. Next, we require `STRESS` to be *greater than or equal to 4 and strictly less than 6*. Negative affect or `NegAff` must be *strictly less than 2*. Notice from Table 4-1 this could have been written just as well with `NegAff < 2`. Often, in `R`, as in life, there is more than one way. Lastly, `SurveyInteger` is required to be precisely `2`:

```
acesData[UserID %like% "19" &
           STRESS  %gel%  c(4, 6) &
           NegAff %l% 2 &
           SurveyInteger == 2]
```

```
##      UserID  SurveyDay SurveyInteger SurveyStartTimec11 Female Age
##  1:      19 2017-03-07             2             0.2080      1  24
##  2:     119 2017-02-25             2             0.2367      1  25
##  3:     190 2017-03-01             2             0.2042      0  23
##  4:     190 2017-03-05             2             0.2152      0  23
##  5:     190 2017-03-07             2             0.2224      0  23
##  6:     190 2017-03-09             2             0.2577      0  23
##  7:     190 2017-03-12             2             0.1990      0  23
##  8:     191 2017-02-25             2             0.1937      1  21
##  9:     191 2017-03-01             2             0.1875      1  21
## 10:     191 2017-03-02             2             0.1875      1  21
```

```
##      BornAUS SES_1 EDU SOLs WASONs STRESS SUPPORT PosAff NegAff
##  1:       1     7   1   NA     NA      5      NA  3.972  1.267
##  2:       1     8   1   NA     NA      4      NA  2.021  1.270
##  3:       0     5   1   NA     NA      5      NA  1.332  1.042
##  4:       0     5   1   NA     NA      5      NA  2.492  1.579
##  5:       0     5   1   NA     NA      4      NA  1.822  1.000
##  6:       0     5   1   NA     NA      5      NA  1.882  1.528
##  7:       0     5   1   NA     NA      5      NA  1.047  1.812
##  8:       0     6   0   NA     NA      5      NA  3.403  1.000
##  9:       0     6   0   NA     NA      4      NA  1.032  1.682
## 10:       0     6   0   NA     NA      4      NA  1.047  1.619
##     COPEPrb COPEPrc COPEExp COPEDis
##  1:      NA      NA      NA      NA
##  2:      NA      NA      NA      NA
##  3:      NA      NA      NA      NA
##  4:      NA      NA      NA      NA
##  5:      NA      NA      NA      NA
##  6:      NA      NA      NA      NA
##  7:      NA      NA      NA      NA
##  8:      NA      NA      NA      NA
##  9:      NA      NA      NA      NA
## 10:      NA      NA      NA      NA
```

Notice in this most recent example, there are several columns having NA. These NAs represent *missing data*. Due to the sometimes complex nature of NAs, we must be careful how we try to find missing data. Based on everything we learned so far, suppose we wanted to check if we have any participants who did **not** report their EDU status. In other words, do we have anyone of our survey participants in acesData who chose to not respond about education level? Based on what we have learned so far, it might seem that one of the two following lines of code should work. After all, we used similar techniques earlier with both `CountyName == "Victoria"` and `SurveyInteger == 2`:

```
acesData[EDU == "NA"]
```

```
## Empty data.table (0 rows and 19 cols): UserID,SurveyDay,SurveyInteger,
SurveyStartTimec11,Female,Age...
```

```
acesData[EDU == NA]

## Empty data.table (0 rows and 19 cols): UserID,SurveyDay,SurveyInteger,
SurveyStartTimec11,Female,Age...
```

Do not be tricked! Despite what the preceding lines of code seem to show, there are in fact many observations that are missing education status! Missing data are complicated. Now, we will not go into the reasons behind that in this text—it is a topic for another time. What you need to remember is missing data are tricky and R has a special function to cope with them. The function to use is the `is.na()` function. Since it is checking every row inside a specific column for missing data, it lives where we would expect, in the row part of our `data.frame`. As you can see, far from no rows with missing data, we have over 200 rows missing educational data!

```
acesData[is.na(EDU)]

##      UserID  SurveyDay SurveyInteger SurveyStartTimec11 Female Age
##    1:      2 2017-02-22             3                 NA     NA  NA
##    2:      2 2017-02-25             3                 NA     NA  NA
##    3:      3 2017-02-23             3                 NA     NA  NA
##    4:      3 2017-03-05             3                 NA     NA  NA
##    5:      4 2017-03-01             3                 NA     NA  NA
#  ---
## 216:    183 2017-02-23             3                 NA     NA  NA
## 217:    184 2017-03-07             3                 NA     NA  NA
## 218:    189 2017-02-22             3                 NA     NA  NA
## 219:    190 2017-02-28             3                 NA     NA  NA
## 220:    190 2017-03-04             3                 NA     NA  NA
##      BornAUS SES_1 EDU   SOLs WASONs STRESS SUPPORT PosAff NegAff
##    1:     NA    NA  NA 23.476      2     NA      NA     NA     NA
##    2:     NA    NA  NA 34.755      0     NA      NA     NA     NA
##    3:     NA    NA  NA  0.000      0     NA      NA     NA     NA
##    4:     NA    NA  NA  0.000      0     NA      NA     NA     NA
##    5:     NA    NA  NA  6.190      1     NA      NA     NA     NA
## ---
## 216:     NA    NA  NA 19.600      1     NA      NA     NA     NA
## 217:     NA    NA  NA 96.670      3     NA      NA     NA     NA
## 218:     NA    NA  NA  9.266      1     NA      NA     NA     NA
```

```
## 219:      NA    NA  NA  0.000      1      NA      NA      NA      NA
## 220:      NA    NA  NA  6.870      1      NA      NA      NA      NA
##       COPEPrb COPEPrc COPEExp COPEDis
##   1:      NA      NA      NA      NA
##   2:      NA      NA      NA      NA
##   3:      NA      NA      NA      NA
##   4:      NA      NA      NA      NA
##   5:      NA      NA      NA      NA
##  ---
## 216:      NA      NA      NA      NA
## 217:      NA      NA      NA      NA
## 218:      NA      NA      NA      NA
## 219:      NA      NA      NA      NA
## 220:      NA      NA      NA      NA
```

While we discussed it separately here for clarity, just like all prior logical operations, we can chain together `is.na()` by using & along with | to create subsets.

Spend some time reviewing what you just learned, to make sure you understand the building blocks used to create complex logical operations to select just the data you need. In the exercises, you have a chance to extend and modify each of the preceding examples to learn more how these work. After your practice is over, we move onward to the next part of `data.table`. In the next section, we will continue to see more logical operations even as we also learn more about `data.table`.

How Do Column Operations Work?

The second area in `DT[i, j, by]` is for *actions* on columns. This includes choosing only some columns and creating new columns.

Choosing Columns

You may have noticed our `acesData` and `texasData` have several columns. What if we were interested in only a few of those columns? We can select just the columns we want by using column operations in the *j*th position. To select just one column is easy enough. Simply type the column name in the *j*th spot. In order to do that, we need to let R know we are *skipping* the first or *i*th position. To skip, simply use a comma. Here, we take a look at only the data in the EDU column. We see there are 0s for *less than a tertiary degree*, 1s for *earned a tertiary education degree*, and NAs for missing data. Notice the comma in our code!

```
acesData[, EDU]
```

```
##   [1]  0  0  0  0  0  0  0  0  0  0  0  0  0  0  0  0  0  0  0  0  0
##  [22]  0  0  0  0  0  0  0  0  0 NA  0  0  0  0  0  0 NA  0  0  0  0
##  [43]  0  0  0  0  0  0  0  0  0  0  0  0  0  0  0  0  0  0  0  0  0
##  [64]  0  0 NA  0  0  0  0  0  0  0  0  0  0  0  0  0  0  0  0  0  0
##  [85]  0  0  0  0  0  0  0  0  0  0 NA  0  0  0  0  0  0 NA  0  0  0
## [106]  0  0  0  0  0  0  0  0  0  0  0  0  0  0  0  0  0  0  0  0  0
## [127]  0  0  0  0  0  0  0  0  0  0  0  0 NA  1  1  1  1  1  1  1  1
## [148]  1  1  1  1  1  1  1  1  1  1  1  1  1  1  1  1  1  1  1  1  1
## [169]  1  1  1  1  1  1  1 NA  0  0 NA  0  0  0  0  0  0  0  0  0  0
## [190]  0  0  0  0  0  0  0  0  0  0  0  0  0  0  0  0  0  0  0  0  0
## [211]  0  0 NA  0  0  0  0  0  0  0  0  0  0  0  0  0  0  0  0  0  0
## [232]  0  0  0  0  0  0  0  0  0  0  0  0 NA  0  0  0  0  0  0  0  0
## [253]  0  0  0  0  0  0  0  0 NA  0  0  0  0  0 NA  0  0  0  0  0  0
## [274]  0  0  0  0  0  0  0  0  0  0  0  0  0  0  0  0  0  0  0  0  0
## [295]  0  0  0  0  0  0  1  1  1  1  1  1  1  1  1  1  1  1  1  1  1
## [316]  1  1  1  1  1  1  1  1  1  1  1  1  1  1  1  1  1  1  0  0  0
## [337]  0  0  0  0  0  0  0  0  0  0 NA  0  0  0  0  0  0  0  0  0  0
## [358]  0  0  0  0  0  0  0  0  0  0  0  0 NA  0  0  0  0  0  0  0  0
## [379]  0  0  0  0  0  0  0  0  0  0  0  0  0  0  0  0  0  0  0  0  0
## [400]  0  0  0  0  0  0  0 NA  1  1  1  1  1  1  1  1  1  1  1  1  1
## [421]  1  1  1  1  1  1  1  1  1  1  1  1  1  1  1  1  1  1  1  1  1
## [442]  1  1 NA  0  0  0  0  0  0  0  0  0  0  0  0  0  0  0  0  0  0
## [463]  0  0  0  0  0  0  0  0  0  0  0  0  0  0 NA  1  1  1  1  1  1
## [484]  1  1  1  1  1  1  1  1  1  1  1  1  1  1  1  1  1  1  1  1  1
## [505]  1  1  1  1  1  1  1  1  1 NA  0  0  0  0  0  0  0  0  0  0  0
## [526] NA  0  0 NA  0  0  0  0  0  0  0  0  0  0  0 NA  0  0  0  0 NA
## [547]  1  1  1  1  1  1  1  1  1  1  1  1  1  1  1  1  1  1  1  1  1
## [568]  1  1  1  1  1  1  1  1  1  1  1  1  1  1 NA  0  0  0  0  0  0
```

```
## [589]  0  0  0  0  0  0  0  0  0  0  0  0  0  0  0  0  0  0  0  0  0
## [610]  0  0  0  0  0  0  0  0  1  1  1  1  1  1  1  1  1  1  1  1  1
## [631]  1  1  1  1  1  1  1  1  1  1  1  1  1  1  1  1  1  1  1  1  1
## [652]  1  1  1 NA  0  0  0  0  0  0  0  0  0  0  0  0  0  0  0  0  0
## [673]  0  0  0  0  0  0  0  0  0  0  0  0  0  0  0  0  0  0  0  0  0
## [694]  0 NA  0  0  0  0  0  0  0  0  0  0  0  0  0  0  0  0  0  0  0
## [715]  0  0  0  0  0  0  0  0  0  0  0  0  0  0  0  0  0 NA  0  0  0
## [736]  0  0  0  0  0  0  0  0  0  0  0  0  0  0  0  0  0  0 NA  0  0
## [757]  0  1  1  1  1  1  1  1  1  1  1  1  1  1  1  1  1  1  1  1  1
## [778]  1  1  1  1  1  1  1  1  1  1  1  1  1  1  1 NA  0  0  0  0  0
## [799]  0  0  0  0  0  0  0  0  0  0  0  0  0  0  0  0  0  0  0  0  0
## [820]  0  0  0  0  0  0  0  0  0  0  0  0  0  0  0  0  0  0  0  0  0
## [841]  0  0  0  0  0  0  0  0  0  0  0  0  0  0  0  0  0  0  0  0  0
## [862]  0  0  0  0  0  0  0  0  0  0  0  0  0  0  0  0  0  0  0  0  0
## [883]  0  0  0  0  0  0  0  0  0  0  0  0  0  0  0  0  0  0  0  0  0
## [904]  0  0  0  0  0  0  0  0  0  0  0  0  0  0  0  0  0  0  0  0  0
## [925] NA  0  0  0  0  0  0  0  0  0  0  0  0  0  0  0  0  0  0  0  0
## [946]  0  0  0  0  0  0  0  0  0  0  0  0  0  0  0  0  0  0  0 NA  1
## [967]  1  1  1  1  1  1  1  1  1  1  1  1  1  1  1  1  1  1  1  1  1
## [988]  1  1  1  1  1  1  1  1  1  1  1  1  1
##  [ reached getOption("max.print") -- omitted 5599 entries ]
```

Note that when selecting a single column, whether we use the brackets or dollar sign yields the same result. To avoid showing so much text, we will show just the unique values, using `unique()`:

```
unique(acesData$EDU)

## [1]  0 NA  1

unique(acesData[, EDU])

## [1]  0 NA  1
```

Of course, we often want more than one column. To collect more than one column, use `.()` to list which columns are needed. Again, remember that column operations belong in the second position, so a leading comma is required to signal the "empty" first space:

```
acesData[, .(UserID, STRESS, NegAff, SurveyInteger, EDU, SES_1)]

##       UserID STRESS NegAff SurveyInteger EDU SES_1
##    1:      1      5  1.669             2   0     5
##    2:      1      1  1.000             3   0     5
##    3:      1      1     NA             1   0     5
##    4:      1      2  1.357             2   0     5
##    5:      1      0  1.000             3   0     5
## ---
## 6595:    191      3  1.767             3   0     6
## 6596:    191      0  1.705             1   0     6
## 6597:    191      3  1.162             2   0     6
## 6598:    191      4  1.153             3   0     6
## 6599:    191      0  1.171             1   0     6
```

Just because we have been leaving our first space empty does not mean it must always be empty. From our work on subsets, recall our fairly complex set of logic operations. Start with the logical operations on `UserID`, `STRESS`, `NegAff`, and `SurveyInteger`. Combine those logical subsets by row with our column selection by using a comma. Notice we include a *comment* in the following code by using #. A code *comment* tells R to ignore the text after the #. At the same time, the comment helps us read our code and see exactly where we are combining row subsets with column selection:

```
acesData[UserID %like% "19" &
           STRESS %gel% c(4, 6) &
           NegAff %l% 2 &
           SurveyInteger == 2, #Pay Attention to the comma!
           .(UserID, STRESS, NegAff, SurveyInteger, EDU, SES_1)]

##      UserID STRESS NegAff SurveyInteger EDU SES_1
##   1:     19      5  1.267             2   1     7
##   2:    119      4  1.270             2   1     8
```

```
##  3:         190        5  1.042              2   1    5
##  4:         190        5  1.579              2   1    5
##  5:         190        4  1.000              2   1    5
##  6:         190        5  1.528              2   1    5
##  7:         190        5  1.812              2   1    5
##  8:         191        5  1.000              2   0    6
##  9:         191        4  1.682              2   0    6
## 10:         191        4  1.619              2   0    6
```

Spend a bit of time studying the preceding code. You are already learning to use quite complex techniques! Between row subsets and column selection, quite a lot can be achieved with these two *i*th and *j*th positions of our data table: `DT[i, j, by]`.

Creating Columns

Notice in the preceding example there are ten observations. These ten observations seem to only have four unique participants (based on `UserIDs`). Sometimes, it can help to count. Of course, this was a restricted enough subset we can do this via inspection. However, that is not always feasible. Creating a count of how many observations or rows is creating a new column. Keep in mind, in your top-right Environment pane, rows are observations and columns are *variables*. So a count of how many rows there are is creating a new variable. Thus, it is a *column* operation. To get a count, we use `.N`. In statistics, *N* usually stands for total population count. It makes sense to use this to count our entire data set. Since it is a column operation, we must use the column or *j*th position. Remember to use a comma to skip over the row operation space!

```
acesData[, .N]
```

```
## [1] 6599
```

Naturally, this gives us the count of *every* row in `acesData`. If we want to automatically count the ten rows in our subset, we need to fill the row operation space with our logical operations. We use our familiar restrictions, swapping out choosing some columns for *counting*:

```
acesData[UserID %like% "19" &
           STRESS %gel% c(4, 6) &
           NegAff %l% 2 &
```

```
          SurveyInteger == 2, #Pay Attention to the comma!
        .N]
```

```
## [1] 10
```

What if, instead of counting the total number of rows, we wanted to count how many unique participants there were who fit these criteria? In that case, we need to use the `uniqueN()` function. Because we are counting unique participants and because each participant has a unique `UserID`, we need to let our column counting function know in which column to check. Again, compare and contrast the following example with the prior example. We swap out `.N` for `uniqueN(UserID)` which answers a different question:

```
acesData[UserID %like% "19" &
          STRESS %gel% c(4, 6) &
          NegAff %l% 2 &
          SurveyInteger == 2, #Pay Attention to the comma!
        uniqueN(UserID)]
```

```
## [1] 4
```

Is this making sense? If it is, great! If not, that is okay too. Learning the R **language** is a lot like learning any other language. It takes time to learn how to write and speak. If you are feeling a bit stuck, go back over these examples a second time. Explore making a change here or there to see what happens. For example, what happens in this latest example when we switched `.N` for `uniqueN(UserID)`? What was the original subset looking like? Go back and see what happened in the last section, when we were looking at all ten rows of data. Never be shy about asking for help. If you are in one of our classes, we enjoy answering students' questions. If you are in someone else's class, we promise most instructors and professors enjoy answering questions (feel free to show them this sentence if you feel shy).

After taking a ten-second break to fortify ourselves, we proceed on to some more column creation. So far, we have not actually built a single new column. We did some counting, yet we did not actually make a new column. It is time to change that by talking about *recoding*. Sometimes, the raw data we collect is not quite what we want. For example, in `acesData`, the `STRESS` scale ranges from 0 to 10. What if we instead wanted to break participants into three groups for low, medium, and high stress? That is an example of *recoding*. To do this requires creating a new column and then assigning values to that column based on our `STRESS` scale. To build a new column, we use the column assignment operator :=. We call our new column STRESS_3. Using our logical row operations, we split them into three ranges: [0, 3] for *low*, (3, 6] for *medium*, and (6, 10] for *high*.

In the following code, take a moment to notice several key features. First, the row selection operation of %gele% stands for "greater than or equal" and "less than or equal." It captures the [0, 3] inclusive range for our low-stress group. If necessary, review Table 4-1. Next, notice the commas, in particular the commas before STRESS_3. They separate the row operation area from the column operation area. Go ahead and run this code:

```
acesData[STRESS %gele% c(0, 3),
         STRESS_3 := "low"]
```

You ran the preceding code, and nothing seems to have happened. As is usual for *assignment*, nothing prints to the console. However, behind the scenes, things have in fact changed. First, in the Environment pane, you should see a circular arrow in the top right. Click that arrow. When you do, the 19 columns of acesData will refresh, showing you that there are now 20 columns. We can view the revised data set by using the View() function, which will bring up the data viewer in RStudio:

```
View(acesData) ## no output, opens RStudio data viewer
```

If we look at the variable STRESS_3, only certain rows have values with the rest missing. Why? We created the new STRESS_3 variable but only assigned the value "low" to it for certain levels of stress. Other levels of stress have no values assigned so are missing. To code the remaining stress scores into categories, we can run the following code:

```
acesData[STRESS %gle% c(3, 6),
         STRESS_3 := "medium"]

acesData[STRESS %gle% c(6, 10),
         STRESS_3 := "high"]
```

Note that we chose to use %gle% and not %gele% like we did when creating the assignment for "low" in STRESS_3. This is important because we need to be careful how boundary values are coded. For example, if we call values that are exactly 3 "low," we would not also want to include them under "medium," or the "low" value will get overwritten. Overall, this creates the following intervals: [0, 3], (3, 6], (6, 10]. This is exactly what we would want to get all values from 0 to 10, the possible range of stress scores, labeled with one and only one term, "low," "medium," or "high." We can view the revised data set by using the `View()` function, which will bring up the data viewer in RStudio. Now we should see `STRESS_3` values for every row:

```
View(acesData) ## no output, opens RStudio data viewer
```

Next, take a look at the following code. We are almost perfectly recycling some prior code we ran. Can you spot the difference? Look closely, and you see we added our new column's name, `STRESS_3`. When you run this code, confirm that our participants all have medium stress. This makes sense, because their numerical stress scores are 4s and 5s. So far, so good!

```
acesData[UserID %like% "19" &
           STRESS %gel% c(4, 6) &
           NegAff %l% 2 &
           SurveyInteger == 2, #Pay Attention to the comma!
         .(UserID, STRESS, STRESS_3, NegAff, SurveyInteger, EDU, SES_1)]

##     UserID STRESS STRESS_3 NegAff SurveyInteger EDU SES_1
##  1:     19      5   medium  1.267             2   1     7
##  2:    119      4   medium  1.270             2   1     8
##  3:    190      5   medium  1.042             2   1     5
##  4:    190      5   medium  1.579             2   1     5
##  5:    190      4   medium  1.000             2   1     5
##  6:    190      5   medium  1.528             2   1     5
##  7:    190      5   medium  1.812             2   1     5
##  8:    191      5   medium  1.000             2   0     6
##  9:    191      4   medium  1.682             2   0     6
## 10:    191      4   medium  1.619             2   0     6
```

Empowered with the ability to select specific rows, choose specific columns, and create or count columns, we are ready to explore the last spot of our data.

How Do by Operations Work?

The third area in `DT[i, j, by]` is for *groups*. This area we are short changing a bit - it can do much more. However, your main goal is to learn statistical methods, not `R` or `data.table`. Thus, for now, forgive us our lack of completeness.

Remember commas are used to separate out each part. Suppose we want to count how many of each level of stress there was? Recall we use `.N` to count items. The new code we use is in the last, third position, and that is `by = STRESS_3`. In this case, our count is going to be by each of our three groups—namely, creating group counts for each of low, medium, and high stress:

```
acesData[, .N, by = STRESS_3]

##    STRESS_3    N
## 1:   medium 1240
## 2:      low 4625
## 3:     high  537
## 4:     <NA>  197
```

Did the output match what you expected? Are you surprised by the `<NA>`? From where did that arrive? Remember `STRESS_3` is a recoded or calculated column we built ourselves. It is based on the `STRESS` column. Remember we want to understand why some of the `STRESS_3` data are missing. To answer this question at scale, we must use each of our three skills. First, we only pick the rows where `STRESS_3` has missing data. This is a row operation and uses `is.na()`. Second, we need to count up what is happening. This is a column operation and we continue to use `.N`. Third, we use `by =` again, only this time on the original `STRESS`:

```
acesData[is.na(STRESS_3),
         .N,
         by = STRESS]

##    STRESS   N
## 1:     NA 197
```

As you can see, the result is our original `STRESS` data had some missing data. To contrast, take a look at what happens when we set `STRESS_3 == "low"`. Now we see what we would expect. Namely, the low stress measure comes from raw `STRESS` scores whose values are greater than or equal to 0 and less than or equal to 3:

```
acesData[STRESS_3 == "low",
         .N,
         by = STRESS]
```

```
##     STRESS    N
## 1:       1 1075
## 2:       2  716
## 3:       0 2146
## 4:       3  688
```

You now have seen the three ways to use and manipulate table data. These building blocks are all the tools you need to use real-world information in your statistical learning. Before you head into the exercises at the end of this chapter, we show some examples of using these table operations.

4.4 Examples

It is often said there is a difference between teaching and learning. Right now, technically, we have *taught* you everything needed. That does not mean it makes sense yet! We are not even sure it could make sense yet. Seeing is different from understanding, and both take practice and experience to master. However, know we are confident you can do this. It just takes some work, and that is okay. Before we send you off to the exercises to get some practice, we want to walk through three examples together. Instead of starting with a *technique*, we start each example with a **question**. Starting with a question helps, because that is how data science usually starts in the real world.

Example 1: Metropolitan Area Counts

Considering the 254 observations in `texasData`, *how many counties are in each type of* `MetroArea`*?* To answer this question, we want to think through which of our three areas of `i/j/by` we may need.

Considering the `i` or row operations, there is no need to subset by row. Each row of data in `texasData` holds precisely one Texas county. There *may* be value in imposing an `order()` on our data. In particular, if we are discussing the types of `MetroAreas`, we may wish to order on that value, rather than the default alphabetical order on `CountyName`.

Considering the `j` or column operations, we must count counties. This can be achieved with either `.N` or perhaps `uniqueN(CountyNumber)`. If we use the first, we are supposing our data does not have duplicated rows. In this case, that is a fair assumption.

However, in the acesData, notice each UserID for our participants shows up multiple times. Often, if there is an identity column of some sort, uniqueN() is safer.

Considering the by or group operations, we must perform our count by MetroArea.

Putting these three parts together gives us a solution to our question. In this case, we did not give a name to our newly created column. Thus, it was automatically named V1 for us. Take some time to experiment with this solution, perhaps by removing the order(MetroArea) piece. What happens?

```
texasData[order(MetroArea),
          uniqueN(CountyNumber),
          by = MetroArea]

##     MetroArea  V1
## 1:      Metro  82
## 2:  Non-Metro 172
```

What happens if we add some more columns to our order() function as a row operation? What happens if we add another column to our by = .() group operation? Notice the list function, .(), works in the third area just as well as in the second area to select more than one column:

```
texasData[order(MetroArea, NCHSUrbanRuralClassification_2013),
          uniqueN(CountyNumber),
          by = .(MetroArea, NCHSUrbanRuralClassification_2013)]

##    MetroArea NCHSUrbanRuralClassification_2013  V1
## 1:     Metro               Large Central Metro   6
## 2:     Metro                Large Fringe Metro  29
## 3:     Metro                      Medium Metro  25
## 4:     Metro                       Small Metro  22
## 5: Non-Metro                      Micropolitan  46
## 6: Non-Metro                          Non-core 126
```

Example 2: Metropolitan Statistical Areas (MSAs)

Thinking about the results from Example 1, we may begin to have more questions about our data. *How many counties tend to be in each "small metro" area?*

First, some background may be in order. A Metropolitan Statistical Area (MSA) is often set by the US Census Bureau. The goal is to group an area together that is predominantly influenced by a specific city or town. In the case of "small metro" areas, these are smaller cities that are still influential enough to be the hub or crossroads of their local community.

Considering the `i` or row operations, there is now a need to subset by row. In this case, we have some choices. We can either explore our data set and confirm we need `NCHSUrbanRuralClassification 2013 == "Small Metro"` or we can use our `%like%` operator to just get those classified with the word `"Small"`. In this case, either way would work. Often, there is more than one right answer.

Considering the `j` or column operations, we must count counties. This can be achieved with either `.N` or perhaps `uniqueN(CountyNumber)`. If we use the first, we are supposing our data does not have duplicated rows. In this case, that is a fair assumption. However, in the `acesData`, notice each `UserID` for our participants shows up multiple times. Often, if there is an identity column of some sort, `uniqueN()` is safer.

Considering the by or group operations, we must perform our count by `MetropolitanStatisticalArea MSA`.

In our solution, we have 11 MSAs that have between 1 and 3 counties in each of them. Keep in mind these are all "small" metro areas:

```
texasData[NCHSUrbanRuralClassification_2013 %like% "Small",
          uniqueN(CountyNumber),
          by = .(MetropolitanStatisticalArea_MSA)]
```

```
##    MetropolitanStatisticalArea_MSA V1
##  1:                  Wichita Falls  3
##  2:                      Texarkana  1
##  3:          College Station-Bryan  3
##  4:                        Abilene  3
##  5:                         Odessa  1
##  6:                       Victoria  2
##  7:                Sherman-Denison  1
##  8:                       Longview  3
```

```
## 9:                      San Angelo  2
## 10:                        Midland  2
## 11:                          Tyler  1
```

Example 3

How many higher or not-low stress (STRESS ¿ 3) reports were made by not yet college and college-educated participants?

Considering the i or row operations, there is a need to subset by row. We must require STRESS > 3 as a row operation.

Considering the j or column operations, we must count each row as each row is a unique report. This requires .N.

Considering the by or group operations, we must perform our count by the EDU level.

Notice our solution is **not** by unique participants. Rather, our solution is by specific instances of not-low stress:

```
acesData[STRESS > 3,
         .N,
         by = .(EDU)]
```

```
##     EDU    N
## 1:    0 1150
## 2:    1  620
## 3:   NA    7
```

4.5 Summary

This chapter explored the three areas of DT[i, j, by]. Understanding and using these areas to perform manipulations of raw data are key to successfully applying statistical methods to real-world data. Table 4-2 should be a good reference of the key ideas you learned in this chapter.

Table 4-2. Chapter Summary

Idea	What It Means
`<-`	New data/variable/item assignment operator.
`colnames()`	Prints all column names of a table.
`unique()`	Prints only unique items (e.g., only once per item).
`head()`	Prints the first six rows of a table.
`tail()`	Prints the last six rows of a table.
DT[**i**, j, by]	The first/ith/row operation area of a `data.table`.
`==`	Are values/vectors equal?
`!=`	Are values/vectors NOT equal?
`%like%`	Are values like the characters in quotes?
`<` **OR** `%l%`	Less than.
`<=` **OR** `%le%`	Less than or equal.
`>` **OR** `%g%`	Greater than.
`>=` **OR** `%ge%`	Greater than or equal.
`%gl%`	Greater than AND less than.
`%gel%`	Greater than or equal AND less than.
`%gele%`	Greater than AND less than or equal.
`%gle%`	Greater than or equal AND less than or equal.
`%in%`	In.
`%nin%`	Not in.
`%c%`	Chain operations on the RHS together.
`%e%`	Set operator, to use set notation.
`is.na()`	A row operation function that selects the rows of a specified column with missing data.
&	Logical *and* operator used to combine two or more of the preceding operations on rows.

(*continued*)

Table 4-2. (*continued*)

Idea	What It Means
|	Logical *or* operator used to combine two or more of the preceding operations on rows.
`order()`	A row operation function that orders the rows of a specified column.
`DT[i, **j**, by]`	The second/*j*th/column operation area of a `data.table`.
`.()`	Used to choose more than one column.
`.N`	Row count function – counts all rows.
`uniqueN()`	Row count function – counts unique rows.
`:=`	New column assignment operator.
`DT[i, j, **by** =]`	The third/by/group operation area of a `data.table`.

4.6 Practice for Mastery

Check your progress and grow through practice by working through some exercises. Comprehension checks ask critical thinking questions that may be best answered with a written or verbal response. Part of the art of statistics is successfully communicating results to your stakeholders or audience. Sometimes that audience is highly technical and other times very much not technical. Exercises are more direct applications of the concepts explored in the chapter.

Comprehension Checks

1. Why is it important that a function such as `order()` keep each row intact when it organizes? Have you ever used a spreadsheet software and accidentally only ordered a single column?

2. How and when do `data.table` operations *permanently* change the table data? Compare and contrast these bits of code: `<-`, `order()`, `:=`, and `uniqueN()`. Do any of these *always* change the table? Do any of these only *sometimes* change the table?

Exercises

1. Which bit of code gives you the correct result?

   A. `acesData[STRESS == NA]`

   B. `acesData[is.na(STRESS)]`

2. In *Example 2*, we used `NCHSUrbanRuralClassification_2013` in our solution. What happens if you change that for `NCHSUrbanRuralClassification_2006`?

3. In the `texasData`, are there any cases where the classification from 2006 to 2013 for the Urban vs. Rural differs? In other words, are there cases where `NCHSUrbanRuralClassification_2006` is not the same as `NCHSUrbanRuralClassification_2013`? Hint: Use `!=`.

CHAPTER 5

Data and Samples

Think about a time when you heard someone talking in a different language. Maybe you listened to the voice tones or the volume or the flow of words. Maybe you almost started thinking you could hear a familiar word or two, only to realize not. Statistics can sometimes feel like you are listening to a different language. In fact, in many ways, it *is* a different language. Just like an English class has nouns and verbs and a dictionary, so too statistics has new ideas that need a dictionary definition. It also has new verbs – new actions – that you must practice. No one is good at playing guitar or violin the first time they try. No one starts off good at football or swimming. Most likely, you, like us, "cooked" more than one meal that was not exactly good eating. Statistics is exactly the same way; no one starts off being good at *thinking statistically*. It is a new language, a new skill set, and a new way of thinking about the world around us. All the same, just like learning all those other skills, we think one of the best ways to learn is by seeing and doing statistics rather than just talking about it.

Remember you want to make sure you are on the right track as you explore statistics. Check and make sure that by the end of this chapter, you are able to

- Apply the definitions and terminology of the language of statistics to familiar data.
- Analyze different types of data.
- Understand how statistical research works.
- Evaluate sample selection methods.
- Create frequency tables from sample data.

M. Wiley and J. F. Wiley, *Beginning R 4*, https://doi.org/10.1007/978-1-4842-6053-1_5

5.1 R Setup

As usual, to continue practicing creating and using projects, we start a new project for this chapter. If you are already feeling comfortable starting projects in R, go ahead and skim this part.

If necessary, review the steps in Chapter 1 to create a new project. After starting RStudio, on the upper-left menu ribbon, select *File* and click *New Project*. Choose *New Directory* ➤ *New Project*, with *Directory name* `ThisChapterTitle`, and select *Create Project*. To create an R script file, on the top ribbon, right under the word *File*, click the small icon with a plus sign on top of a blank bit of paper, and select the *R Script* menu option. Click the floppy disk–shaped *save* icon, name this file `PracticingToLearn_XX.R` (where XX is the number of this chapter), and click *Save*.

In the lower-right pane, you should see your project's two files, and right under the *Files* tab, click the button titled *New Folder*. In the *New Folder* pop-up, type `data` and select *OK*. In the lower-right pane, click your new *data* folder. Repeat the folder creation process, making a new folder titled `ch05`.

Remember all packages used in this chapter were already installed on your local computing machine in Chapter 2. There is no need to re-install. However, this is a new project, and we are running this set of code for the first time. Therefore, you need to run the following `library()` calls:

```
library(data.table)

## data.table 1.13.0 using 6 threads (see ?getDTthreads). Latest news:
r-datatable.com

library(palmerpenguins)
library(JWileymisc)
library(extraoperators)
```

You are now free to explore statistics!

5.2 Populations and Samples

Thinking statistically is all about organizing and sorting a messy world around us into some sort of meaning. Sure, we use numbers (often) to make this happen. However, thinking statistically does **not** require numbers at all!

Take a moment and think of all the families in the world. Now take a moment and think of your family. Now take a moment and think of the family of someone you know. While the specifics of all families vs. one or two families can be somewhat fuzzy, can you think of some common similarities that all families might share? Can you think of at least one difference that your family has? Notice that we have two *levels* of family. One is big, and it is "all the families in the world." One is smaller, and it is your family or the family of someone you know.

This relationship between a large and comprehensive group (e.g., "all the families") and a small and specific group (e.g., "one or two families") happens often when we think statistically.

Consider some other examples of this type of large, total group vs. small, specific group. We could talk about all the people in your country vs. all the people in your hometown. We could talk about all the dogs vs. only Australian shepherds.

In statistics, we call the large, comprehensive set of things or group of people a **population**. The population is the real idea or group we want to understand better and study. However, it can be tough to get access to an entire population set. Thus, we often talk about a smaller slice or subset and call that a **sample**. Your family is a sample of all the families in the world. By studying your family – easier to do since it is smaller than the whole world – you can learn something about all the families in the world.

Congratulations! You just learned your first two words in the language of statistics! When thinking about the real world statistically, we first identify the larger *population* we want to understand better. Then, once the population idea is locked down, we start to think about what sorts of *samples* we might be able to actually reach and use and study.

5.3 Variables and Data

Of course, when we want to study something, we need some way to record and think about key features. For families, we might be interested in the names of each family member. We might also be interested in the family hierarchy: Who gets their way most often? Who is the next most influential? What is the temperature where the family lives? Is it hot or cold? Lastly, we might be interested in the heights of the family members. Who is tallest, and who is shortest?

Each of these measurements we just described can *change*. It will be different in different families – and maybe even different for the same family from one day or year to the next. We call these types of changing measurements **variables**. In general, statisticians talk about two types of variables: **qualitative** and **quantitative**.

Qua*lit*ative variables are categories such as our family members' names. Because of this, qualitative data are also sometimes called **categorical** or **nominal** data. Eye color, skin tone, gender, biological sex, personality – these types of variables all have categories described by names or labels. While it may be easy (in some cases) to determine clear differences (e.g., two siblings may have clearly different eye colors), none of these differences give any sort of order. Green eyes are not "first" or "second" compared to blue. One way to help yourself remember this type of data is that the word *nominal* means "name" and names are found in *lit*erature which is one way to think of qua*lit*ative data.

Qua*n*titative is *n*umeric (and the "n" is how we remember this). There are three types of quantitative data we discuss.

First child and second child are an example of **ordinal** data. Ordinal data has a clear numeric *ord*er to it. The "oldest college in the western hemisphere" vs. the "youngest college." The first dollar a company ever earned. These data have obvious numbers attached to them, and yet, there is no deeper meaning to the numbers. Being first or second in a race does not allow us to calculate much of anything. Silver medalists are not "twice" as slow as gold medalists, after all. So while ordinal data has a number order, that is about it.

Another example of ordinal data is the popular-in-surveys Likert (*lick*-urt) scale of strongly disagree, disagree, agree, and strongly agree. This may not quite seem numeric, yet is often coded in data as -2, -1, 0, 1, and 2.

Interval data on the other hand not only have a numeric order; they have a clear difference between numbers. Centigrade temperature is an example of interval where 100 C is the boiling point of water and 0 C is the freezing point of water. In this temperature system, 23 C is a great room temperature, and 45 C would be horrid. The name "interval" is because it makes sense to perform interval or difference calculations such as 45 C – 23 C = 22 C (in other words, the interval between 23 C and 45 C is 22 C). Contrast this meaningful interval against the ordinal data (where the "interval" between -2 and -1 in the Likert scale is meaningless). A feature of interval data is that "0" is also not of any major significance. For example, 0 F on the Fahrenheit temperature measure has no mathematical significance. Granted, 0 C is the freezing point of water at one standard atmosphere; however, that is not *universally* significant.

The last type of quantitative data is **ratio** data which has numeric order, has a clear difference between numbers, and has a true zero. Examples of ratio data include blood pressure, height, age, and money.

We summarize these four types of data in Table 5-1. Take a moment, consider some data you already know, and figure out where it lives in the table. Is it qualitative or quantitative? Can you place it more precisely?

Table 5-1. *Qualitative or Quantitative Data*

Type	Examples
Nominal	Names, eye color, skin tone, gender, biological sex, personality "types," generation, categories, and other qua*lit*ative data.
Ordinal	First/second children, oldest college, gold vs. silver medalist, most populous nation on earth, and Likert scale are all examples of quantitative, numeric data.
Interval	Fahrenheit/Celsius temperature, time of day, and IQ scores are also examples of quantitative data.
Ratio	Blood pressure, height, age, and money are also examples of quantitative data.

Example

Let us take a closer look at the `aces_daily` data we met in Chapter 4. Go ahead and convert it to the `data.table` style using the same process we used last time, `as.data.table()`. Now, however, we are interested in what types of data our various columns are. To see the *structure* of our data, we use the `str()` function:

```
acesData <- as.data.table(aces_daily)
str(acesData)
```

```
## Classes 'data.table'  and 'data.frame':      6599 obs. of 19 variables:
##  $ UserID             : int  1 1 1 1 1 1 1 1 1 1 ...
##  $ SurveyDay          : Date, format: "2017-02-24" ...
##  $ SurveyInteger      : int  2 3 1 2 3 1 2 3 1 2 ...
##  $ SurveyStartTimec11 : num  1.93e-01 4.86e-01 1.16e-05 1.93e-01 4.06e-
01 ...
##  $ Female             : int  0 0 0 0 0 0 0 0 0 0 ...
##  $ Age                : num  21 21 21 21 21 21 21 21 21 21 ...
##  $ BornAUS            : int  0 0 0 0 0 0 0 0 0 0 ...
##  $ SES_1              : num  5 5 5 5 5 5 5 5 5 5 ...
##  $ EDU                : int  0 0 0 0 0 0 0 0 0 0 ...
##  $ SOLs               : num  NA 0 NA NA 6.92 ...
##  $ WASONs             : num  NA 0 NA NA 0 NA NA 1 NA NA ...
##  $ STRESS             : num  5 1 1 2 0 0 3 1 0 3 ...
##  $ SUPPORT            : num  NA 7.02 NA NA 6.15 ...
```

```
## $ PosAff               : num  1.52 1.51 1.56 1.56 1.13 ...
## $ NegAff               : num  1.67 1 NA 1.36 1 ...
## $ COPEPrb              : num  NA 2.26 NA NA NA ...
## $ COPEPrc              : num  NA 2.38 NA NA NA ...
## $ COPEExp              : num  NA 2.41 NA NA 2.03 ...
## $ COPEDis              : num  NA 2.18 NA NA NA ...
## - attr(*, ".internal.selfref")=<externalptr>
```

While `UserID` may *look* quantitative, it is in fact qualitative! That ID, while an integer, is a proxy for the individual study participants' *names*. Other nominal data elements in this set include the biological sex flag `Female`, the born in Australia flag `BornAUS`, and the education level marker `EDU`. These are all examples of using the number 1 to indicate "yes" and the number 0 to indicate "no." Because the number 1 means the category is true (and because in electric computer design the value 1 means the circuit is "live"), we call this type of data storage "one-hot" coding.

It might seem a bit confusing at first. One way to see these are all qualitative, nominal data is to see the `UserID` column is essentially the *names* of the study participants. The `Female` variable could just as well be a column titled "Biological Sex" where each row was either "Female," "Male," or "Declined to Respond." Similarly, `BornAUS` could simply be "Yes," "No," or "Declined to Respond."

Ordinal variables in this data set include `SurveyInteger` telling us which of three daily surveys is being recorded and socioeconomic status `SES_1` which ranges from 4 to 8. While they do help us rank surveys or statuses in some sort of order, there are no sensible mathematical calculations possible based on these quantitative variables:

```
unique(acesData$SurveyInteger)

## [1] 2 3 1

unique(acesData$SES_1)

## [1]  5 NA  7  8  6  4
```

Interval variables include the date value `SurveyDay` as well as the time variable `SurveyStartTimec11`:

```
unique(acesData$SurveyDay)

##  [1] "2017-02-24" "2017-02-25" "2017-02-26" "2017-02-27" "2017-02-28"
##  [6] "2017-03-01" "2017-03-02" "2017-03-03" "2017-03-04" "2017-03-05"
```

```
## [11] "2017-03-06" "2017-03-07" "2017-02-22" "2017-02-23" "2017-03-08"
## [16] "2017-03-09" "2017-03-10" "2017-03-11" "2017-03-12" "2017-03-13"
## [21] "2017-03-14"
```

The rest of the data in this set are considered ratio data. It may be clear enough that `Age` is going to be ratio data. It has a real zero, and it makes sense to say that someone who is 40 years of age has twice the years of a 20-year-old. The self-reported sleep onset `SOLs` variable measures time in minutes before a participant fell asleep. This stopwatch-like measurement thus has a true zero for "fell directly asleep," and 5 minutes is half the time as 10 minutes. Similarly, the self-reported number of awakenings after sleep onset, `WASONs`, is a quantitative ratio data point.

While you are already familiar with the `unique()` function in `R`, we introduce a new function, `summary()`. We discuss summary more at a later point in time. Of interest right now is the range which stretches from a minimum of 0 minute until falling asleep all the way to 180 minutes (3 hours). It may also be somewhat interesting to notice that of the 6,599 observations in our data set, there are missing values – `NAs` – for 4,502. It is tough to self-report how long it takes to fall asleep!

```
unique(acesData$Age)

##  [1] 21 NA 23 24 25 22 20 18 26 19

summary(acesData$SOLs)

##    Min. 1st Qu.  Median    Mean 3rd Qu.    Max.    NA's
##       0       7      16      27      32     180    4502

unique(acesData$WASONs)

## [1] NA  0  1  2  4  3
```

Note

The last set of measures in our list are the psychometric data of the stress rating `STRESS`, social support scale `SUPPORT`, positive affect rating `PosAff`, negative affect rating `NegAff`, problem-focused coping scale `COPEPrb`, emotional process coping `COPEPrc`, emotional expression coping `COPEExp`, and mental disengagement coping `COPEDis`. These are often treated as continuous data, although this is a bit of a nuance and somewhat beyond the scope of this course.

Example

Recall the mtcars car data set. Definitely the type of car is going to be a nominal, qualitative variable. However, both vs (whether the engine is v shaped) and am (transmission automatic or manual) are also nominal data. These are one-hot encodings of categories.

All the rest of the values are ratio data:

```
head(mtcars)
```

```
##                    mpg cyl disp  hp drat    wt  qsec vs am gear carb
## Mazda RX4         21.0   6  160 110 3.90 2.620 16.46  0  1    4    4
## Mazda RX4 Wag     21.0   6  160 110 3.90 2.875 17.02  0  1    4    4
## Datsun 710        22.8   4  108  93 3.85 2.320 18.61  1  1    4    1
## Hornet 4 Drive    21.4   6  258 110 3.08 3.215 19.44  1  0    3    1
## Hornet Sportabout 18.7   8  360 175 3.15 3.440 17.02  0  0    3    2
## Valiant           18.1   6  225 105 2.76 3.460 20.22  1  0    3    1
```

Example

Our last example for this section is the penguins data set. The species, island, and sex columns are qualitative, nominal data. However, when using the str() function, notice these are shown as Factor data. Factor data are a multilevel version of one-hot encoding. While R has kept the text of the species names (e.g., "Gentoo") visible for our eyes, behind the scenes there are numeric values of 1, 2, and 3 for each of Adelie, Chinstrap, and Gentoo.

The length, depth, and mass measurements are ratio data:

```
str(penguins)
```

```
## tibble [344 x 8] (S3: tbl_df/tbl/data.frame)
##  $ species          : Factor w/ 3 levels "Adelie","Chinstrap",..: 1 1 1
1 1 1 1 1 1 1 ...
##  $ island           : Factor w/ 3 levels "Biscoe","Dream",..: 3 3 3 3 3
3 3 3 3 3 ...
##  $ bill_length_mm   : num [1:344] 39.1 39.5 40.3 NA 36.7 39.3 38.9 39.2
34.1 42 ...
```

```
## $ bill_depth_mm    : num [1:344] 18.7 17.4 18 NA 19.3 20.6 17.8 19.6
18.1 20.2 ...
## $ flipper_length_mm: int [1:344] 181 186 195 NA 193 190 181 195 193 190
...
## $ body_mass_g      : int [1:344] 3750 3800 3250 NA 3450 3650 3625 4675
3475 4250 ...
## $ sex              : Factor w/ 2 levels "female","male": 2 1 1 NA 1 2 1
2 NA NA ...
## $ year             : int [1:344] 2007 2007 2007 2007 2007 2007 2007
2007 2007 2007 ...
```

Thoughts on Variables and Data

The reason it is important to understand the types of data you have is because some statistical techniques will only work on certain types of data. Suppose you have both a checking account and a savings account – those ratio data are easy to sum together and get a grand total. On the other hand, we cannot sum together your name and our names – nominal data do not work that way. More importantly, there are certain statistical methods that may *seem* to work mathematically and yet, on the wrong kinds of data, will simply not be true. For example, consider degree Celsius temperature. Mathematically, we *can* take survey integer 3 less survey integer 1 and get a "2" (because 3 – 1 = 2). However, that has no meaning about the contents or use of survey integer 2.

Keep this idea in your mind: There is a difference between something working *mechanically* and something working *successfully*. A coffee grinder might well work on wood chips; it takes some skill (especially in the early morning!) to have the sense about one to not put those wood chips into the coffee maker.

One risk of using a statistical language like R is that if the data work mechanically, R will be happy to give you an output. It may even look and feel "about right." Taking time at the beginning to make sure your data are the correct *type* to perform a particular calculation may seem silly four times out of five. However, on the fifth time, you will be glad that is part of your regular mental process.

5.4 Thinking Statistically

The universe around us is complicated; there are many ways to try to simplify that complexity into a usable model. Remember what we said about families? The *population* of all families on planet earth is huge – perhaps too huge to easily understand. If we can find a *sample* of families – just a few – we may be able to study those more closely and then understand something about all families (or at least most families).

This idea of picking a subset or sample from our larger population is statistics. In fact, we often use the word **statistics** when talking about data or information related to *samples*. In the (more rare) cases where we have *population*-level data, we often use **parameter**.

As with most philosophies, there are pros and cons to this approach. A *good* sample ought to give us fairly reasonable insights into the population. However, picking a good sample is tough! Our family experiences are likely not quite the same as yours. There are many kinds of families in the world; even deciding which types of people to include on your list of family differs.

Suppose we are studying ideal keyboard layouts; the Roman alphabet is only a sample of all the population of written scripts. Or we could study university and college students in geographic places labelled "Victoria"; yet Victoria, Australia, may well be different from Victoria, Texas. Some samples ask people about political opinions; yet the political beliefs of "random adult citizens" often differ (by ratio) from the beliefs of "likely voters" or even "actual voters."

This tension between sample and population is both a strength and a weakness of statistics. Select a good sample, and you have a smaller (and more manageable) group to study and research. However, the reduction in complexity often removes some key features – making it tougher to identify universal trends. In this book, we occasionally generate a large "population" of data and pull a smaller sample from that population. By comparing the two sets of data (courtesy of using a programming language like R), you are gaining highly useful insight into the power *and* the limitations of this discipline.

5.5 Evaluating Studies

There are two types of research methods, generally. An **observational** study has no direct *interventions* from the researcher. Political polling is an example of an observational study. A sample is determined, questions are asked, and the results are totaled. Indeed, most polling tends to be observational. While the study of polling questions and such *psychometrics* is a field in its own right, there are some easy questions you are already very able to answer.

Are the questions fair to ask? Do the questions "lead" people in one direction or another? Consider the following unfair question: "Have you stopped hurting other people yet?" There are few good answers to this question, because it presupposes the answerer has been hurting other people all along. While a "yes" is certainly better than a "no," it is not a great answer really. If you are a researcher for a company, an unfair question might include "Would you prefer to pay $10 or $5 for our widget?" The average consumer might well answer "$5." However, if the question instead was "Assuming you are in the market for widgets, would you prefer a premium, full-featured widget at $10 or an entry-level, basic widget at $5?" that question would get a little closer to getting an actionable answer from your prospective clients.

Are the right people being observed? In a way, this gets back to sample selection. Are there any features that make the group under observation different from the population? Were enough people observed?

Contrastingly, **experimental** studies have direct interventions. The group of people being studied are split into two groups. The **control** group is asked the same questions or given the same tests, yet there is no intervention. The other group is given the intervention or treatment and also asked the same questions or given the same tests as the control group.

This can be done on websites, where it is sometimes called A/B testing (version A and version B of a website or sales pitch). This is done in medical research trials, where the control group is given a **placebo** (a harmless drug that traditionally was only a "sugar pill"), while the experimental group is given the new drug under study.

Using experimental groups raises an important question of **ethics**. In the case of website testing, most likely the differences between websites are not overly harmful to the average consumer. On the other hand, in a medical trial, two equally sick groups of patients are given two very different treatments. The severity of the illness, the disparity between treatments (if there is a traditional treatment already in place that can be swapped out for a true placebo), and the danger of the experimental intervention (what if the new drug turns out halfway through trials to not be safe) all increase the risk of experimental research.

While experimental research has ethical risks that need to be carefully and thoughtfully considered, when done with informed consent of participants, there can be great benefits to such research. The benefits start with the researcher's ability to fully control the sample of participants given the treatment.

5.6 Evaluating Samples

By now, you are seeing that getting the right sample may well make a significant difference in how we understand the population. So how does one "get the right sample"? What is a good sample? Can a sample be "good enough"?

The answer is, as is often the case, "it depends." Key to a good sample is that, despite being smaller than the population, the sample somehow still has all the characteristics of the population. In other words, we want the sample to be a "true" representation of the population. In statistics, when a sample "fits" the population well, we say the sample is **unbiased**. To flip that around, a **biased** sample will not match the population well on one or more potentially key variables. In some cases, even though a sample might appear to match on the variables we initially considered for our research, there may be some other variable, one we did not consider, which has an outsized influence on our measures. We call this type of unexpected influence from unplanned variables **confounding** variables.

As we explore the various types of samples together, keep in mind the pros and cons of each. Keep a close watch for sample bias and confounding variables. As we work through each sampling method, you will learn some `R` code about how to pull a sample from a "population" using that method. Keep in mind that, during "regular" research (contrasted with teaching or learning about research), we need samples because the population is too big or too expensive to collect full data. However, when using a programming language, we can *simulate* population data, pull smaller samples from that larger mix, and hands-on learn how each sample works. For *just this section*, we will *pretend* that `acesData` is a *population*:

```
#this comment reminds us we pretend acesData is a population
colnames(acesData)
```

```
## [1] "UserID"              "SurveyDay"           "SurveyInteger"
## [4] "SurveyStartTimec11"  "Female"              "Age"
```

```
##  [7] "BornAUS"            "SES_1"              "EDU"
## [10] "SOLs"               "WASONs"             "STRESS"
## [13] "SUPPORT"            "PosAff"             "NegAff"
## [16] "COPEPrb"            "COPEPrc"            "COPEExp"
## [19] "COPEDis"
```

Convenience Samples

The easiest sample method is called a **convenience** sample. As the name suggests, rather than any formal method of reducing a population to a selected sample, we simply grab a subset that is convenient. Our family would be an example of a convenience sample of families. The students in one of our classrooms would be a convenience sample of college students. Putting a poll on the front page of a company website would be a convenience sample of likely customers or clients. These samples are easy and fast to access, and results can be quickly tabulated and processed – a powerful pro. On the other hand, we are at the mercy of our sample. Our primary teaching in higher education involves research and statistical methods courses. Choosing to use our classrooms as a convenience sample for *all* college students would be biased. Only rarely do art students or music majors or computer engineers take our courses; whole segments of students are not part of our courses. On the other hand, our classrooms might make a good-enough sample for various science majors.

Example

Convenience samples are all about quick and easy. Suppose we wanted to quickly understand the age ranges of our survey participants. We might just take the first 20 participants. Remember, in `data.table`, the ith position allows for row selection. Here, we want *less than or equal* to `20`. As we are only interested in ages for this example, we go ahead and choose only the `Age` column. Participants actually take many surveys in this study, so we use the `unique()` function on our result to only show us each age once (we are currently only interested in age *range*):

```
convenienceAge <-  acesData[UserID <= 20, Age]
unique(convenienceAge)
```

```
## [1] 21 NA 23 24 25 22 20 18
```

While we can see the ages range from 18 to 25, this is not the most organized data output ever. Writing more effective code often takes several rounds of trial and error. As you refine your code through each edit, it can help to *comment* your code. R uses the pound sign or hashtag # to indicate words typed after that symbol are not meant to be processed. Code comments help future you and other readers understand your goal. It can help your research be *reproducible*:

```
#convenience sample selection of first 20 participants
convenienceAge <- acesData[UserID <= 20, Age]

#each participant takes many surveys, just need age once
convenienceAge <- unique(convenienceAge)

# sort the ages for better readability
sort(convenienceAge)

## [1] 18 20 21 22 23 24 25
```

Did you notice our `sort()` function removed the NA? The NAs are placeholders for participants who choose to not report their Age. By default, `sort()` does not return NAs. If you want to include the NAs in a sort, you must specify whether you want them last or first in your list by using another of `sort()`'s arguments:

```
sort(convenienceAge, na.last = TRUE)

## [1] 18 20 21 22 23 24 25 NA

sort(convenienceAge, na.last = FALSE)

## [1] NA 18 20 21 22 23 24 25
```

Remember, in R and RStudio, you can always learn more about a function by typing ? in front of a function:

```
?sort()
```

Example

What if we tried this on our penguinsData data to check flipper_length_mm? We could reuse almost all the code from the prior example. Of course, we would have to convert our data to a data.table format.

Something not so nice though about our first 20 penguins: They all seem to be the same species. Might this cause some bias?

```
penguinsData <- as.data.table(penguins)

#convenience sample selection of first 20 penguins
convenienceFlippers <- penguinsData[1:20, flipper_length_mm]

#Just need each length once each; only need range
convenienceFlippers <- unique(convenienceFlippers)

# sort the lengths for better readability
sort(convenienceFlippers, na.last = TRUE)

## [1] 180 181 182 184 185 186 190 191 193 194 195 197 198 NA
```

Kth Samples

Another common sample method is **K** or **Kth** sampling. The "K" stands in for an ordinal such as "2nd" or "10th." It is a systematic sample method where every Kth item or person is sampled. Every 100th customer to visit our company's website might be given a questionnaire, or every 3rd customer in line in the store might be asked a question. Kth sampling is often used in quality control, because it can ensure a wider coverage by requiring spacing between items. However, in the case of automated processes (e.g., manufacturing or compute efforts), be sure there is not an accidental pattern! For example, suppose you have a factory that sews shirts. If there are 14 sewing machines in total and you always pick the 13th shirt, you might accidentally only be sampling the 13th machine ever.

Example

Going back to our Age range, we need to edit our method a bit to successfully capture every 13th participant. Think back to long division and *remainders*. A number such as 6 divided by 2 would have remainder 0 (because 2 divides into 6). On the other hand, the number 5 divided by 2 would be 2 with a remainder of 1. That remainder is sometimes called the *modulo*, and the R operator for modulo is %%:

```
#the modulo output of 0 is the 'remainder'
6 %% 2

## [1] 0

8 %% 2

## [1] 0

#the modulo output of 1 is the 'remainder'
5 %% 2

## [1] 1

7 %% 2

## [1] 1
```

Notice the output of the modulo for 2 would always be 0 for even numbers and 1 for odd numbers.

We can exploit the fact that when the output of modulo is 0, we have found numbers that are perfect modulos for our Kth value. In our case, we want to find numbers that are in multiples of 13:

```
13 %% 13

## [1] 0

26 %% 13

## [1] 0
```

Remember that `UserID` is one of the columns inside `acesData`. Thus, we want all the rows where the `UserID` is a multiple of 13 (because we want every 13th user for our Kth sample). That means we want all the rows where modulo 13 is 0. To check that we are using this technique correctly, before we use our code to capture Ages, we assign the IDs to a new variable we call idCheck. Notice we want to select every row where the UserID is a perfect multiple of 13. So we use modulo 13 to test that and require that output to be exactly 0 (that is done with the code snippet UserID %% 13 == 0 in the ith position for row selection of our data.table):

```
# Assign every 13th UserID to idCheck
idCheck <- acesData[UserID %%  13 == 0, UserID]
# Make sure we got the correct UserIDs
unique(idCheck)
```

```
##  [1]  13  26  39  52  65  78  91 104 117 130 143 156 169 182
```

We are now sure our modulo idea works for Kth sample selection (which in this case happens to be every 13th). Notice we first experimented with our idea and are now moving our idea forward. This is not just a technique used to help you see each bit of code in isolation (although it works well for that too). In real-life data science, exploring the data and figuring out how to capture a sample are often best done with a bunch of small steps. A challenge for a new learner can be trying to do too many things at once, which makes it tougher to figure out where a code failure point might live. Debugging shorter code is easier than debugging longer code, so always try to start small and run your code often!

With modulo working, we are ready to mostly copy and paste our convenience sample code. Go back and review that code, and highlight the changes we made from there to here. Not much difference! This is one way that R makes it so much easier to do statistics at scale - your research is reproducible across methods:

```
#kth sample selection of every 13 participants
kthAge <- acesData[UserID %% 13 == 0, Age]

#each participant takes many surveys, just need age once
kthAge <- unique(kthAge)

# sort the ages for better readability
sort(kthAge, na.last = TRUE)
```

```
## [1] 18 19 20 21 23 24 25 NA
```

Example

What if we tried Kth sampling on our penguins data to check flipper_length_mm range? Again, you could reuse much code. The penguinsData does not have IDs, however. Instead, it simply has rows. In this case, the seq() function will create a sequence for us. This function takes three inputs, a starting from value, an ending to value, and a by value telling the sequence how many entries to skip:

```
#observation-rows in penguins
nrow(penguinsData)
```

```
## [1] 344
```

```
#sequence of every 13th row needed
seq(13, nrow(penguinsData), 13)
```

```
##  [1]  13  26  39  52  65  78  91 104 117 130 143 156 169 182 195 208
## [17] 221 234 247 260 273 286 299 312 325 338
```

Before we actually check the range of the flipper_length_mm using recycled code, go ahead and take a look at what we get out of selecting every 13th row of penguins data. Using our sequence list, we use the ith position of data.table to capture every 13th row. In this case, we leave the jth position of data.table blank to easily see that now we have a better mix of species. Even though we now have examples of all three species, notice it is not perfect. Can you spot the one species rather underrepresented in our sample? We only have five Chinstrap. Also notice R renumbered our rows in this new data set. Unlike our participant data with fixed IDs, we have

```
#kth sample selection of every 13th penguin
penguinsData[seq(13, nrow(penguinsData), 13)]
```

```
##       species    island bill_length_mm bill_depth_mm
##  1:    Adelie Torgersen           41.1          17.6
##  2:    Adelie    Biscoe           35.3          18.9
##  3:    Adelie     Dream           37.6          19.3
##  4:    Adelie    Biscoe           40.1          18.9
##  5:    Adelie    Biscoe           36.4          17.1
## ---
```

```
## 22: Chinstrap     Dream            51.3            19.9
## 23: Chinstrap     Dream            43.2            16.6
## 24: Chinstrap     Dream            47.5            16.8
## 25: Chinstrap     Dream            51.5            18.7
## 26: Chinstrap     Dream            46.8            16.5
##     flipper_length_mm body_mass_g    sex year
##  1:               182        3200 female 2007
##  2:               187        3800 female 2007
##  3:               181        3300 female 2007
##  4:               188        4300   male 2008
##  5:               184        2850 female 2008
## ---
## 22:               198        3700   male 2007
## 23:               187        2900 female 2007
## 24:               199        3900 female 2008
## 25:               187        3250   male 2009
## 26:               189        3650 female 2009
```

Putting together everything we know and using as much of our old code as possible, we assign our penguinsData to a kthFlipper variable. We continue to focus on range, and now we see our range is longer with all three species in the sample:

```
#assign every 13th sample to kthFlipper variable
kthFlipper <- penguinsData[seq(13, nrow(penguinsData), 13), flipper_length_mm]

#Just need each length once each
kthFlipper <- unique(kthFlipper)

# the lengths for better readability
sort(kthFlipper, na.last = TRUE)

##  [1] 181 182 184 187 188 189 190 198 199 202 210 214 215 218 219 220
## [17] 221
```

When thinking about Kth sampling, there are some things to consider. In the case of our participants' age ranges, we seem to have gotten lucky with our first 20 in the convenience sample vs. every 13th. In both cases, the range is 18–25. However, with penguin flipper lengths, definitely the kth sample, while a little harder to code, gave us a more complete understanding.

Cluster Samples

Sometimes, it makes sense to choose a group or a **cluster** sample. To survey students, we might choose 20 classrooms and assign a grade for a survey. In the grocery store, we might walk down aisle 10 and ask each customer if they are finding everything they wanted. Choosing a cluster makes sense when the clusters have some sort of natural group behavior; this is different from convenience sampling because we are either willing or able to include even inconvenient clusters. In some student surveys, for example, the institution's research office tends to choose classrooms rather than individual students. This is because giving surveys during class time leads to better survey completion rates. Cluster sampling makes more sense over Kth sampling because that way every student in that class is still on the same schedule and lesson plan.

Cluster sampling also happens more globally – some medical research might choose one hospital in a handful of countries. Some educational research might take place in one grade school in every district. Cluster sampling might also occur with shipping containers – a handful of trucks might be stopped at a border and then every item in the truck checked for contraband.

While cluster sampling is often geographic in some sense, it also has a time component. Additionally, a feature of cluster data is that every (or at least most items) item inside the cluster is chosen. In other words, what is being sampled is an entire cluster, not one individual at a time.

Example

Suppose we have a visiting researcher who wants to ask our survey participants some additional questions. One example of a cluster sample might be "all the participants on a specified date." This requires us to select a specified date, and we use the `as.Date()` function. This is required, in this case, because `SurveyDay` has a `Date` data format. We also use the `%in%` logical operator. These are used, as now is more familiar, in the ith row selection position of our `data.table`. We use the jth column selection position to inspect the `UserID` and `Age` of our participants. We can see we have several, and they are duplicated (participants are meant to complete three surveys each day after all).

Notice that, rather than place the date in-line in the ith position, we assigned the date(s) its (their) own variable, `clusterDate`. This helps our code be more human readable. If we are sharing results, later, with colleagues who are not familiar with `R`, it still makes sense to say, "In the survey data set, we selected only the IDs and ages of our participants on the date we did the cluster survey":

```
#assign to a well-named variable
clusterDate <- c(as.Date("2017-02-23"))

acesData[SurveyDay %in% clusterDate,
         .(UserID, Age)]

##      UserID  Age
##   1:      2   23
##   2:      2   23
##   3:      2   23
##   4:      3   NA
##   5:      6   NA
##  ---
## 103:    187   19
## 104:    187   19
## 105:    189   19
## 106:    189   19
## 107:    189   19
```

Remember to compare and contrast this recycled code against what we have used earlier. Overall, however, there are no major changes:

```
#cluster sample of all participants on 27 Feb 2017
clusterAge <- acesData[SurveyDay %in% clusterDate, Age]

#each participant takes many surveys, just need age once
clusterAge <- unique(clusterAge)

# sort the ages for better readability
sort(clusterAge, na.last = TRUE)

##  [1] 18 19 20 21 22 23 24 25 26 NA
```

There is one difference, however; we have an age of 26 for the first time.

Example

Using mtcars is quite easy to do a cluster sample. We could collect our clusters by manufacturer. Suppose we wish to use cluster sampling to understand the ranges of horsepower possible. We might choose Mercedes and Toyota as our two manufacturers.

To get started, we assign mtcars as a data table and keep the row names. This creates a data.table column named rn that has our car row names. Using the logical operator %like% as well as the logical or operator |, we are able to collect only those rows that are Mercedes or Toyota:

```
mtcarsData <- as.data.table(mtcars, keep.rownames = TRUE)

mtcarsData[rn %like% "Merc" |
                    rn %like% "Toyota"]

##                rn  mpg cyl  disp  hp drat    wt  qsec vs am gear carb
## 1:      Merc 240D 24.4   4 146.7  62 3.69 3.190 20.00  1  0    4    2
## 2:       Merc 230 22.8   4 140.8  95 3.92 3.150 22.90  1  0    4    2
## 3:       Merc 280 19.2   6 167.6 123 3.92 3.440 18.30  1  0    4    4
## 4:      Merc 280C 17.8   6 167.6 123 3.92 3.440 18.90  1  0    4    4
## 5:     Merc 450SE 16.4   8 275.8 180 3.07 4.070 17.40  0  0    3    3
## 6:     Merc 450SL 17.3   8 275.8 180 3.07 3.730 17.60  0  0    3    3
## 7:    Merc 450SLC 15.2   8 275.8 180 3.07 3.780 18.00  0  0    3    3
## 8: Toyota Corolla 33.9   4  71.1  65 4.22 1.835 19.90  1  1    4    1
## 9:  Toyota Corona 21.5   4 120.1  97 3.70 2.465 20.01  1  0    3    1
```

However, once our data set is ready and selected, the same process continues to work. In this case, we do use the hp column selection for horsepower:

```
#cluster sample of hp
clusterCar <- mtcarsData[rn %like% "Merc" |
                    rn %like% "Toyota", hp]

#some cars may share hp, we still are focused on range
clusterCar <- unique(clusterCar)

# sort the hp for better readability
sort(clusterCar, na.last = TRUE)

## [1]  62  65  95  97 123 180
```

Stratified Samples

Stratified samples ensure some of each type or demographic are selected. In the case of `penguins`, a stratified sample might be best, to select some of each species of penguin. Strata can be collected across biological sex, ethnicity or race, age brackets, income levels, a socioeconomic indicator, or other variables. One feature of strata is that not every element inside a particular strata needs to be collected. So while the strata groups might seem somewhat similar to cluster groups (and there is admittedly some conceptual overlap between the two methods), they are distinct in that one typically expects to see some difference between strata, the strata are more thoughtfully chosen, and not every element/item/individual in the strata is usually selected into the sample.

Example

In the case of our survey participants, we may wish to sample by socioeconomic status and biological sex (`SES_1` and `Female`). Again, for now, we are only interested in the age ranges. Our goal is to pick some data from each strata. Unlike cluster data, we do not need to pick all the data in each strata. Our method of selecting the participants inside each strata can be any method we like. We could use a convenience sample of the first one or two, or we could use every Kth participant. It is our choice.

For simplicity, we simply select the first row in each strata. To do that, we use `data.table`'s *S*ubset *D*ata operator, `.SD[]`. The subset operator is used to signal `data.table` we are about to do *row* operations with our `by` operation.

Remember our goal is to use a convenience sample of the first row in each of our strata. In this case, we have six possibilities for socioeconomic status and three possibilities for biological sex. There is a maximum total of `6*3 = 18` strata (if all possible combinations occurred):

```
unique(acesData$SES_1)

## [1]  5 NA  7  8  6  4

unique(acesData$Female)

## [1]  0 NA  1
```

If we simply did what might seem natural, we would get a warning message. That is because `data.table` processes commands from left to right, and since a `1` in the ith row selection position means row 1, the by is never processed:

```
acesData[1, , by = .(SES_1, Female)]

## Warning in '[.data.table'(acesData, 1, , by = .(SES_1, Female)):
Ignoring by= because j= is not supplied

##    UserID  SurveyDay SurveyInteger SurveyStartTimec11 Female Age
## 1:      1 2017-02-24             2             0.1927      0  21
##    BornAUS SES_1 EDU SOLs WASONs STRESS SUPPORT PosAff NegAff COPEPrb
## 1:       0     5   0   NA     NA      5      NA  1.519  1.669      NA
##    COPEPrc COPEExp COPEDis
## 1:      NA      NA      NA
```

Instead, what we want is to take *all* rows, do a subset by socioeconomic status and biological sex, and take only the first row for each part of that subset. Because we want to start with *all* rows, we must leave the ith row position blank. We signal this to `data.table` by leaving a blank before our first comma.

Next, we use the subset operation, `.SD[]`. This signals `data.table` we are subsetting our data. Because subsets happen by row, the subset operation is expecting an entry telling it which row of each subset. The `1` tells it we want the first row. If we wanted to do a Kth sample, we could use our `seq()` function to make that happen. For now, we keep it simple.

Lastly, subsetting can include a by, and that is where our strata belong. We only need to tell `data.table` which columns hold our strata; the rest will follow automatically. Notice we do not quite get 18 rows; not all possible combinations occurred. For example, there is no socioeconomic status of 8 with biological sex `1` in our participant list:

```
acesData[,
         .SD[1],
         by = .(SES_1, Female) ]

##     SES_1 Female UserID  SurveyDay SurveyInteger SurveyStartTimec11
##  1:     5      0      1 2017-02-24             2            0.19272
##  2:    NA     NA      2 2017-02-22             3                 NA
##  3:     7      1      2 2017-02-23             1            0.14466
```

```
##  4:     8       1       3 2017-02-24             1              0.09568
##  5:     5       1       6 2017-02-24             1              0.08155
## ---
##  9:     6       1      10 2017-02-24             3              0.40065
## 10:     4       1      21 2017-02-24             1              0.09767
## 11:     8       0      28 2017-02-23             1              0.12850
## 12:    NA       1      70 2017-03-02             2              0.21267
## 13:    NA       0     107 2017-02-24             2              0.18896
##      Age BornAUS EDU  SOLs WASONs STRESS SUPPORT PosAff NegAff COPEPrb
##  1:  21        0   0    NA     NA      5      NA  1.519  1.669      NA
##  2:  NA       NA  NA 23.48      2     NA      NA     NA     NA      NA
##  3:  23        0   0    NA     NA      2      NA  3.099  1.037      NA
##  4:  21        1   0    NA     NA      3      NA  3.597  1.127      NA
##  5:  22        1   0    NA     NA      9      NA  1.396  1.966      NA
## ---
##  9:  25        1   1    NA     NA      2   7.345  3.571 1.059   2.135
## 10:  20        0   0    NA     NA      0      NA  3.053 1.076      NA
## 11:  19        1   0    NA     NA      0      NA  3.261 1.099      NA
## 12:  19        1   0    NA     NA      3      NA  4.189 1.093      NA
## 13:  NA       NA  NA    NA     NA      7      NA  1.886 4.452      NA
##      COPEPrc COPEExp COPEDis
##  1:       NA      NA      NA
##  2:       NA      NA      NA
##  3:       NA      NA      NA
##  4:       NA      NA      NA
##  5:       NA      NA      NA
## ---
##  9:    1.613   2.389   2.121
## 10:       NA      NA      NA
## 11:       NA      NA      NA
## 12:       NA      NA      NA
## 13:       NA      NA      NA
```

Do not let the subset operator be too much of a holdup. First of all, it can be a bit confusing at first. Second, we will slowly work through any additional uses of the subset operator. For now, we take our new bit of sample code and plug it into our process to check for the range of the age data. In this case, it is about what we have come to expect, although we lost those of age 18 and 26:

```
#strata sample of age by SES and sex
strataAge <- acesData[ ,
          .SD[1],
          by = .(SES_1, Female) ][, Age]

#each participant takes many surveys, just need age once
strataAge <- unique(strataAge)

# sort the ages for better readability
sort(strataAge, na.last = TRUE)

## [1] 19 20 21 22 23 25 NA
```

Example

In the case of penguins, a natural stratification would be `species`. Rather than only select the first entry, we might choose to select the first three rows of each species. Other than that one change, this is very much the same subset call using `.SD[]` we just saw:

```
# recall : can be used as a shortcut for sequence
1:3

## [1] 1 2 3

# make one tiny change to .SD to get first three rows.
penguinsData[,
         .SD[1:3],
         by = species]

##      species    island bill_length_mm bill_depth_mm flipper_length_mm
## 1:    Adelie Torgersen           39.1          18.7               181
## 2:    Adelie Torgersen           39.5          17.4               186
## 3:    Adelie Torgersen           40.3          18.0               195
## 4:    Gentoo    Biscoe           46.1          13.2               211
## 5:    Gentoo    Biscoe           50.0          16.3               230
## 6:    Gentoo    Biscoe           48.7          14.1               210
## 7: Chinstrap     Dream           46.5          17.9               192
```

```
## 8: Chinstrap     Dream              50.0          19.5                196
## 9: Chinstrap     Dream               51.3         19.2                193
##    body_mass_g    sex year
## 1:        3750   male 2007
## 2:        3800 female 2007
## 3:        3250 female 2007
## 4:        4500 female 2007
## 5:        5700   male 2007
## 6:        4450 female 2007
## 7:        3500 female 2007
## 8:        3900   male 2007
## 9:        3650   male 2007
```

We are only interested in the lengths of the flippers, and thus we only want the flipper column. Because we are already using the jth column position for the subset data function, we use a new set of brackets and put `flipper_length_mm` in the column position of those brackets. It may look rather odd; all the same, what we are left with in the variable `strataPenguins` is only flipper lengths. Other than that difference, the majority of this code should look very familiar:

```
#strata sample of penguin flipper lengths by species
strataFlippers <- penguinsData[,
         .SD[1:3],
         by = species][, flipper_length_mm]

#Just need each length once each
strataFlippers <- unique(strataFlippers)

#sort the lengths for better readability
sort(strataFlippers, na.last = TRUE)

## [1] 181 186 192 193 195 196 210 211 230
```

Random Samples

There is one last type of sample left, and it is considered the gold standard of sampling. **Random** samples are usually considered the safest method of sampling to prevent bias. This is because we never have a chance to let our own preconceived notions as researchers come into play. Instead, there is a purely random selection of participants or items.

There are two main cons with random sampling. Technically, at random, eventually, all possible samples will be selected – even samples that are only one species or only one socioeconomic status. While that is a risk for any single sample, across all samples (especially taken over a lifetime of research), there will be less sample bias. Furthermore, mathematically, you can't simply keep taking random samples until you get what you want in terms of demographics that defeats the meaning of random. The other con is that in real life, it is often difficult to take a purely random sample. Imagine a random sample of families – even just a hundred families might require several trips to visit! For medical research, a pure random sample would likely not be ethical (testing a new drug on a perfectly healthy person who simply got their name drawn out of a hat is not humane). That said, the more randomness that can be built into sample selection, the less bias from us can leak into the data collection.

Example

As you might expect by now, we pull a sample from our participant data set to check the range of the ages. To do this, we need to explore the `sample()` function. The `sample()` is quite versatile; however, for our purpose now, we only need two formal arguments. The first argument is a set of values (these could be names, IDs, or numbers). The second is how many of these we want to pull for our sample.

The `sample()` function works *randomly*, and because of that randomness, by default, your sample and ours would be different. For your first time using this function, it would be tough to know if everything was working correctly if you could not match our results. To allow us to match, `R` has a function called `set.seed()` that takes a numeric argument that makes sure your "random" matches our random. When initially writing code that involves samples, it can help to use `set.seed()` to keep everything consistent while you make sure your code works. However, it is important to eventually **stop** using the function – otherwise, you do not actually have a random sample!

To see how this works, we populate our random number generator with the number `1234`. While any number would work, this one is easy, and as long as you use the same number, you will get the same result as we do. Then, using the numbers 1–191 (this is the range of the `UserID`s), we take a sample of size 20. Notice the results are not pulled in order – remember this is meant to be random (or, in this case, pseudo-random):

```
set.seed(1234)
sample(x = 1:191, size = 20)

##  [1]  28  80 150 101 111 137 133 166 144 132  98 103  90  70  79 116
## [17]  14 126  62   4
```

As long as we select those UserIDs, we will have a random sample of our participants. Notice we use `set.seed()` again to ensure we get the same numbers in our sample that we did earlier. Again, note that in real-life random sampling, we need to **not** use `set.seed()`!

```
#remember, do not use this in general
set.seed(1234)

#random sample of age
randomIDs <- sample(x = 1:191, size = 20)
randomAge <- acesData[UserID %in% randomIDs,
                      Age]

#each participant takes many surveys, just need age once
randomAge <- unique(randomAge)

# sort the ages for better readability
sort(randomAge, na.last = TRUE)

## [1] 18 19 20 21 23 24 25 NA
```

Our sample ages seem to range from 18 to 25, which again is about what we have been seeing.

Example

For the last time (at least in this chapter), we take a sample of penguins to see the range of flipper lengths. From our `penguinsData`, we select *rows* rather than ID numbers. This makes sense because these data have only one row for each unique penguin. Thus, each row is an unduplicated penguin. Our participant data, on the other hand, has the same participant on many rows. That is because those rows are really unique surveys the participant takes, not unique participants.

To ensure your results agree with ours, we once again use `set.seed(1234)`. We assign a sample of 20 values randomly (in this case pseudo-randomly) pulled from the numbers 1 through the 344 rows of penguin data. Notice we could have just as well used `1:344` rather than `1:nrow(penguinsData)`. In programming, it is often a good idea to make code more human readable. Now, you know and we know the penguin data set has 344 observations. So it might be shorter to simply write `1:344`. However, other people may look over this code someday, and `1:nrow(penguinsData)` is much easier to understand in English as "the numbers 1 through the number of rows/observations in the penguin data set."

After that, our code becomes quite familiar:

```
#remember, do not use this in general
set.seed(1234)

#random sample for penguins
randomRows <- sample(x = 1:nrow(penguinsData), size = 20)

#random sample of penguins
randomFlippers <- penguinsData[randomRows, flipper_length_mm]

#Just need each length once each
randomFlippers <- unique(randomFlippers)

#sort the lengths for better readability
sort(randomFlippers, na.last = TRUE)

##  [1] 183 184 185 187 190 192 193 194 195 196 197 198 209 210 221 222
## [17] 224  NA
```

Compare this back to the ranges you already have seen. While it seems the stratified sample had the widest range, the random sample was certainly better than the convenience sample. In general, random samples are usually best. That said, a well-considered stratified sample can be superior.

Sample Recap

This has been a lot to take in so far, and it is worth having a bit of a recap. Populations are often quite large entities – studying everyone in a population might either be too expensive or too time intensive or involve much duplication of effort. By reducing the entire population down to a manageable sample, one can often gain insight into the whole population at a fraction of the cost, time, and effort.

However, the smaller number of observations in a sample necessarily means some quantity of information has been lost. Different sample methods can be used to find levels of compromise between faster and accurate results.

Convenience samples tend to be cheap and may yield the fastest results, yet can be the least accurate. We certainly saw that with the penguin data which, in the convenience sample, ranged only from 181 to 194. An alternative to a convenience sample is to systematically pick every Kth member of a population (where *K* is a large enough integer that the sample is small enough to be manageable). As long as there is not some cyclical nature to our data that interferes with Kth sampling, this can be a great way to get data in real time while avoiding the worst risks of convenience sampling. Cluster sampling picks convenient groups – yet because (generally) more than one group is picked, the hope is to average out any isolated data oddities. Stratified sampling goes even further, with researchers carefully thinking through what sorts of features/ demographics might change the results of the data. By ensuring the sample is made up of representatives from every strata, there can be a more robust sample that more closely matches the population. The risk of stratified sampling comes from the researcher's own biases or lack of understanding making it possible to inadvertently leave out a useful strata. Random sampling is typically the most robust in terms of accurately matching the population; it is the most time intensive of real-world sampling methods.

To all this, we must add a consideration of the ethics of asking participants into a study. In the case of medical research, a true random sample of the overall population could be unethical – giving risky medicine to a perfectly healthy person is rarely good.

Before you meet us in the next section, take a moment to let this all settle. Get a snack, stretch, and join us to learn more about just what types of data are in our samples:

- Convenience sample simply polls whoever/whatever can be quickly found – most bias.
- Kth sample takes every Kth person/item (e.g., on the production line).
- Cluster sample is often geographic (e.g., entire warehouse/school).
- Stratified sample controls for "layers" to avoid bias.
- Random sample is generally least biased.

5.7 Frequency Tables

One of the goals behind taking a sample is to understand population data better by reducing the complexity of the population. Another way to reduce complexity is to summarize. Just as an executive summary quickly shares key takeaways of an action report or the abstract shares key ideas of a research publication, so too frequency tables share key features of a large quantity of data.

Take a quick look at this frequency table – it is the `acesData` by `Age`. The first column shows the category, and the second column shows the *frequency* of participants who self-identified their age.

```
##      Age   N
##  1:   18  11
##  2:   19  27
##  3:   20  26
##  4:   21  38
##  5:   22  24
##  6:   23  17
##  7:   24  12
##  8:   25  31
##  9:   26   4
## 10:   NA 142
```

In general, frequency tables have two columns of data. The first column, like the one you just saw, will contain information about a category or feature of the data. The second column will show the frequency – how often the first column category happens in the data set. In the case of our participant data, this helps us see many participants did not care to identify their age.

Thus, the frequency table is faster at helping us see a quick summary of our data set compared to scrolling through 6,599 observations!

Example

Thinking about the `acesData` participants, let us consider the frequency by age. Recall the `i`-rows, `j`-columns, and `by` of the data table structure from our prior studies. In this case, we do want to summarize all our observations. Thus, we have no row restrictions or logic. Next, since our participants fill out many surveys for us, we know that `UserID` is duplicated across rows. If we just want to count participants once each, we will need to use the uniqueN() function on UserID. Lastly, we wish the frequency by age. The column for that data element is Age:

```
acesData[,
         uniqueN(UserID),
         by = Age]
```

```
##      Age  V1
##  1:   21  38
##  2:   NA 142
##  3:   23  17
##  4:   24  12
##  5:   25  31
##  6:   22  24
##  7:   20  26
##  8:   18  11
##  9:   26   4
## 10:   19  27
```

This does not look quite like our initial frequency table. It has all the right pieces; it is just not "finished." Firstly, while you know R has simply invented a column for the new values (V1 is "value 1"), that does not mean much to most people. It is much more common to see a capital N for **N**umber in data summaries. To achieve this, we use the column selection mechanism of .(). Compare and contrast the prior code with the following code:

```
acesData[,
         .( N = uniqueN(UserID)),
         by = Age]
```

```
##      Age   N
##  1:   21  38
##  2:   NA 142
##  3:   23  17
##  4:   24  12
##  5:   25  31
##  6:   22  24
##  7:   20  26
```

```
## 8:  18  11
## 9:  26   4
## 10: 19  27
```

The two bits of code are almost the same. We have just added a bit of code that selects our column unique count calculation; and, while selecting that column, we name it `N`. Notice we have not used any assignment operators. In particular, we did not assign our results back to any variable (such as using `freqTable1 <-`). You can confirm this by seeing no new variables show in our global environment pane. We also have not used the column assignment operator (such as using `acesData[, N := uniqueN(UserID)]`). You can confirm this by clicking the refresh button in the global environment pane and confirming `acesData` has not gained a new column/variable.

In real-life coding, often, frequency tables are simply a way for a data scientist – you and us – to understand their data a bit better. Thus, it is not always required to assign data to a stored variable. There is no right or wrong answer, just a trade-off between speed (it takes a bit longer to type out the assignment) and longevity (there is nothing like saving something to have it later). In practice, it can help to start with not assigning and deciding later if you need to add some code.

However, our table also does not quite look right yet. We need to `order()` the results. Ordering data is ordering all our *rows*, and that function belongs in the `i` part of our data table. Since the `i` portion is currently free, we go ahead and add one more bit to our code:

```
acesData[order(Age),
         .( N = uniqueN(UserID)),
         by = Age]
```

```
##      Age   N
##  1:   18  11
##  2:   19  27
##  3:   20  26
##  4:   21  38
##  5:   22  24
##  6:   23  17
##  7:   24  12
##  8:   25  31
##  9:   26   4
## 10:   NA 142
```

Example

Sometimes, in a frequency table, we want not only the single category frequencies; we want a **cumulative** frequency. The cumulative frequency is simply the running sum or *cumulative* sum of that frequency column, N.

To make it a bit easier (and echoing back to our observation, it is always easy to decide later to assign data to a variable), we go ahead and assign that prior bit of frequency table code to a new variable, acesAgeFrequency. We make one change (besides assignment) to our prior code; we swap out the name N in favor of Freq:

```
acesAgeFrequency <- acesData[order(Age),
                             .(Freq = uniqueN(UserID)),
                             by = Age]
```

The cumulative frequency is going to be built entirely off the Freq column and is going to be a new column in our new data table. As such, it is a column operation, and this time we go ahead and assign our new data to a column named CumulaFreq. Remember, since we are assigning our value to a new column inside the data table, we use the column assignment operator :=. To get the cumulative sum, we use the aptly named cumsum() function. Because that function sums up across each row, there is no need for a by part of our data table.

Since we are using a column assignment, that operation does not return a result to the console. Thus, we also directly call our data table to see the results. Notice, in the CumulaFreq column, that second entry of 38 = 11 + 27. The cumsum() function is just adding up each frequency:

```
acesAgeFrequency[, CumulaFreq := cumsum(Freq)]
acesAgeFrequency

##      Age Freq CumulaFreq
##  1:   18   11         11
##  2:   19   27         38
##  3:   20   26         64
##  4:   21   38        102
##  5:   22   24        126
##  6:   23   17        143
##  7:   24   12        155
```

```
## 8:  25   31          186
## 9:  26    4          190
## 10: NA  142          332
```

Cumulative frequency can help us understand how much of the total data has certain characteristics. In this case, 191 participants choose to self-identify their biological sex. This contrasts with the 142 who choose to not respond to that question. By having the cumulative frequency in the table alongside the frequency, we can see that most participants choose to respond (although it is rather close).

If you were paying careful attention, you might have noticed we only have 191 participants in this study. How did the cumulative frequency get so high? The most likely answer is that participants (who fill out many surveys each remember) do not always enter their age. Thus, the 142 `NA` are counts of participants/`UserID`s who sometimes responded with their age and other times did not. We will explore this more later (and in the exercises). For now, keep in mind traditional statistics classes tended to teach with toy examples (to avoid this sort of data weirdness). You are learning about data and statistics in a hands-on way, so always keep your eye out for results that do not make sense. Often, such results help us spot features (or flaws) in our data sets.

Example

Not all frequency tables have such nice categories. Thinking to the `penguins` data set and flipper lengths, those have a lot of options. A frequency table by length would get messy. While 56 rows is smaller than 344 observations, this is not exactly the greatest summary ever:

```
penguinsData[order(flipper_length_mm),
         .N,
         by = flipper_length_mm]
```

```
##     flipper_length_mm N
##  1:               172 1
##  2:               174 1
##  3:               176 1
##  4:               178 4
##  5:               179 1
## ---
```

```
## 52:                228 4
## 53:                229 2
## 54:                230 7
## 55:                231 1
## 56:                 NA 2
```

This is where the idea of *bins* comes into play. In essence, what we want is to place each penguin in a particular category. One way to do this is to create numerical cut-offs by length to determine category placement. We already used the `seq()` function to create a list for us; let us recycle that idea to create an evenly spaced set of cut-offs that break our `flipper_length_mm` range into even segments. Notice that by having `seq()` do this for us, we are assured of evenly spaced segments. It would give us a weird view of the data, if our segments were not evenly spaced. It would be a not-so-great summary that could be easily misinterpreted.

However, we need a bit of setup first to use sequence this way. Our sequence needs to start at the smallest or `min()` value of `flipper_length_mm`. It also needs to end at the largest or `max()` value. Additionally, we need to decide how many bins we want in total. We will hard-code this example to be five bins. Lastly, we need to take the total length or range of `flipper_length_mm` and break that into our number of bins. That will be the width of each bin. Notice some lengths were not recorded. Those can be removed via `na.rm = TRUE`:

```
minLength <- min(penguinsData$flipper_length_mm, na.rm = TRUE)
maxLength <- max(penguinsData$flipper_length_mm, na.rm = TRUE)

numberOfBins <- 5

binWidth <- (maxLength - minLength)/numberOfBins
```

As you can see in your global environment, the minimum length is 172 mm, the maximum length is 231 mm, and we are splitting our range of 231 – 172 = 11.8 into five bins of widths 11.8 mm:

```
minLength

## [1] 172

maxLength

## [1] 231
```

```
numberOfBins
```

```
## [1] 5
```

```
binWidth
```

```
## [1] 11.8
```

Using these data, we can create a sequence having the correct cut-offs for us:

```
cutOffs <- seq(from = minLength,
               to = maxLength,
               by = binWidth)
cutOffs
```

```
## [1] 172.0 183.8 195.6 207.4 219.2 231.0
```

Next, we use the `findInterval()` function. This function takes an input, compares it to a set of cut-offs, and then outputs an integer telling us in which interval or *bin* that input belongs. In our case, since we have five bins, it will look at each flipper length and place it in bins 1–5 based on those cutoffs.

We use the column assignment operator to store this ordinal value in a new column named `BinOrdinal`. We also look at the first few rows of `penguinsData` and visually confirm that `findInterval()` correctly sorted each `flipper_length_mm`. Remember, done correctly, lengths from 172 to 172 + 11.8 = 183.8 should be in the first bin. As you see, the inspection seems to show we are correctly placed. Notice `findInterval()` naturally copies `NA` over (like it should):

```
penguinsData[ , BinOrdinal := findInterval(flipper_length_mm, cutOffs)]
head(penguinsData)
```

```
##    species    island bill_length_mm bill_depth_mm flipper_length_mm
## 1:  Adelie Torgersen           39.1          18.7               181
## 2:  Adelie Torgersen           39.5          17.4               186
## 3:  Adelie Torgersen           40.3          18.0               195
## 4:  Adelie Torgersen             NA            NA                NA
## 5:  Adelie Torgersen           36.7          19.3               193
## 6:  Adelie Torgersen           39.3          20.6               190
##    body_mass_g    sex year BinOrdinal
## 1:        3750   male 2007          1
## 2:        3800 female 2007          2
## 3:        3250 female 2007          2
```

```
## 4:          NA   <NA> 2007          NA
## 5:        3450 female 2007           2
## 6:        3650   male 2007           2
```

Now that every penguin has been fit into an ordinal rank, we are ready to build our frequency table! We want our table ordered by shortest to longest flipper length, which is a row operation, so `order(BinOrdinal)` goes first. Next, we want a number count, so `.N` will do, which is a new column and thus is a column operation. Lastly, we want those counts by the ordinal rank, so `by = BinOrdinal`:

```
flipperLengthFreq <- penguinsData[order(BinOrdinal) ,
                              .(Freq = .N),
                              by = BinOrdinal]
flipperLengthFreq
```

```
##    BinOrdinal Freq
## 1:          1   25
## 2:          2  131
## 3:          3   59
## 4:          4   84
## 5:          5   42
## 6:          6    1
## 7:         NA    2
```

Again, should a cumulative frequency be required or useful, we can readily assign `cumsum()` to a new column:

```
flipperLengthFreq[, cumulaFreq := cumsum(Freq)]
flipperLengthFreq
```

```
##    BinOrdinal Freq cumulaFreq
## 1:          1   25         25
## 2:          2  131        156
## 3:          3   59        215
## 4:          4   84        299
## 5:          5   42        341
## 6:          6    1        342
## 7:         NA    2        344
```

5.8 Summary

This chapter explored ways to understand populations and to think through different ways to reduce complexity by pulling a sample. Different sample methods give us the ability to choose what is most important to our research. If we are designing a new landing page for our website, a convenience sample of the first 100 people to click our site will give us fast results that may be "good enough." On the other hand, if we are doing equity research, stratified or random samples may take more time to pull together, yet can reduce bias.

You also applied programming techniques from prior sections to explore real-world (and sometimes messy) data to get a feel for what these sampling methods look like. Along the way, you learned some new functions especially suited to sampling and frequency tables. Practice makes perfect, so as you work through the examples, checks, and exercises, be sure to reference Table 5-2 for key ideas you learned in this chapter.

Table 5-2. *Chapter Summary*

Idea	What It Means
Population	The entire group under study (often large).
Parameter	Population-level or "total" data.
Sample	A subset of the population under study (small enough to manage).
Statistic	Sample-level or "partial" data.
Qualitative	Text or literary data – not meaningfully numeric.
Quantitative	Numeric data.
Nominal	Qualitative or categorical data.
Ordinal	Quantitative, ordered data (e.g., "first").
Interval	Quantitative, "decimal" data (e.g., temperature or time).
Ratio	Quantitative, data with a meaningful zero (e.g., age, money).
Observational study	Researchers only watch – do not intervene.
Experimental study	Researchers deploy interventions, likely a control group.
Placebo	A harmless "intervention" to ensure control group does not know it is control group.

(*continued*)

Table 5-2. *(continued)*

Idea	What It Means
Ethics	Hugely important in good research – participants must give informed consent.
Bias	A "bad" sample may not reflect the population well.
Confounding variable	An unexpected feature in a sample that biases results.
Convenience sample	Simply poll whoever/whatever can be quickly found. Most bias.
Kth sample	Every Kth person/item is sampled (e.g., on the production line).
Cluster sample	Often geographic (e.g., entire warehouse/school).
Stratified sample	Researcher controls for "layers" to avoid bias.
Random sample	Takes more time yet is generally least biased.
`sort()`	Sorts a set of numbers/letters. Similar to `order()`.
`order()`	Used to order rows **inside** a `data.table`.
%%	Modulo (aka long division remainder).
`seq()`	Creates sequence using from/to/by.
`nrow()`	Counts number of rows/observations in a dataset.
`unique()`	Reduces data down to no duplication.
`uniqueN()`	Counts data reduced to no duplication.
`as.Date()`	Converts "YYYY-MM-DD" text string to a data format in R.
`as.data.table()`	Converts to `data.table` format (can use `keep.rownames = TRUE`).
`.SD[]`	`data.table` Subset Data operator; does row ops AFTER by operation.
`set.seed()`	Ensures we share same fake "random" sample; do not use in real life.
`sample()`	Pulls a random sample of given size.
`.()`	`data.table` column selection function.
`cumsum()`	Cumulative sum adds each row to a running total.
`min()`	Returns minimum number in a data set.
`max()`	Returns maximum number in a data set.
`findInterval()`	Checks a value against cut-offs and returns ordinal.

5.9 Practice for Mastery

Check your progress and grow through practice by working through some exercises. Comprehension checks ask critical thinking questions that may be best answered with a written or verbal response. Part of the art of statistics is successfully communicating results to your stakeholders or audience. Sometimes that audience is highly technical and other times very much not technical. Exercises are more direct applications of the concepts explored in the chapter.

Comprehension Checks

1. In your own words, what is the difference between a population and a sample?
2. Can you think of one additional type of data for each of our four types (i.e. nominal, ordinal, interval, ratio)?
3. A **confounding variable** exerts an unexpected influence. For example, suppose you are researching the range of human temperatures. Your population is "all humans." You happen to know that, on entry to a hospital, patient temperature data are recorded. Thus, we might imagine using those temperatures to find the usual range of temperatures. You even (somehow) have access to all hospital data worldwide for the last 20 years. That is a large (cluster) sample! A confounding variable, in this case, would be "hospital patient." We might expect patients to have different temperatures than all the humans who were not hospital patients. Now think of the last poll or survey you were spammed with. What might a confounding variable be for those data?
4. A researcher at a college is curious what might happen if all the students are required to take courses online. The population is thus "all students at the college." Rather than take a sample, a poll is emailed to all students asking if they have regular access to email and computer networks. After three days, the poll is closed, and two out of every five students responded. Even before looking at the data, what might be a confounding variable?

5. In the preceding example, you are called in to consult. You explain that even though it might look impressive to email the whole population, sometimes sampling is a better way to research. What method of sampling (convenience, Kth, cluster, stratified, random) do you recommend? How would you go about capturing a sample of the method you recommend?

Exercises

1. What is the difference between `sort(penguinData$flipper_length_mm)` and `penguinData[order(flipper_length_mm), flipper_length_mm]`? What happens if you switch those functions around accidentally?

2. In one of this chapter's examples, you saw a frequency table on `penguinsData` by `flipper_length_mm`. Change just three parts of this code to move from penguins data to `mtcarsData` and from `flipper length mm` to `cyl`:

```
penguinsData[order(flipper_length_mm),
           .N,
           by = flipper_length_mm]

##     flipper_length_mm N
##  1:               172 1
##  2:               174 1
##  3:               176 1
##  4:               178 4
##  5:               179 1
## ---
## 52:               228 4
## 53:               229 2
## 54:               230 7
## 55:               231 1
## 56:                NA 2
```

3. Now that you have a functioning frequency count of the cylinders of `mtcarsData`, go ahead and rename that column (using the methods used earlier in the chapter) to `Freq` and add a cumulative frequency column named `CumulaFreq`. Bonus points if you copy, paste, and modify code from earlier in the chapter. When writing code in real life, one often copies and pastes from the Web. In fact, the functions you use (such as `cumsum()`) are essentially easy ways for you to copy and paste code from other programmers like you. We all stand on the shoulders of giants to look a little further ahead.

4. Explore the idea of using `set.seed(1234)` a bit more. We used the function `sample(x = stuff, size = n)` to let `R` take care of the random selection of sample data. To visually see and understand the use of `set.seed()`, run the following bits of code on your computer several (three or more) times:

```
sample(x = 1:10, size = 1)

## [1] 10

sample(x = 1:10, size = 1)

## [1] 5
```

Do your results on your computer always match ours? Are your results always the same? Now, go ahead and use `set.seed()`. Do your results match ours now?

```
set.seed(1234)
sample(x = 1:10, size = 1)

## [1] 10

set.seed(NULL) #this cancels the use of set.seed()
```

CHAPTER 6

Descriptive Statistics

Statistics is the mathematics of understanding sample data. The larger goal is to use our understanding of a sample to better understand a population. The smaller, more immediate goal is to develop a regular set of tools and workflow to help us understand sample data. Because understanding often requires *exploring* a data set, **e**xploratory **d**ata **a**nalysis (EDA) is a modern term given to this stage. As you explore a sample, you will understand the sample data well enough to *describe* those data. Thus, this process is also called **descriptive** statistics.

By the end of this chapter, you should be able to

- Evaluate a data set for types of qualitative/quantitative data.
- Create plots and other visual aids to understand large quantities of data quickly.
- Apply central tendency, position, and turbulence calculations to summarize data.
- Remember your short-term goal is understanding the sample, with a long-term goal to project onto the population.

6.1 R Setup

As usual, to continue practicing creating and using projects, we start a new project for this chapter.

If necessary, review the steps in Chapter 1 to create a new project. After starting RStudio, on the upper-left menu ribbon, select *File* and click *New Project*. Choose *New Directory* ➤ *New Project*, with *Directory name* `ThisChapterTitle`, and select *Create Project*. To create an `R` script file, on the top ribbon, right under the word *File*, click the small icon with a plus sign on top of a blank bit of paper, and select the *R Script* menu option. Click the floppy disk–shaped *save* icon, name this file `PracticingToLearn_XX.R` (where XX is the number of this chapter), and click *Save*.

M. Wiley and J. F. Wiley, *Beginning R 4*, https://doi.org/10.1007/978-1-4842-6053-1_6

In the lower-right pane, you should see your project's two files, and right under the *Files* tab, click the button titled *New Folder*. In the *New Folder* pop-up, type `data` and select *OK*. In the lower-right pane, click your new *data* folder. Repeat the folder creation process, making a new folder titled `ch06`.

Remember all packages used in this chapter were already installed on your local computing machine in Chapter 2. There is no need to re-install. However, this is a new project, and we are running this set of code for the first time. Therefore, you need to run the following `library()` calls:

```
library(data.table)

## data.table 1.13.0 using 6 threads (see ?getDTthreads). Latest news:
r-datatable.com

library(ggplot2)
library(palmerpenguins)

library(JWileymisc)
library(extraoperators)
```

In addition, go ahead and set our three usual data sets into our familiar variables and `data.table` format. Something to think about, in terms of real-life use of statistics and `R`, are your "usual" data sets. Having a nice structure and process for getting those data into action (like we do in the following lines of code) can make a daily workflow much easier to start:

```
acesData <- as.data.table(aces_daily)
penguinsData <- as.data.table(penguins)
mtcarsData <- as.data.table(mtcars, keep.rownames = TRUE)
```

You are now free to learn more about statistics!

6.2 Visualization

One amazing feature of `R` is the ability to quickly generate pictures and graphs to visualize entire data sets. Much like frequency tables, graphs provide an easy way to understand features of our data set. Remember the types of qualitative and quantitative data we studied? Different graphs work with different types of data.

Often, when doing exploratory graphing, you will find misconceptions or concerns about your data come to light. In that case, often the answer is to perform data manipulation of some type to *clean* or convert your data into a usable format (this is sometimes called data *munging*).

Histograms

Histograms are very similar to frequency tables. They are designed to work with quantitative, interval, or ratio-style data. The data should be set up in such a way that each copy of the data represents a single item (unduplicated data). Contrast this with our `acesData` column `Age`, which has many copies (due to each participant taking multiple surveys over multiple days).

As you explore the histogram examples, be on the watch for the types of data input and think about whether the data are duplicated.

Example

Recall, from Chapter 5, the `penguinData` variable of `flipper_length_mm` was binned and then counted into a frequency table. You may recall we chose to bin our data in that frequency table into five categories. While there are histogram graphs where we can control the number of bins precisely, the `hist()` function by default uses an *algorithm* to decide a sensible number of bins. Algorithms are logic programming that make choices using some process. While one could make a study of its own just on methods of deciding bins for frequency tables or histograms, for this book, we simply go with the default choice.

Because our data for `flipper_length_mm` are lengths, which are quantitative, ratio data, it makes sense to use a histogram. It also makes sense to use a histogram because the `penguin` data set is unduplicated - every row is a unique observation of a specific penguin:

```
hist(x = penguinsData$flipper_length_mm)
```

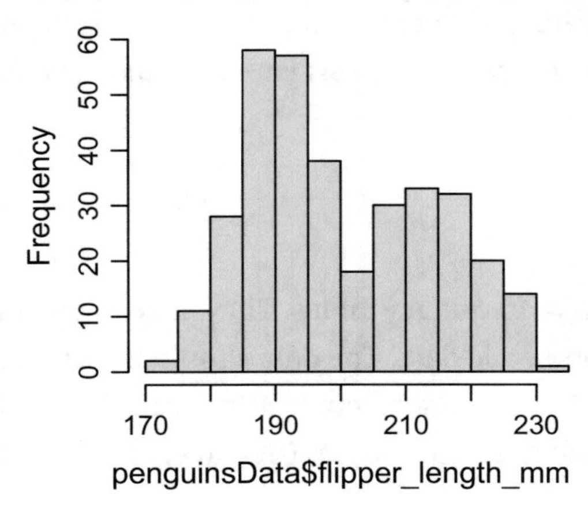

Figure 6-1. *penguinsData histogram*

In Figure 6-1, the horizontal x-axis is labeled with the approximate lengths of the flippers. In other words, that first, leftmost bin contains flippers of lengths 170–175 mm. The vertical y-axis shows the frequency – how often a particular bin of flipper lengths occurs. In the case of that leftmost bin, you can see not many penguins have flippers with lengths between 170 and 175 mm.

What we see is the lengths of flippers are not *uniform*. Instead, some bins have more flippers than other bins. We knew this on some level already – some of our sampling methods in Chapter 5 gave us too narrow a range.

Take a moment to look at the histogram. Pretend the histogram was on a wall and you had a dart in your hand. You close your eyes and throw your dart randomly at the histogram. It is perhaps easy to see your dart is not likely to hit the 230-and-up bin. That is a very small bin indeed – and not much area to hit. On the other hand, you might well hit somewhere between 180 and 200 – it is narrow, yet tall. Or you might hit somewhere between 205 and 220. Those are likely places to randomly get.

Why do we talk about random? Well, the random sample method is R's way of chucking a dart at the histogram. When we say the random sample method is *unbiased*, what we mean is randomly selecting data into our sample ought to get us a sample that looks somewhat close to our actual data.

Take a moment, and go back to Chapter 5 where you worked through an example to build the `randomFlippers` variable. Did that random pull mostly return values either between 180 and 200 or 205 and 220? If it did, we might start to trust random samples do a pretty good job, in general, of helping us avoid sample bias.

Example

We spent a fair bit of time exploring Age from our aces_daily data set. Remember histograms are designed to work with interval or ratio data that is unduplicated. So what is wrong with just using our age data as it is given? After all, Age is definitely qualitative, ratio data, right?

We start by just using the raw data in the Age column in the hist() function. The first thing to notice about Figure 6-2 is the frequencies are quite high. This data set has only 191 unique participants, yet each participant fills out the survey many times. Recall in Chapter 5 we had some challenges with that in a frequency table. The next thing to notice is that hist() completely ignored and dropped the NA entries. There is no sign of those data points in this plot:

```
hist(x = acesData$Age)
```

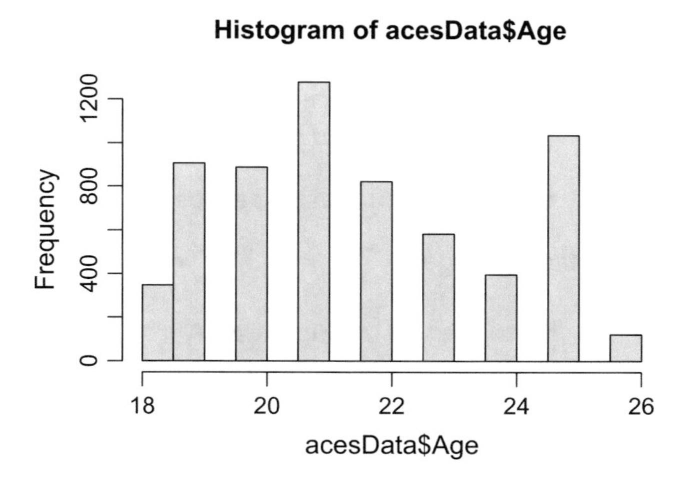

Figure 6-2. *Duplicated aces Age histogram*

So we now see some challenges to simply depending on hist() alone to understand data. While ages in general are qualitative, ratio data, our Age column is not quite ratio-level data. The presence of NAs (which stand for *not available*) is a sign of a *category*. The histogram knew it could not cope with that sort of data and simply dropped it.

Another challenge with what we see in Figure 6-2 is that taking the surveys is optional. Thus, it looks like we have many more 21-year-old participants than we do 18-year-old participants (look at the height of the first and fourth bars). What if 21-year-old participants are simply more conscientious at filling out surveys for some reason?

We are trying to understand just the ages of our participants. Thus, we have a bit of data cleanup to achieve. Remember the function `unique()` removes duplicates. Additionally, using the `.()` column selection function in `data.table`, we can select only specific columns. Now, we for sure want the Age column. However, we want the frequency of the ages to match up to each user/participant. If we blindly hit the age column, by itself, with `unique()`, we would only have one copy of each age (and get a very boring histogram).

Armed with this mental idea of what needs to happen, we can consider the data.table process. First, we know histograms only work on numerical data; thus, we have no need to pick rows that have missing values. That is a row selection problem. The function `is.na()` would give us all the rows in a particular column that *are* NAs. We want the rows that are *not*. Thus, you want to use the logical negation sign `!`. As for columns, by including both the `userID` and the Age columns, you can be sure each user will have each age only represented once.

Study the code that follows, see how each goal is achieved, and consider the resulting histogram in Figure 6-3.

```
unduplicatedAcesAge <- acesData[! is.na(Age),.(UserID, Age)]

unduplicatedAcesAge <- unique(unduplicatedAcesAge)

hist(x = unduplicatedAcesAge$Age)
```

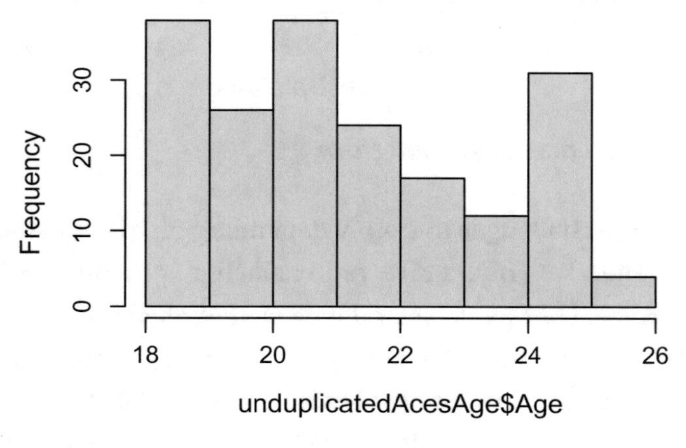

Figure 6-3. *Unduplicated aces Age histogram*

Right away we see some visual differences. The height of the frequencies is much lower. Notice 18 years and 21 years show up at the same frequency. There was indeed a difference between the age groups in survey completion! As a sidenote, it may well not have anything to do with age directly – there may be a *confounding* variable that motivates the 21-year-old participants to complete more surveys. In fact, there was such a variable in that study. Participants accrued a token amount on a gift card to a local grocery store for each survey completed. It may be those of 21 years are more likely to value grocery vouchers.

Did we lose any participants when we dropped all the NAs? Based on the frequencies (and the number of rows in unduplicatedAcesAge), it would seem the answer is not many. Rather than visually inspect UserID for the missing value, we ask R to find it for us using the setdiff() function.

This function takes an input x and compares it to a second group y. If any elements of x are missing from y, it will return those data points. Despite the number of NAs in our frequency table, it seems participants simply get tired of repeating their age three times a day. In fact, only one participant fully declined to report their age, and that participant has UserID = 107:

```
setdiff(x = 1:191,
        y = unduplicatedAcesAge$UserID)

## [1] 107
```

Notice quite a bit is going on in this code. While it is nice to see results right away, it can sometimes help to explore an idea more after seeing it in use. Let us take a moment together to see all the pieces and logic to the preceding code.

First, setdiff() simply finds the difference between two groups or sets of data. In particular, it only returns the items in the x set which are not in the y set. Notice in both the examples that follow, only the number 107 is returned in both cases. Even though the 500 in the y field is not in the first group, setdiff() **only** gives us "things in x not found in y":

```
setdiff(x = c(107, 108),
        y = c(108))

## [1] 107

setdiff(x = c(107, 108),
        y = c(108, 500))

## [1] 107
```

Because `setdiff()` works this way, to find any missing users (the user(s) who might have been dropped when we did `!is.na(Age)`), we must make `x` be all possible users. R is happy to give us a whole set of numbers (or letters). To get all the integers between two values, simply put a colon in the middle (e.g., `1:191`). To get the lower or upper letters, use either `letters` or `LETTERS`:

```
1:191

##   [1]   1   2   3   4   5   6   7   8   9  10  11  12  13  14  15  16
##  [17]  17  18  19  20  21  22  23  24  25  26  27  28  29  30  31  32
##  [33]  33  34  35  36  37  38  39  40  41  42  43  44  45  46  47  48
##  [49]  49  50  51  52  53  54  55  56  57  58  59  60  61  62  63  64
##  [65]  65  66  67  68  69  70  71  72  73  74  75  76  77  78  79  80
##  [81]  81  82  83  84  85  86  87  88  89  90  91  92  93  94  95  96
##  [97]  97  98  99 100 101 102 103 104 105 106 107 108 109 110 111 112
## [113] 113 114 115 116 117 118 119 120 121 122 123 124 125 126 127 128
## [129] 129 130 131 132 133 134 135 136 137 138 139 140 141 142 143 144
## [145] 145 146 147 148 149 150 151 152 153 154 155 156 157 158 159 160
## [161] 161 162 163 164 165 166 167 168 169 170 171 172 173 174 175 176
## [177] 177 178 179 180 181 182 183 184 185 186 187 188 189 190 191

letters

##  [1] "a" "b" "c" "d" "e" "f" "g" "h" "i" "j" "k" "l" "m" "n" "o" "p"
## [17] "q" "r" "s" "t" "u" "v" "w" "x" "y" "z"

LETTERS

##  [1] "A" "B" "C" "D" "E" "F" "G" "H" "I" "J" "K" "L" "M" "N" "O" "P"
## [17] "Q" "R" "S" "T" "U" "V" "W" "X" "Y" "Z"
```

With that, we have all the pieces we need to quickly spot any missing user:

```
setdiff(x = 1:191,
        y = unduplicatedAcesAge$UserID)

## [1] 107
```

Dot Plots/Charts

Frequency tables and histograms are very good at providing quick summaries of data. By fitting data into categories or bins and providing a count, large quantities of data can be rapidly read by information consumers and data researchers like you. As such, they are useful tools in exploratory data analysis. However, the choice of bin can sometimes hide important information – sometimes seeing all the data can help reveal a trend or bias.

Dot plots or **dot charts** are one way to see data. The graph in Figure 6-4 shows one circle for every penguin flipper length. The horizontal x-axis shows the length. Notice the vertical height is not a measure in this case. Instead, every flipper is getting its own vertical position – this allows us to see all the dots (with a bit of overlap). Compare and contrast Figure 6-4 with Figure 6-1. Both are entirely showing the same underlying data. The histogram is a good fast-to-understand summary, while the dot plot shows much more detail (and is still easier to glance over than the raw data):

```
dotchart(x = penguinsData$flipper_length_mm)
```

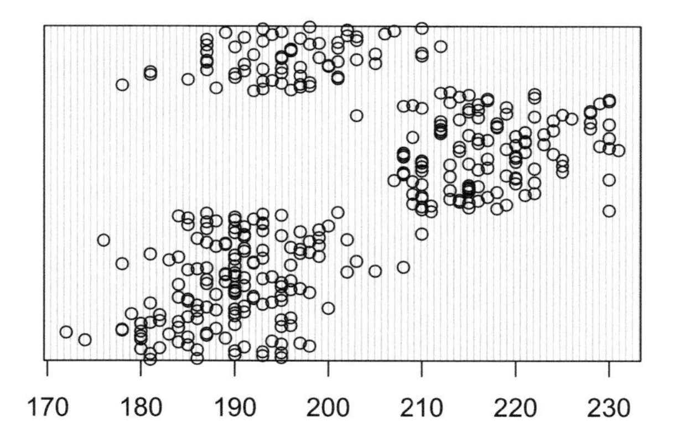

Figure 6-4. *Penguin dot plot*

Descriptive statistics – exploring our data – usually involves using several different methods or tools to help us understand what types of data we have. Unpacking the code behind Figure 6-4 adds another skill to your "usual process" for understanding data and thinking statistically.

Example

The R code creating a dot chart is quite simple. The function is `dotchart()`, and we are interested in two variables it takes (while it has more variables, we will explore more advanced visualization later). The first variable, `x`, takes the raw, numeric data that will create the horizontal x-axis. Thus, like histograms, dot charts assume our data are quantitative and either interval or ratio. The second variable, `groups`, is optional. This variable takes either nominal or ordinal data.

In Figure 6-4, we only used the first variable `x = penguinsData$flipper_length_mm`. In Figure 6-5, we keep that first variable and add in `group = penguinsData$species`:

```
dotchart(x = penguinsData$flipper_length_mm,
         groups = penguinsData$species)
```

This group feature is telling, because it shows that some species of penguin seem to be distinguishable based on the length of their flippers.

What does this tell us about sample methods based on species clusters or convenience samples? Is it easy to see how our understanding of what penguin flipper lengths are changes if we only had one species? While we discussed bias and confounding variables in Chapter 5, Figure 6-5 is our first solid visual of how sample bias might occur.

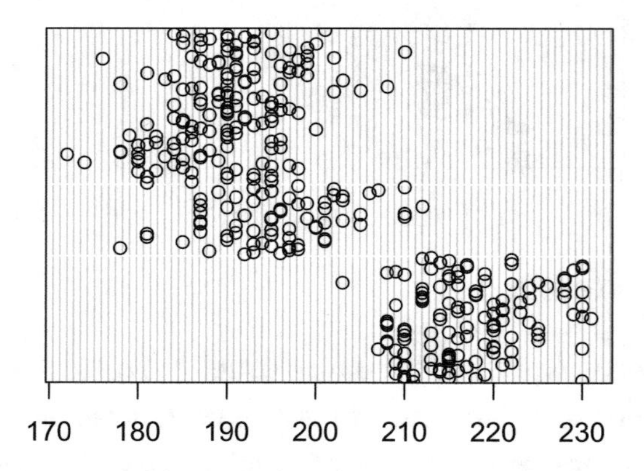

Figure 6-5. *Penguin dot plot by species*

ggplot2

Both hist() and dotchart() are useful functions for exploratory data analysis. That said, both graphs are not very customizable. This can make it tough to change perspective. What if you wanted to group by more than one category (e.g., viewing age by socioeconomic status and education attainment)? Often, early explorations lead to further questions and ideas. Now, please do not misunderstand – we (and many other data scientists) use hist() and dotchart() early on *because* they are not customizable. They are really fast to get an initial view of the data! It is only on the second pass that more detail may become necessary.

The ggplot2 package [17] develops a more comprehensive visualization process. In fact, *gg* stands for ***g**rammar of **g**raphics*. For now, we need only scratch the surface of this framework. It will help us see more clearly and give you some solid, marketable skills you can leverage. Being able to visualize using ggplot2 in RStudio may be something to mention during job interviews.

The power of this package lies in something you already know. Both Figure 6-1 and Figure 6-4 are built on exactly the same data. Yet, we had to consider them in two different ways. Conceptually, however, it was the same data on the same coordinate system – just two different ways of organizing the plot.

At the heart lies the ggplot() function, which takes two variables. The first variable, data =, holds our data.table element. This saves us from having to constantly repeat ourselves as we call data from every column. The second variable is mapping = which connects or *maps* columns in our data set to our visuals or *aesthetics*. Aesthetics are visual choices, such as choosing the x-axis to be flipper_length_mm or a group to be filled in with a certain color.

An example of this base is shown in Figure 6-6. By setting data = penguinsData, there is no need to reference a column by using the penguinsData$flipper_length_mm we used earlier. Instead, we simply set x = flipper_length_mm directly.

In this grammar, the data in the flipper_length_mm column is mapped to the x-axis *aesthetic* (or visual). While the following code creates the base of our graph, we have not specified the *geometry*. In ggplot(), *geometries* are the visual shapes one sees (e.g., a histogram is a geometry). Thus, while you do see the x-axis has the correct range of points (from 1 to 7) and is properly labeled, everything is presently blank:

```
ggplot(data = penguinsData,
       mapping = aes(x = flipper_length_mm))
```

That is all about to change. From this building block, additions can be made to ggplot() via literal use of the + operator. These additions are usually specific geometric specifications. Our earlier graphs can be recreated using this grammar.

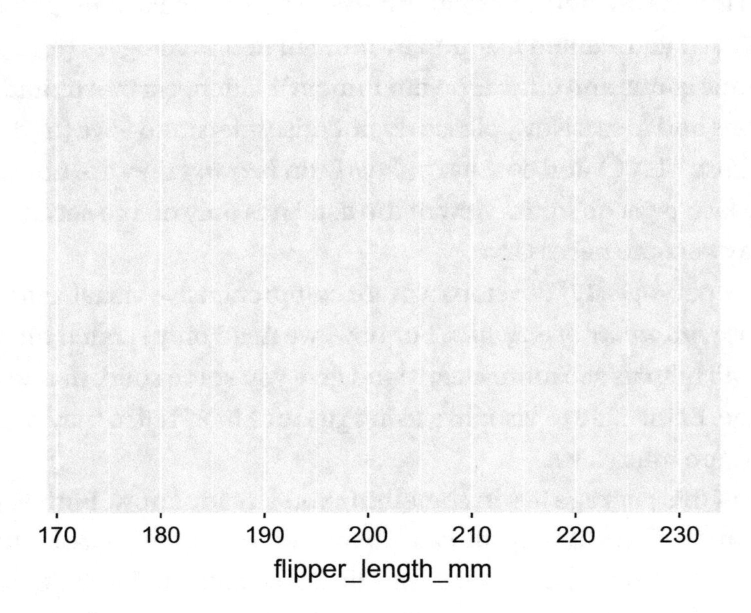

***Figure 6-6.** ggplot() base settings, no geometry*

To create a histogram, we simply add on to our base graph the geometry of histogram. This is achieved with the geom_histogram() function. The resulting visual is shown in Figure :

```
ggplot(data = penguinsData,
      mapping = aes(x = flipper_length_mm))+
  geom_histogram()
```

```
## 'stat_bin()' using 'bins = 30'. Pick better value with 'binwidth'.
```

```
## Warning: Removed 2 rows containing non-finite values (stat_bin).
```

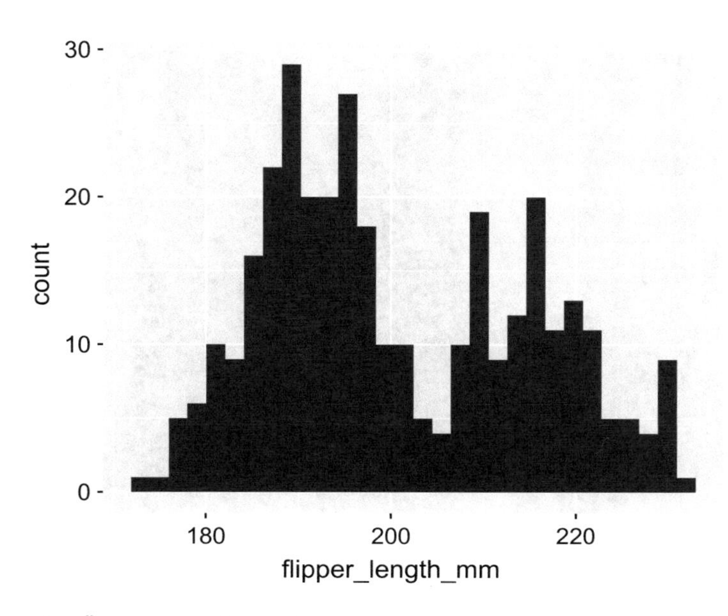

Figure 6-7. *ggplot() base settings, histogram*

On the other hand, if we wish our dot plot to show us distinct data points, that can be approximated as well using a different geometry. Keep in mind, in this grammar, our base stays the same as long as we are using the same data and wanting the same x-axis aesthetic. Instead, only the layout or geometry is changing; so the only change is to use the geom_dotplot() as the addition to achieve Figure 6-8.

```
ggplot(data = penguinsData,
       mapping = aes(x = flipper_length_mm))+
  geom_dotplot()
```

```
## 'stat_bindot()' using 'bins = 30'. Pick better value with 'binwidth'.
```

```
## Warning: Removed 2 rows containing non-finite values (stat_bindot).
```

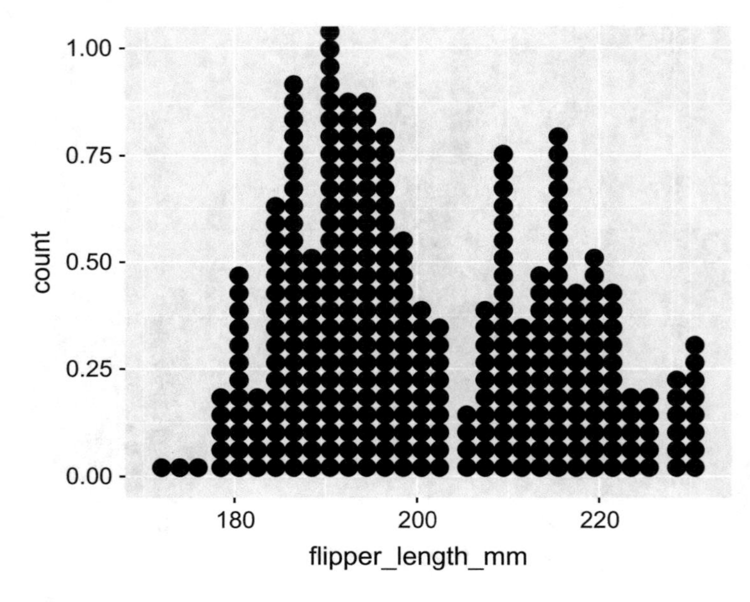

Figure 6-8. *ggplot() base settings, dot*

For now, ignore the wonky numbers on the y-axis. Because each dot represents a genuine point in the data, you can visually see the actual count. The y values are actually representing physical position of each dot (and we will learn later on how to hide them when needed). We ignore the many frills of `ggplot()` to focus on the grammar of these graphical sentences and consider three highly instructive examples. These examples will take us deeper in our exploration of the data.

Example

Recall our penguins come in three distinct species. Now, back in Figure 6-5 we tried to visualize that feature. It was not exactly a success. We did not even have labels for our species!

Our data set column is `species`, and that column must be mapped to an aesthetic choice before we can use it. Labeling each dot could get messy. The solution is to `fill` each dot with a color mapped to each species. This is achieved via the *fill* aesthetic:

```
ggplot(data = penguinsData,
       mapping = aes(x = flipper_length_mm,
                     fill = species))+
  geom_dotplot()
```

```
## 'stat_bindot()' using 'bins = 30'. Pick better value with 'binwidth'.

## Warning: Removed 2 rows containing non-finite values (stat_bindot).
```

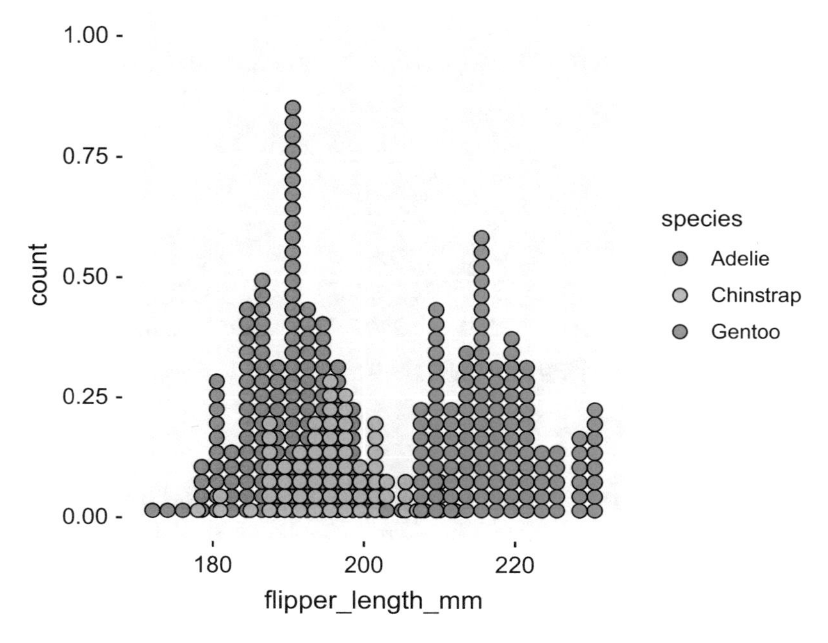

***Figure 6-9.** Penguin dot plot with fill = species*

For the first time, you see something very interesting in the penguin data in Figure 6-9. Notice each species tends to have flippers of similar lengths. In fact, notice for each single species, only a handful of penguins have the shortest or longest flipper lengths for that species. The dots pile up tallest in the *middle* of the lengths for each species.

Take a closer look at this phenomenon in Figure 6-10, and forget about the R code for a moment. For each species, the most common flipper lengths are in the middle:

```
## 'stat_bindot()' using 'bins = 30'. Pick better value with 'binwidth'.

## Warning: Removed 2 rows containing non-finite values (stat_bindot).
```

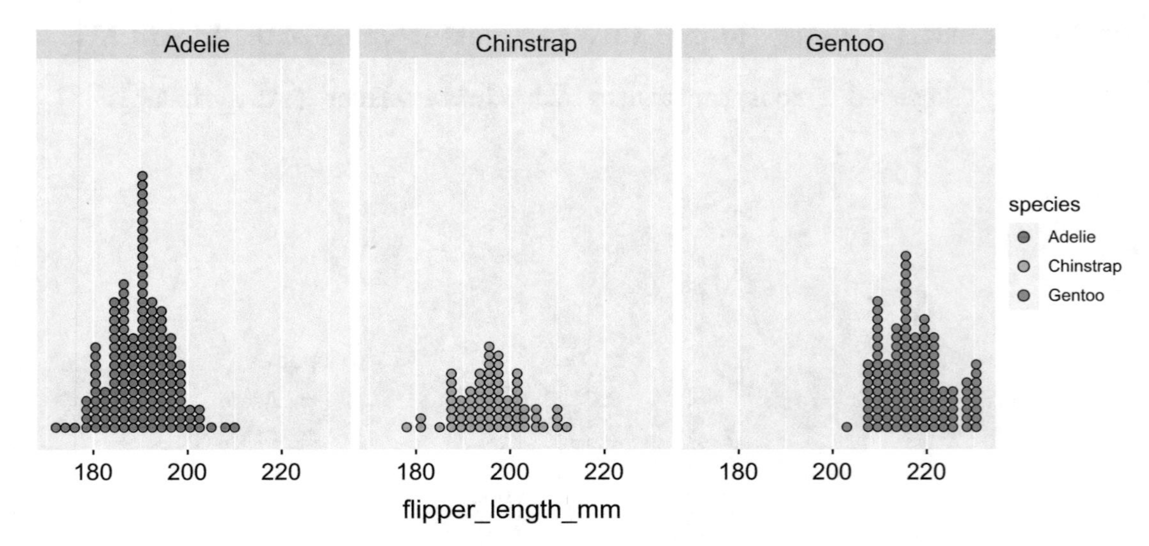

Figure 6-10. *Penguin dot plot by species*

This is almost a perfect reflection for Chinstrap (this is called **normal**). However, Adelie and Gentoo both have a bit of a tail on the right (this is called **right skew**). If the data trailed off the other direction, this would be called **left skew**.

Example

When trying to understand what a sample looks like, it can help to have a comparison. Take a look at Figure 6-11. In the prior example, you learned the word **normal** for a set of data where the most common values are in the middle and the least common are on the edges. That is the shape you see on the left side of Figure 6-11. On the right side, you are seeing what is called **uniform** data; it looks rather like a rectangle.

Both data mostly seem to range from 0 to 5 (give or take). In fact, both these shapes have the same *area* shaded! They both represent 10,000 data points (generated via computer simulation to show you nearly perfect examples of normal and uniform).

When you look at a histogram from sample data, you want to compare it to normal or uniform. Does it look closer to one or the other of these shapes? Remember it is okay if it is *almost* normal. Depending on which side the data trail off, that just makes it a left- or right-skewed version of normal.

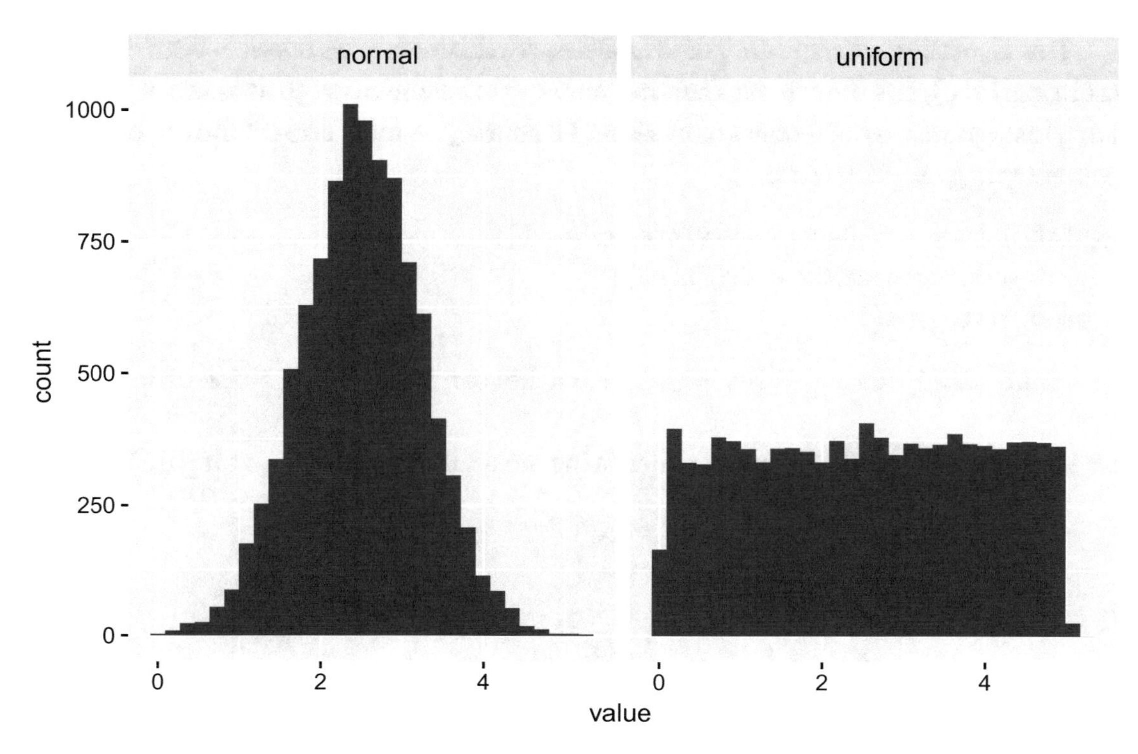

Figure 6-11. *Nearly perfect normal and uniform histograms*

Now, the penguin data you saw were really quite normal (with just a bit of skew – Gentoo had the most). What about some of our participant data from the aces daily surveys? You are now armed with an idea of both normal and uniform data, and you know how to make `ggplot()` work. Thus, you can safely venture a little further into the `acesData`. Perhaps we can explore `COPEExp`, which is an emotional expression coping measure on a 1–4 scale that is done on the evening survey.

The evening survey is the third survey of the day in this data set; thus, it makes sense to restrict our plot data to only those rows in the data table which are part of the third survey. That is a row operation on our `data.table`, so set that ith row selection via `SurveyInteger == 3`.

These data are repeated often; unlike age which is a one-time thing, the participants are meant to record their emotional expression coping every evening of the study. Thus, there is no need to perform any `unique()` work on these variables.

This is just some exploratory data analysis, so a histogram works well. With the data set in `ggplot()`, it is time to map column name(s) to aesthetics such as x-axis or y-axis. For a histogram, we need only set an x-axis. Of course, we must also set the display geometry to `geom_histogram()`:

```
ggplot(data = acesData[SurveyInteger == 3],
       mapping = aes(x = COPEExp))+
  geom_histogram()
```

```
## 'stat_bin()' using 'bins = 30'. Pick better value with 'binwidth'.

## Warning: Removed 264 rows containing non-finite values (stat_bin).
```

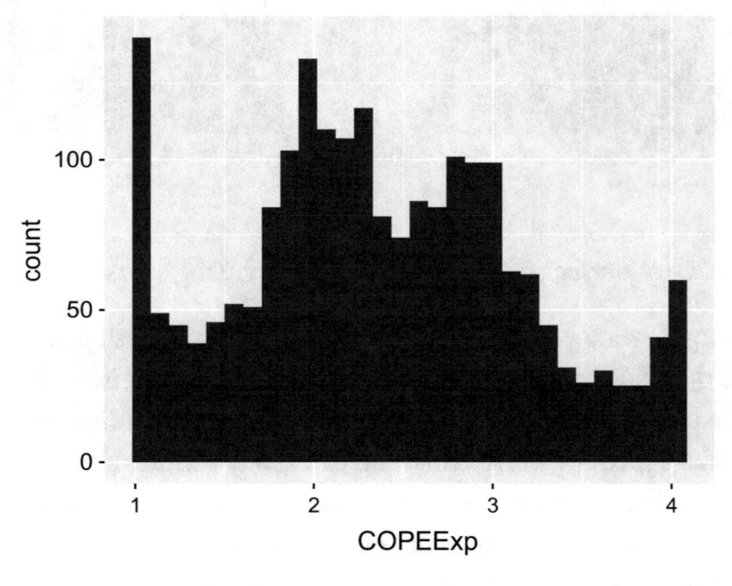

Figure 6-12. *Histogram of COPEExp*

Looking at Figure 6-12, are these data normal? Normal yet skewed? Uniform? Perhaps none of these?

Think back to some recent surveys you took. If you were bored with the survey, did your answers get more extreme? Sometimes, the end points of a data set get confusing. One technique is to **trim** the data set to see if the middle of the data fits a pattern. Now, this must be done cautiously, and with good reason. An arbitrary trim of the data just to force some normal or uniform features on the sample would be silly (and lead to wrong conclusions).

All the same, just to see what a trim looks like, go ahead and remove the extreme values of 1 and 4. To do this is a row selection operation. We only want the COPEExp values that are greater than 1 and less than 4. Recall our logical operator for that – %gl%. Other than that one change, our code is the same:

```
ggplot(data = acesData[SurveyInteger == 3 &
                        COPEExp %gl% c(1,4)],
       mapping = aes(x = COPEExp))+
  geom_histogram()

## 'stat_bin()' using 'bins = 30'. Pick better value with 'binwidth'.
```

With that change, Figure 6-13 looks almost normal. In fact, it looks rather like our penguin species did – it has two spots in the data that are common. These trimmed data are close to normal, perhaps with a bit of a right skew.

The final thought for this section is for you to keep in mind that while this book will walk you through many different patterns and models, most real-world data does not look like Figure 6-11. Thus, when working with data, one must be somewhat flexible. All the same, the techniques you learn and practice in this book earn you two useful skills.

The first is a greater fluency with the foundations of statistics and data science. Handling messy, real-world data can require fancier models than normal or uniform. That does not mean you cannot get quite close (or even close enough) using these two. The other skill is simply experience. You have seen how easy it is to switch ggplot() from histograms to dot plots. With the experience you earn here, it will not take much work to go a further step or two when you encounter wild data that needs some extra care.

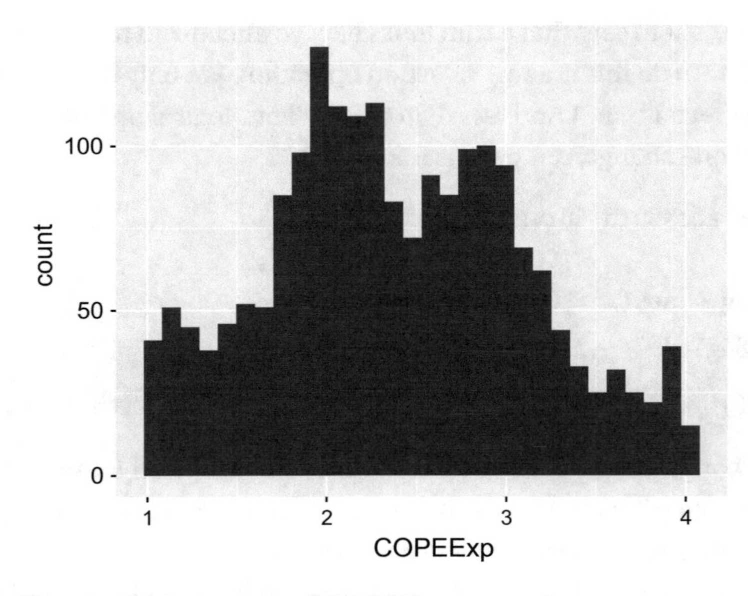

Figure 6-13. *Trimmed histogram of COPEExp*

6.3 Central Tendency

The word "middle" has come up quite a bit in the preceding sections. However, there are some synonyms for middle we quite painstakingly avoided, even though these are the parts of statistics most people are already familiar. We have not said the word "average" or "mean" or "median." Now is the time to fix that and talk about a single number that can collapse an entire column of data down to one central tendency.

Remembering stories is often easier than memorizing a thousand seemingly different facts. Your goal is to understand populations by understanding easier-to-access, manageably sized samples. Even so, samples often are quite large. You already know some good visuals to summarize the numeric behavior of a sample. Central tendency is all about finding some usual, middle, or common numbers to describe the "most likely" values of our sample.

In everyday life, people often use the word "average" to describe some of these ideas. Because we are getting mathematically precise here, we will mostly avoid that word. Instead, we will describe three different ways to compute an "average" that will often yield different answers. By comparing and contrasting these differences, you not only will learn some good ways to summarize a sample; you will also learn more about a sample.

Arithmetic Mean

Most commonly, when people say "average" in a mathematical sense (e.g., the average income in Victoria), they are referring to the **arithmetic mean**. Suppose you have two people and one person earns $0 and the other earns $100,000. Sum the two numbers together, divide by the two people, and you have an average of $50,000 salary.

While we could do the calculation by hand, as you can see, that is not very informative. Tougher for some random reader to see what we are doing:

```
(0 + 100000)/ 2
```

```
## [1] 50000
```

A better way would be to use the R function for arithmetic mean, `mean()`, on a variable we create that holds those two salaries:

```
salaries <- c(0, 100000)
mean(salaries)
```

```
## [1] 50000
```

What if we had three people in our sample? Suppose we added one more person into the mix who had a salary of $57,364. Well, you are now summing three incomes together and dividing by 3, not 2:

```
(0 + 100000 + 57364)/3
```

```
## [1] 52455
```

```
salaries <- c(0, 100000, 57364)
mean(salaries)
```

```
## [1] 52455
```

In mathematics and statistics, we often use symbols to represent ideas. While the symbols can sometimes feel a bit arcane, they let researchers quickly share ideas. There are two ways to talk about the arithmetic mean equation. This is useful to know simply so you can follow along in other texts; in this book, we use `mean()`.

The first way uses ellipsis notation (...) to show "carry on in like fashion." To show the arithmetic mean (denoted by $\overline{x}$ or "x bar" symbol) is the sum of all the variables divided by the *count* of all the variables; the equation is $\overline{x} = (x_1 + \ldots + x_n) / n$. You saw this just now for n = 2 and n = 3.

A more traditional or formal way to say exactly the same thing is to use a capital Greek *sigma* (think sigma for ***sum***). In this case, the *i* shows we iterate across all the values in our set from beginning to end.

$$\overline{x} = \frac{x_1 + \ldots + x_n}{n} = \frac{\sum x_i}{n}$$

Notice we used lowercase *Latin* characters for our sample variables (e.g., *x*, *n*, and $\overline{x}$). Of course, we did use a capital *Greek* character in place of the ellipsis.

Populations are large, their data values and summaries are called *parameters*, and most often we use lowercase *Greek* or capital *Latin* letters for populations. These are what we want to study.

Samples are smaller subsets of a population, their data values and summaries are called *statistics*, and most often we use lowercase *Latin* letters for populations. Because it is usually tough to study a whole population, we "settle" for statistics and use them to *estimate* or *infer* information about populations and population parameters.

If you are curious, even though the mathematics works out the same, there is a formal population parameter equation. If one had the luxury of knowing the value of every salary in an entire population, that would most properly be noted via the Greek letter μ (roughly pronounced "mew") and a capital Latin N for the number of people in the entire population. The Greek μ corresponds to *mean* (see rough pronunciation):

$$\mu = \frac{\sum x_i}{N}$$

Now, we stress again all these symbols are to give you a good ability to engage with data and researchers around the world. Whether you use R (`mean()`) or ellipses or capital sigma (Σ) or lowercase mu (μ), in the case of arithmetic mean, there is no computational difference in the end.

Example

Calculating the arithmetic mean (so called because one uses arithmetic to sum up each value before dividing by the count) uses the `mean()` function. By default it takes two arguments. The first is quantitative data, such as might be found in a column. The second is optional and only required if the data have NAs; it is `na.rm = TRUE`.

Here, you see the results of running `mean()` on the `penguinsData$flipper_length_mm` with and without removal of missing data. Either way you gain some useful information:

```
mean(penguinsData$flipper_length_mm)

## [1] NA

mean(penguinsData$flipper_length_mm, na.rm = TRUE)

## [1] 200.9
```

Visually, we can add a vertical line to our graph using the function `geom_vline()`. However, we have no data mapped to the aesthetic appearance of a vertical line. As vertical lines are determined by a single point on the x-intercept, we must provide that information in a secondary call to the mapping argument. Do not worry about the code too much in this case. What you want to focus on is that the arithmetic mean in Figure 6-14, while clearly in the middle of all the points, does not actually cut through the tallest parts of the histogram:

```
ggplot(data = penguinsData,
       mapping = aes(x = flipper_length_mm))+
  geom_histogram()+
  geom_vline(mapping = aes(xintercept = mean(flipper_length_mm,
  na.rm = TRUE)))

## 'stat_bin()' using 'bins = 30'. Pick better value with 'binwidth'.

## Warning: Removed 2 rows containing non-finite values (stat_bin).
```

So while 201 might be a great "average" number to say when asked about the lengths of penguin flippers, you can see we may want other ways of discussing central tendency. More on that soon!

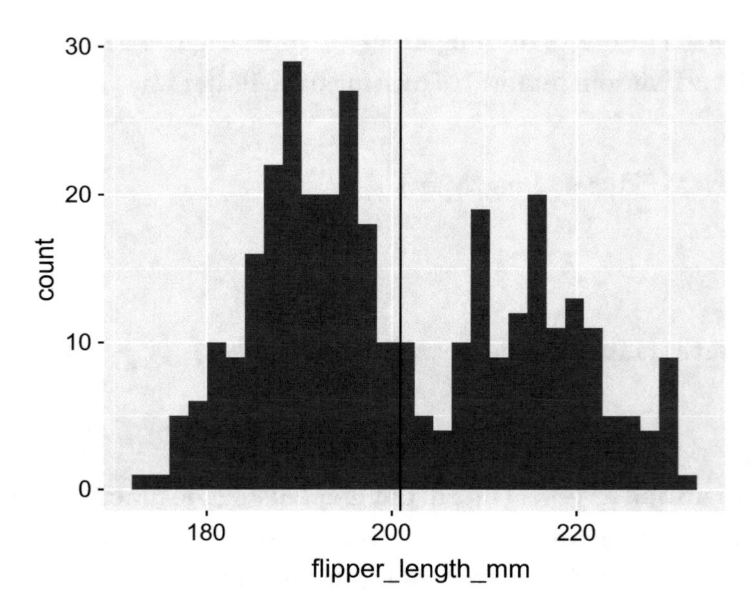

***Figure 6-14.** Penguin histogram with arithmetic mean*

Example

Similarly, the mean() can be computed for our acesData:

```
mean(acesData[SurveyInteger == 3]$COPEExp)

## [1] NA

mean(acesData[SurveyInteger == 3]$COPEExp, na.rm = TRUE)

## [1] 2.37
```

This time, the vertical line in Figure 6-15 seems to not only be in the middle of our data: it is also in "thick" of the data:

```
ggplot(data = acesData[SurveyInteger == 3],
       mapping = aes(x = COPEExp))+
  geom_histogram()+
  geom_vline(mapping = aes(xintercept = mean(COPEExp, na.rm = TRUE)))

## 'stat_bin()' using 'bins = 30'. Pick better value with 'binwidth'.

## Warning: Removed 264 rows containing non-finite values (stat_bin).
```

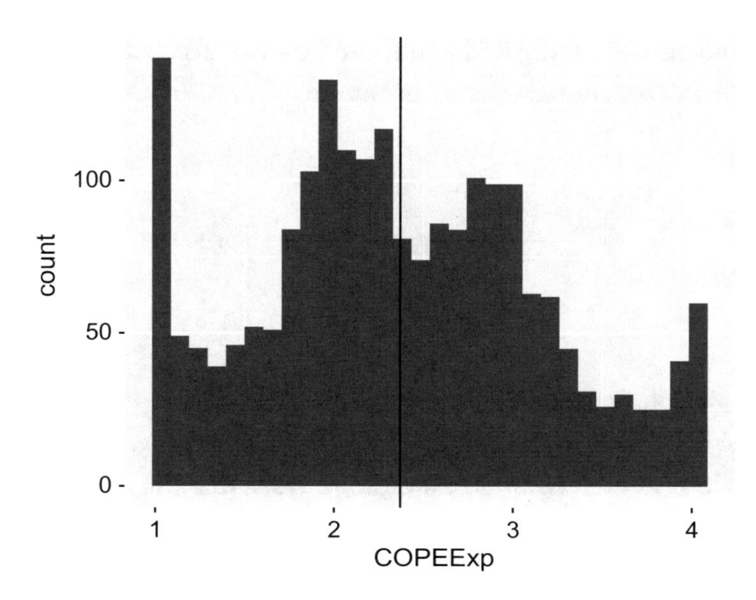

Figure 6-15. *Histogram of COPEExp with arithmetic mean*

Compare and contrast Figure 6-14 with Figure 6-15. Does the arithmetic mean for sure tell you what the "average" data look like?

Median

Because arithmetic mean does not always yield an ideal summary for data, it can help to have other measures. One possibility is to find the data point that is in the exact center of our data point list.

Recall our three salaries mini-example:

```
salaries
```

```
## [1]      0 100000   57364
```

Of the three people in that example, *two* earned *more* than the arithmetic mean. The salary in the middle, \$57364, would be the `median()` salary. In that case, it might be a better example of what an "average" person makes. It is a real salary and is the middle of those three people. As before, the R function can take two arguments, a data entry and the na.rm = TRUE as an optional second argument:

```
mean(salaries)

## [1] 52455

median(salaries)

## [1] 57364
```

One important note is that salaries had an odd count. Thus, the middle value was reported precisely. In the case of an *even* count of values, the *two* middle values are averaged. Take a look at this extended example, and see that the R median() function spotted the middle two values and computed their mean():

```
salariesEven <- c(0, 100000, 57364, 20000)

sort(salariesEven)

## [1]      0  20000  57364 100000

median(salariesEven)

## [1] 38682

mean(c(57364, 20000))

## [1] 38682
```

Example

Compare and contrast the mean with the median for the penguin data. What can you learn from the mean being larger?

```
mean(penguinsData$flipper_length_mm, na.rm = TRUE)

## [1] 200.9

median(penguinsData$flipper_length_mm, na.rm = TRUE)

## [1] 197
```

Take a look at the histogram (again do not spend too much time thinking about the ggplot() code). Here, the median is added to Figure 6-16 as a dashed line. As you can see, in this case, median might be a better measure of central tendency for these data – it cuts through a fuller part of the graph:

```
ggplot(data = penguinsData,
       mapping = aes(x = flipper_length_mm))+
  geom_histogram()+
  geom_vline(mapping = aes(xintercept = mean(flipper_length_mm,
  na.rm = TRUE)))+
  geom_vline(mapping = aes(xintercept = median(flipper_length_mm,
  na.rm = TRUE)),
                                  linetype = "dashed")

## 'stat_bin()' using 'bins = 30'. Pick better value with 'binwidth'.

## Warning: Removed 2 rows containing non-finite values (stat_bin).
```

Example

For the COPEExp data, the median and mean are very close. Recall that earlier, in the trimmed example from Figure 6-13, these data were very close to *normal*. One feature of normal data is the median and mean are the same. Of course, these data were not perfectly normal – we would not expect these two to necessarily be exactly the same!

```
mean(acesData[SurveyInteger == 3]$COPEExp,
     na.rm = TRUE)

## [1] 2.37

median(acesData[SurveyInteger == 3]$COPEExp,
     na.rm = TRUE)

## [1] 2.312
```

The two values are quite close in Figure 6-17 too. Notice the median is *smaller* than the mean. Looking at the graph, do you see how there are a lot of values right at 1? That is the tallest count on the histogram. While a value of 1 is low, it does not "drag down" the mean too much. On the other hand, the count there is quite high, and since median is all about the *count*, it is pulling that dashed median line just a bit to the left:

```
ggplot(data = acesData[SurveyInteger == 3],
       mapping = aes(x = COPEExp))+
  geom_histogram()+
  geom_vline(mapping = aes(xintercept = mean(COPEExp, na.rm = TRUE)))+
  geom_vline(mapping = aes(xintercept = median(COPEExp, na.rm = TRUE)),
             linetype = "dashed")
```

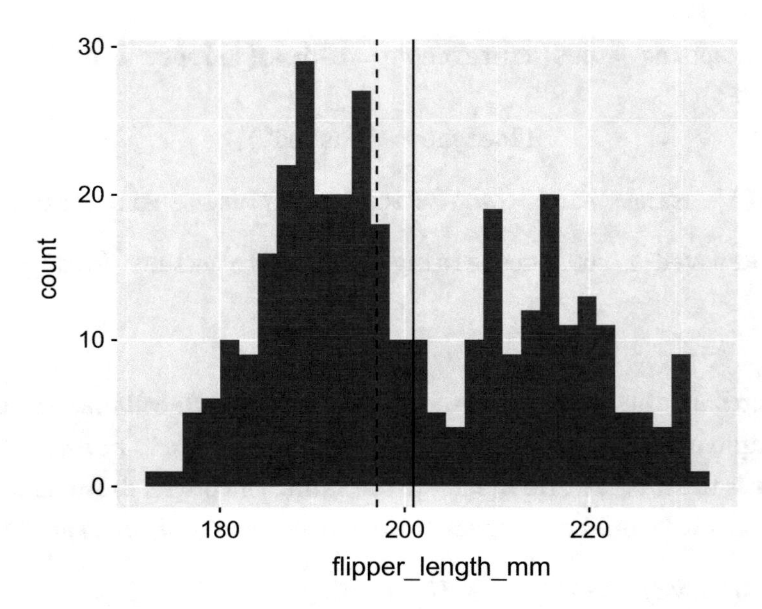

Figure 6-16. *Penguin histogram with arithmetic mean and dashed line median*

```
## 'stat_bin()' using 'bins = 30'. Pick better value with 'binwidth'.

## Warning: Removed 264 rows containing non-finite values (stat_bin).
```

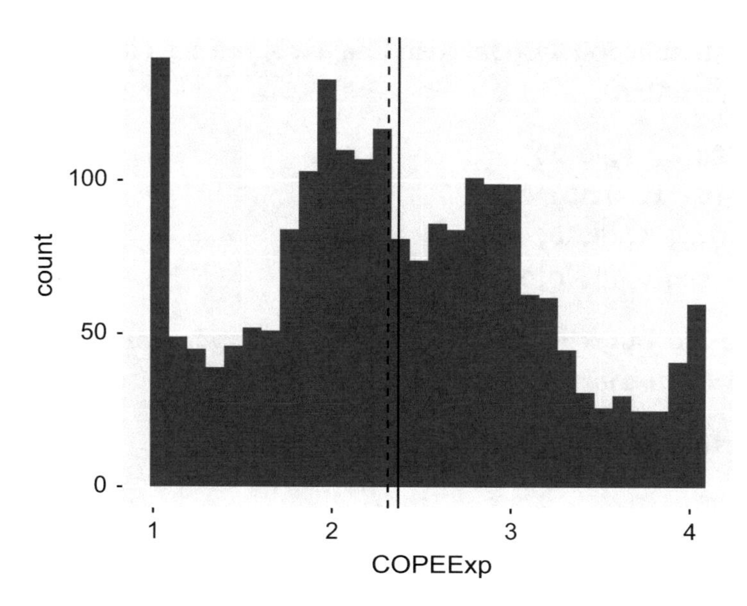

***Figure 6-17.** Histogram of COPEExp with arithmetic mean*

The more you explore data, the more you can learn by comparing and contrasting the central tendency values.

6.4 Position

Having explored the center of the data, let us consider the overall position of the data. You already know the median is the data point exactly in the middle of the data. In other words, 50% of the data points are less than or equal to the median, and the other 50% of the data are greater than or equal to the median.

Sometimes, it can be useful to know the cut-off for other positions besides the exact middle. These are called **quantiles**. Common quantiles you may be familiar with are *quartiles* (quartering data at the 25%, 50%, 75%, and 100% points) and *percentiles* (slicing data by percent from 1% to 100%).

Common uses of quantiles include per capita income or household income (such as we briefly explored in the `salaries` "data"). In fact, such quantiles can lead to categorizing data (such as the `acesData$SES_1` which is in part based on wealth).

The function that computes quantiles is aptly named `quantile()` and takes two arguments. The first is the data, and the second is a number list showing the break points or probabilities.

Using the seq() function introduced in Chapter 5, you can create the cut-offs for some common quantiles:

```
quartiles <- seq(0, 1, 0.25)
deciles <- seq(0, 1, 0.10)
grades <- seq(0.6, 1, 0.10)
percentiles <- seq(0, 1, 0.01)
```

Here, we consider our salaries data by quartiles. As you can see, the median() function is a special case for the 50th percentile:

```
quantile(salaries, quartiles)

##      0%    25%    50%    75%   100%
##       0  28682  57364  78682 100000

median(salaries)

## [1] 57364

quantile(salariesEven, quartiles)

##      0%    25%    50%    75%   100%
##       0  15000  38682  68023 100000

median(salariesEven)

## [1] 38682
```

Now, let us explore this quantile categorization idea with larger data.

Example

Recall the dot plot in Figure 6-8. Rather than line up the data in histogram shape, we could simply view the raw data and make our own determinations.

To do this, we use almost the usual ggplot() code. However, for the first time, we not only map flipper_length_mm to our horizontal x value, we also set y = 0. Then, instead of geom_dotplot(), we use geom_jitter(). The result in Figure 6-18 shows all 344 penguin flipper lengths (well, 344 less the missing values):

```
ggplot(data = penguinsData,
       mapping = aes(x = flipper_length_mm, y = 0))+
  geom_jitter()

## Warning: Removed 2 rows containing missing values (geom_point).
```

Notice on the right side of the graph are the longer Gentoo flippers (recall Figure 6-9) with that bit of a gap separating them out in the middle into their own cluster.

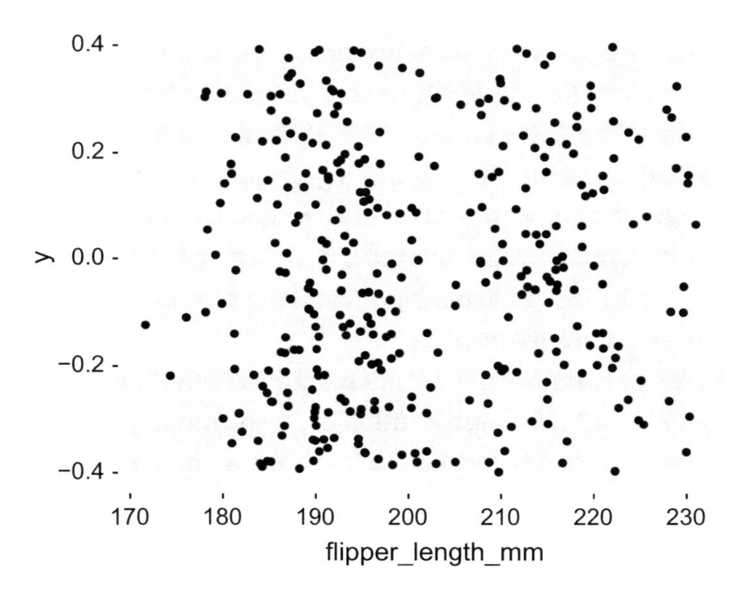

Figure 6-18. *Penguin jitter plot*

What does this have to do with quantiles? The dots in Figure 6-18 can be thought of in quantile chunks. Look closely at those dots, consider the `quartiles`, and see if you can group the various lengths into the shortest 25%, the longest 50%, and so on. As you may have guessed, because there are some missing values in the flipper lengths, we must use the trusty code phrase `na.rm = TRUE`. Notice we do not start by using that. While not all error messages in R are helpful, often they are. Additionally, even obscure error codes often yield useful clues by copying and pasting the exact phrase into an Internet search engine:

```
quantile(penguinsData$flipper_length_mm, quartiles)

## Error in quantile.default(penguinsData$flipper_length_mm, quartiles):
missing values and NaN's not allowed if 'na.rm' is FALSE
```

```
quantile(penguinsData$flipper_length_mm, quartiles, na.rm = TRUE)
```

```
##   0%  25%  50%  75% 100%
##  172  190  197  213  231
```

This is a common way to segment a data set, and it is so common there is a special graph that helps us visualize this. The graph is the **boxplot** or *box and whisker plot*. In Figure 6-19, you can see the faded boxplot drawn over the dots of the penguin flipper lengths.

On either side of the boxplot are the whiskers. These show where the bottom 25% and the top 25% of the data live. *Inside* the box is the *middle* 50% of the data (it ranges from the 25% quartile to the 75% quartile). A solid, thick, vertical line shows the *median*.

Notice that, in the case of flipper lengths, to get one-fourth of the dots into the 25%–50% range creates a fairly narrow portion of the boxplot. On the other hand, from 50% to 75% of the data points is much wider. Compare and contrast this information with the histogram from Figure 6-1. Both the histogram and the boxplot show not only the full *range* of data – they also show *frequency*.

Remember, so far, our goal is only to understand, explore, and describe data. We are not yet evaluating these data. Having a handful of go-to graphs (such as histograms, dot plots, jitter plots, and boxplots) is very useful to see things from slightly different points of view (if you will pardon our pun).

As always, `ggplot()` makes it easy to add `geom_boxplot()`. Now, in real life, one does not always include both the jitter dots and the boxplot. This is because the dots in the jitter can be distracting. If the goal is to summarize data, less is often more in terms of communication. However, boxplots are one of the trickier graphs to understand in this book. So we will tend to show both. Because we are showing both, there is one new argument you see in the code that follows. To make the boxplot transparent enough to let those dots show through, we reduce the `alpha` down from 1 to 0.5:

```
ggplot(data = penguinsData,
       mapping = aes(x = flipper_length_mm, y = 0))+
  geom_jitter()+
  geom_boxplot(alpha = 0.5)
```

```
## Warning: Removed 2 rows containing non-finite values (stat_boxplot).
```

```
## Warning: Removed 2 rows containing missing values (geom_point).
```

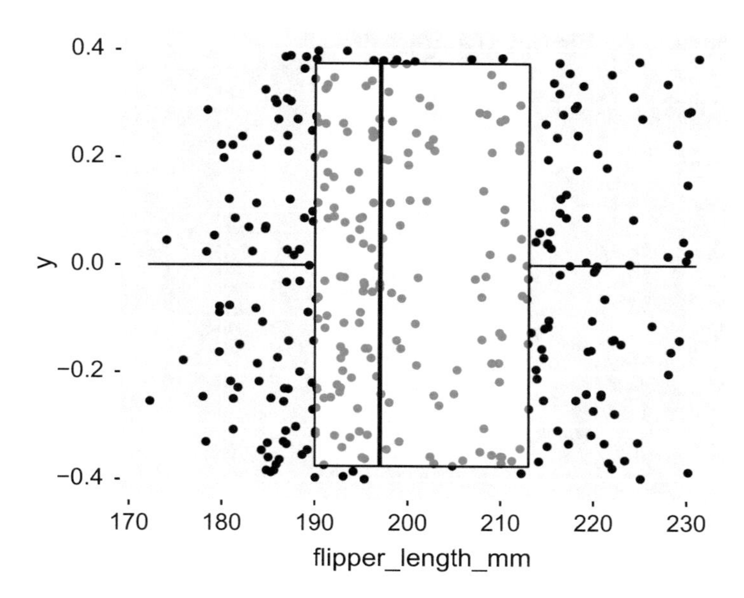

Figure 6-19. *Penguin boxplot*

Example

Suppose you needed to better understand how people of different ages respond to surveys in the `acesData`. It might be worth viewing the entire survey set ages (this could help detect patterns before doing some analysis).

However, remember that participants often do not reenter their age - many surveys have missing ages. Thus, before we can even explore the data, we must clean or *munge* the data.

This is *common* in real-world data. In fact, over half the work on any data set is usually munging. This is important to remember when you are using the techniques you are learning in the real world. It is also important to communicate to stakeholders (e.g., managers, directors, clients, etc.) that while the actual analysis might seem simple, data quality is usually a limitation.

To clean up the `Ages` column, all we need to do is fill in the missing values with the usual age for each participant. Despite our scary intro to this, `data.table` makes this fast!

Recall the frequency table from Chapter 5:

```
acesData[order(Age), .N, by = Age]
```

```
##      Age    N
##  1:   18  347
##  2:   19  908
##  3:   20  889
##  4:   21 1278
##  5:   22  821
##  6:   23  583
##  7:   24  396
##  8:   25 1036
##  9:   26  121
## 10:   NA  220
```

We have 220 missing ages which need to be fixed. One way to do this is to use the `mean()` function we just learned. If we set the `Age` of each `UserID` to that person's average age (and `rm.na=TRUE`), we should be set!

We want to do this for all rows, so our row operations in the `i`th position are left blank. We are assigning new values to `Age`, so that gets our new assignment.

Now is a key moment in data munging: Do we use up memory by adding a new column? If we add a new column in the `j`th position, we get to keep the original or raw `Age` column. This may be helpful for other analyses in the future. On the other hand, adding a new column gives us 6599 new bits of data that are mostly copied (other than the 220 missing values). Plus, we have to remember on *which* age column to do the analysis. Most decisions like this are trade-offs. In this case, we create a new age column titled `meanAge`.

Lastly, we need the average age *for each participant*. Thus, the `by = UserID` in the last space of the data table will ensure that we do not get the overall average age:

```
acesData[, meanAge := mean(Age, na.rm = TRUE), by = UserID]
```

Taking one last look at our frequency table, now with `meanAge`, we are left with only 26 rows of missing data. If you explored those (on your own) a bit further, you would see they are from the one participant who never reported their age:

```
acesData[order(meanAge), .N, by = meanAge]
```

```
##      meanAge    N
##  1:       18  359
##  2:       19  932
##  3:       20  919
##  4:       21 1310
##  5:       22  844
##  6:       23  606
##  7:       24  413
##  8:       25 1064
##  9:       26  126
## 10:      NaN   26
```

Having cleaned up your data set, you are now ready to visualize how participants enter survey data. There are now over 6500 dots on Figure 6-20. Looking at all of those, it would be tough to see patterns. However, you can see how the boxplot gives a good summary of what is happening. Half the survey responses are from younger participants.

The code to achieve this figure is about what you would expect from the preceding example. The only difference here is adding transparency to `geom_jitter` by reducing alpha to 0.4:

```
ggplot(data = acesData,
       mapping = aes(x = meanAge, y = 0))+
  geom_jitter(alpha = 0.4)+
  geom_boxplot(alpha = 0.85)

## Warning: Removed 26 rows containing non-finite values (stat_boxplot).

## Warning: Removed 26 rows containing missing values (geom_point).
```

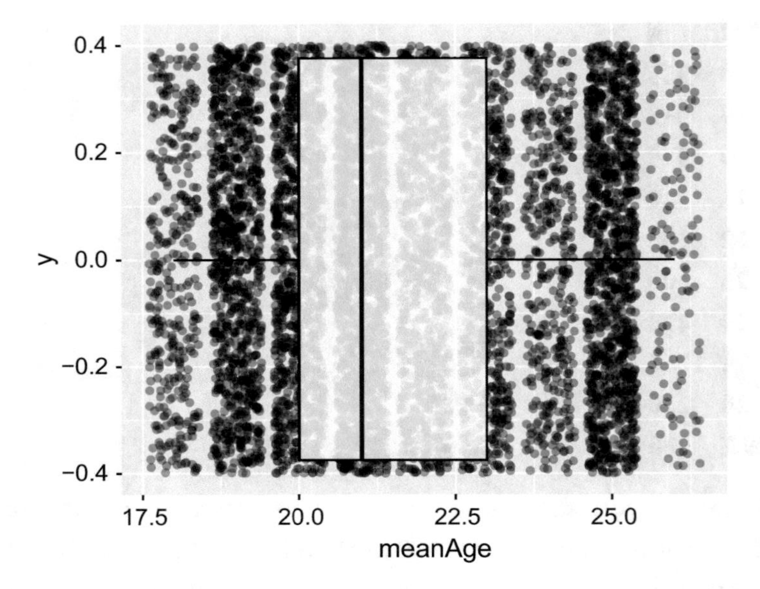

Figure 6-20. *meanAge boxplot*

Example

For the last boxplot example, we explore the idea of **outliers**. Recall the "perfect" data set from Figure 6-11. Take a look at the normal data on the left side of that chart. Now, take a look at Figure 6-21 on the left (normal) side. Notice the middle 50% of the boxplot is quite narrow. That makes sense because that is where the histogram was tallest. Then, the whiskers on either side are comparatively longer; that also makes sense. Those longer areas are the long tails on either side of that histogram.

Now, take a look at those oversized dots in Figure 6-21. Where those dots are on the boxplot corresponds to the faintest scattering of dots from the jitter plot. Those sparse dots (both the large dots which are part of the boxplot summary and the small dots which are the actual raw data values for that data set) are **outliers**.

In the solar system, the outlying asteroids are very far from the middle. So too, outliers in data are points very far from center.

Something else to notice with these two boxplots in Figure 6-21: The normal boxplot is very crunched in the middle, has longer whiskers, and has outliers (the big dots). The uniform boxplot has perfect even spacing of the four parts of the boxplot and has no outliers. Despite this, the median and indeed mean of these data are the same.

These data share medians and modes. They share ranges from about 0 to about 5 (give or take). And yet their boxplots look different. How might we contrast these two data sets? Would such contrasts have use?

6.5 Turbulence

The answer is yes. The language we use to describe the difference in Figure 6-21 is the language of data *turbulence* or *variance*. Normal data cluster about the mean, while uniform data do not.

In general, all else equal, normal data, because of their cluster-at-the-mean behavior, will have *smaller* **variance** or **standard deviation** than other data.

Keeping our "nearly perfect" data sets in Figure 6-22 in mind, consider the following output (again, for this "perfect" example, we suppress the code and only show the output). Notice the difference in standard deviations between the normal and uniform data:

```
##      variable Variance StandardDeviation ArithmeticMean Median
## 1:     normal   0.5486            0.7406          2.505  2.503
## 2:    uniform   2.0830            1.4433          2.522  2.545
```

You have likely met standard deviation before now. Have you ever seen a poll on the news that said "...+/-3%"? That flex or flux or turbulence in the sample estimate is related to variance and standard deviation.

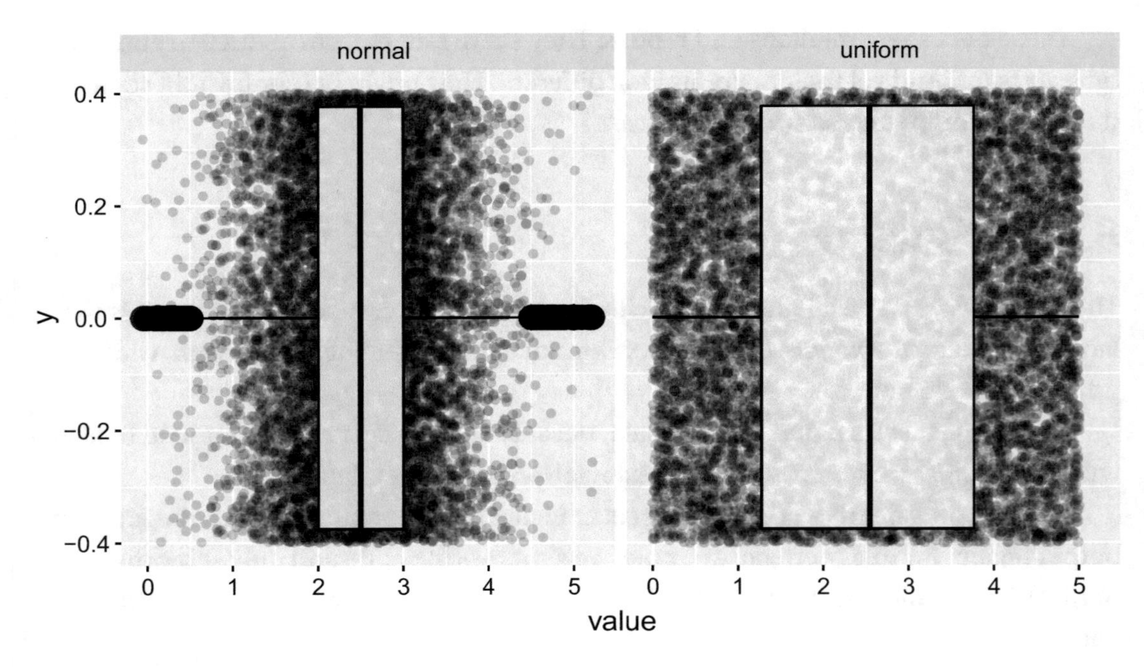

Figure 6-21. *Outliers in boxplot*

Formally, variance is the square of standard deviation. Another way to say this is that the square root of variance is standard deviation.

Recall our earlier discussion about capital Latin and lowercase Greek letters being used for populations, while lowercase Latin is used for samples. Of course, we use R for our calculations. However, it is worth chatting through the formal equations. This will help if you ever choose to explore statistics more deeply, even though this book is a methods book.

For all four equations (population variance, population standard deviation, sample variance, and sample standard deviation), the overall philosophy is the same. The goal is to compare *every* individual row element (e.g., x_i) to the arithmetic mean. In mathematics, differences are found by subtraction. Thus, at the heart of all four equations is $x_i - MEAN$.

Do not let the symbols fool you! These are all really just one formula, not four. Remember population mean is μ, while sample mean is $\overline{x}$. Then of course we have N for population count and n for sample count. The only real difference in these equations is that in population one divides by the whole count. However, in sample, the division is by $n - 1$.

Keep in mind samples are smaller than the population. In real life, we almost never have population-level parametric data. Instead, we take samples and then project what we learn from the smaller sample onto the larger population. $n - 1$ is an adjustment to the equation to make the sample variance and standard deviation larger than they would be without that adjustment. We will see in following chapters how this gives us more conservative estimates for populations from the samples.

Think of it this way: in the normal data, because it clusters near the center, you could be fairly certain that if you drew a point out random off the graph, you most likely would pick one near the middle. Contrastingly, the uniform data is spread across a wider area. In a way, standard deviation tells you the range around the median you would most likely get from a random draw. So, by making the standard deviation wider for samples, statisticians ensure we tell folks a wider possible range for "most" results. Again, we will come back to this idea in depth later.

The one last observation we make about variance and deviation is that R is focused on using samples to consider populations. Thus, the usual R functions for variance (`var()`) and standard deviation (`sd()`) automatically use the sample versions of these equations.

Population variance:

$$\sigma^2 = \frac{\Sigma(x_i - \mu)^2}{N}$$

Population standard deviation:

$$\sigma = \sqrt{\frac{\Sigma(x_i - \mu)^2}{N}}$$

Sample variance - `var()`:

$$s^2 = \frac{\Sigma(x_i - \bar{x})^2}{n-1}$$

Sample standard deviation - `sd()`:

$$s = \sqrt{\frac{\Sigma(x_i - \bar{x})^2}{n-1}}$$

There can be value in working through the sample standard deviation equation by hand. One of the exercises has you do this. However, most such manual effort is done on very small, toy datasets. For sure it is not done on larger data (even comparatively small "larger data" such as the 344 penguins). Modern data are simply too large to calculate by hand.

More often, in today's larger data sets, we are focused on using standard deviation to help summarize and describe data. One way to do this is to describe the "usual ranges" of our data in terms of standard deviations. Just like with the "60% approval rating +/- 3%" you might see on a public opinion survey, we can look at data as the *mean* plus or minus the standard deviation.

To visualize this use of sample standard deviation *s*, see how in Figure 6-22 our "perfect" normal curve has line segments capturing bins around the mean. Notice the histogram bins intersected by `1s` (for *one standard deviation*) represent the majority of the counts (remember height in histogram means we have more data there). Then, if we extend ourselves a distance of `2s`, we get almost all the bins.

In fact, in statistics, there is a rule called the **empirical rule** that states 68% of your data will live in the first standard deviation area. The rule goes on to say 95% of data live in the two standard deviations area and 99.7% of the data live in the three standard deviations area. In fact, this rule is sometimes called the *68-95-99.7* rule. While you will learn ways using code to calculate any exact ratio you need in this book, this rule can be helpful to understand how knowing standard deviation and the mean gives a good idea on how a particular data set may behave.

If you have spent time in the world of business, you may have heard of "six sigma." The name of that management philosophy comes from the idea that precision in work – even with the usual amount of *variance* in quality – should be such that even at *six* standard deviations from the mean, the product will still be of usable quality.

As you might imagine, keeping something usable even at such extremes of standard variation requires having the standard deviation be very small so that $\overline{x} \pm 6s$ stays in a very narrow range.

Another way to look at this same idea can be seen in Figure 6-23 [6]. This version of the normal distribution shows the standard deviations in the σ notation and they go out to ±4σ. Now, do not let all the information in Figure 6-23 overwhelm you; we discuss this image in more detail in Chapter 9. For now, focus on the line near the bottom of the curve that reads "Standard Deviations from The Mean." See how on either side of 0 there are a -1σ and $+1\sigma$? Notice the area between those two is labeled as 0.3413? Of course, if we add those two areas together, 0.3413 + 0.3413 = 0.6826. This rounds to 68 and is where the 68% of the empirical rule gets its name. Similarly, if we shade in the area between -3σ and $+3\sigma$, we get 0.0214 + 0.1359 + 0.3413 + 0.3413 + 0.1359 + 0.0214 = 0.9972. This is where the 99.7% of the empirical rule gets its name.

These two figures, Figure 6-22 which comes from a *sample* generated in R and Figure 6-23 which comes from a (theoretical) *population*, help us understand how thinking about the sample (and the probability of data points) informs our understanding of a population.

As we turn our attention to some examples of standard deviation using our familiar data sets, keep in mind you have learned and thought about many things in this chapter. It may not yet seem that standard deviation summarizes thousands of data points neatly. It takes time to use all these new descriptions you are learning enough that they become useful to you. Keep your focus on practicing using these ideas and perspectives while exploring data. Over time, by comparing and contrasting more data sets, you will develop your sense of what these various descriptors explore.

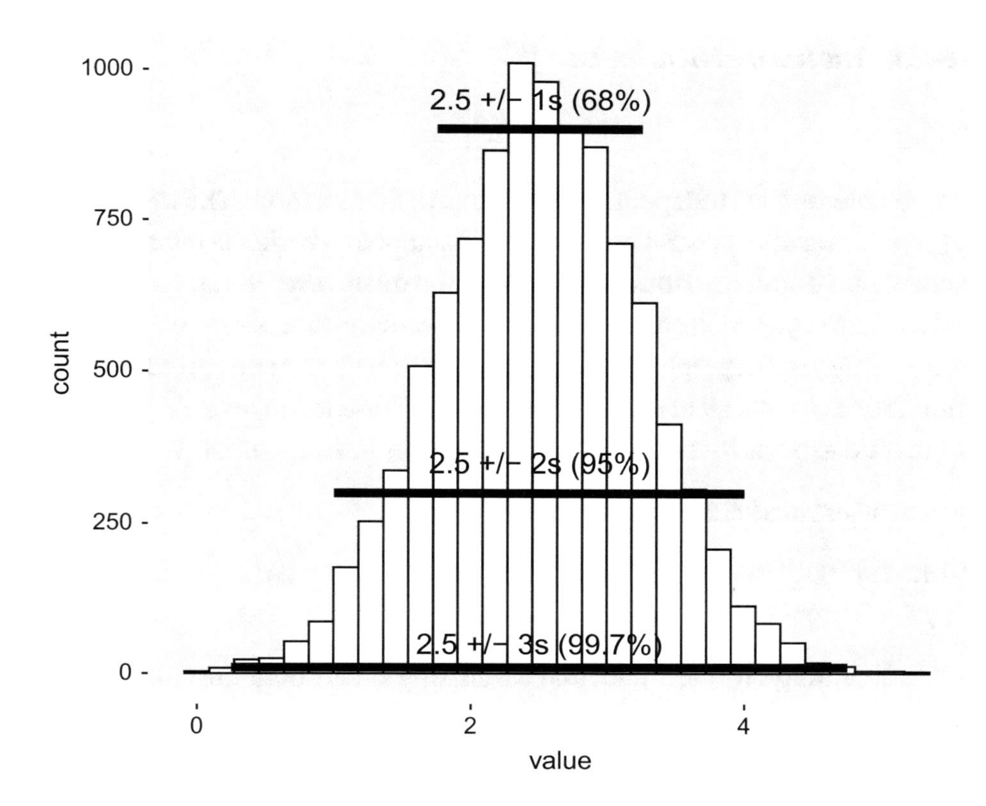

***Figure 6-22.** Bars captured by the standard deviation ranges*

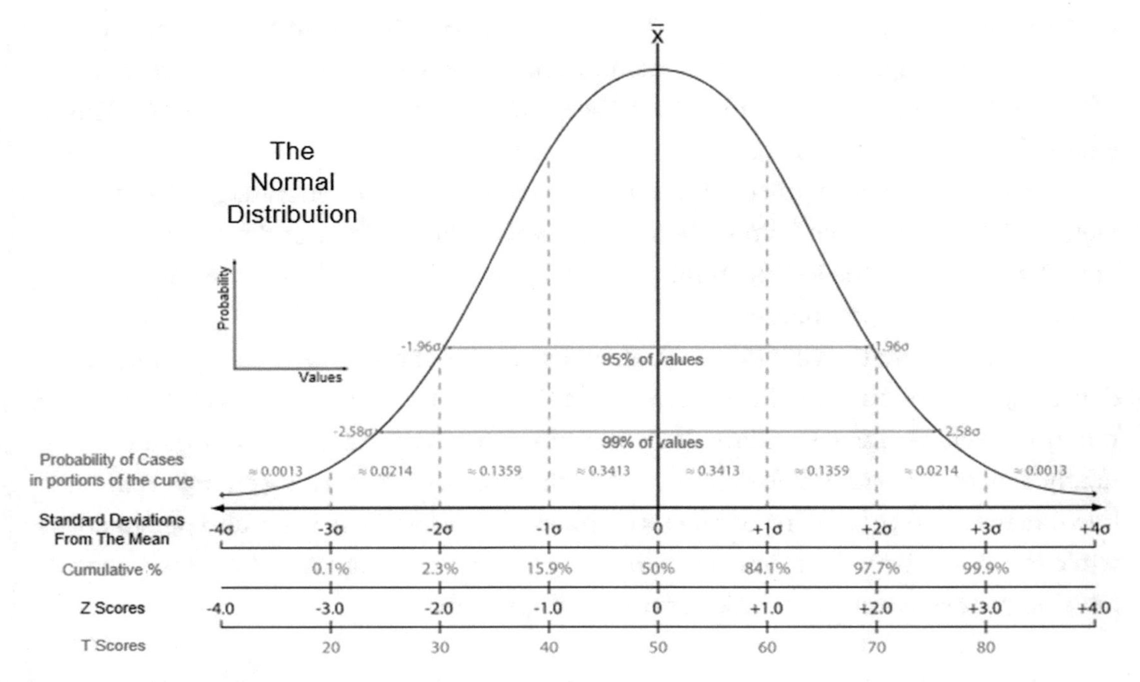

Figure 6-23. *The normal distribution [6]*

Example

We use three different, yet independent functions in this example. The first function, `summary()`, pulls together several of the pieces discussed in earlier sections. In particular, it gives mean, range, and quartiles. If you ignore the mean, everything else is sometimes described as the *five-figure summary*. Notice the five-figure summary are the five points on a boxplot. The boxplot graph was explicitly built to help us understand many five-figure summaries visually all at once. The `summary()` function takes an input of a single data column and expects that column to be interval or ratio, quantitative data:

```
summary(penguinsData$flipper_length_mm)

##    Min. 1st Qu.  Median    Mean 3rd Qu.    Max.    NA's
##     172     190     197     201     213     231       2
```

From the summary, we see much that we already knew about the lengths of flippers. They have a **range** from a minimum of 172 mm to a maximum of 231 mm. Because the median of 197 is smaller than the mean of 201, you know there is a large cluster of data closer to the maximum than the minimum (check Figure 6-9 to see this is the Gentoo species). Lastly, we know that half of all flipper lengths must live between the first quartile at 190 and the third quartile at 213.

We already mentioned `var()` and `sd()` for sample variance and sample standard deviation, respectively. They also take interval or ratio, quantitative data. They also take `rm.na = TRUE` as a secondary argument (which is only needed for data that include missing values):

```
var(penguinsData$flipper_length_mm, na.rm = TRUE)
```

```
## [1] 197.7
```

```
sd(penguinsData$flipper_length_mm, na.rm = TRUE)
```

```
## [1] 14.06
```

From our discussion of the *empirical rule*, we might imagine 68% of the data should live between 201 ± 14. Does this hold true? Subtracting and adding gives us a one standard deviation range from 187 to 215. Based on our quartile data, that looks *about* the right distance to be a bit larger than the middle 50%.

If we want to check more closely, a quick count can help. While we could do something fancy, this is exploratory data analysis. Keeping in mind that the penguin data set has 344 observations, the fact that there are 227/344 = 0.66 tells us these data do not quite follow that empirical rule. Of course, you are not surprised by this. The empirical rule was for *normal* data; the penguin data set does not really fit that pattern as you have already seen. Still, it was a close estimate, quite close in fact:

```
count <- penguinsData[flipper_length_mm %between% c(187, 215),.N]
count
```

```
## [1] 227
```

```
count/nrow(penguinsData)
```

```
## [1] 0.6599
```

The empirical rule holds true for normal data. The further the standard deviation ranges are from matching that rule, the further your data are from normal. You will learn other tests for normal (besides the easiest test of visual inspection that you already know). Still, this is one more exploratory tool in your skill set.

Lastly, recall in our formal exploration of variance and standard deviation that those two formulae are related. Variance is the square of standard deviation, and thus the square root (`sqrt()`) of variance should be the same as standard deviation. For a bit of fun (and to make sure you stay fresh on your logical operators), we include a check of that truth:

```
sqrt(var(penguinsData$flipper_length_mm, na.rm = TRUE)) ==
  sd(penguinsData$flipper_length_mm, na.rm = TRUE)

##  [1] TRUE
```

Example

While the `penguinsData` are always nice to explore because they are fairly short and easy to understand, the `acesData` are simulated off human research. Thus, it always helps to try what we have learned on those data too. The more data sets you can find and practice these techniques on, the better and faster you will train your compare and contrast ability. Comparing and contrasting, in turn, is part of the *artistic* side of data science. It can help you spot the story behind the numbers and amplify important truths to your stakeholders, consumers, managers, or community.

We start with the `summary()` function, and notice there are `NA`s in these data:

```
summary(acesData$meanAge)

##     Min. 1st Qu.  Median    Mean 3rd Qu.    Max.    NA's
##     18.0    20.0    21.0    21.7    23.0    26.0      26
```

Because there are missing data, we include the second argument option in `var()` and `sd()` of `na.rm = TRUE`. This is a good moment to mention something rather key in research. Even in a perfect random sample, participants are free to not respond (this is required to be ethical and is part of informed consent). If participants choose to not respond randomly, then all is well. However, if participants do not respond according to some pattern, bias returns to your data despite all your efforts. Traditionally, statistical equations had no choice except to drop the `NA`s and accept that bias had returned. While it is beyond the scope of this book, you may wish to check one of our statistical programming and data models book [22] to learn more about "solving" *missing data* (and reducing the bias).

For now though, we simply remove the missing variables:

```
var(acesData$meanAge, na.rm = TRUE)

## [1] 4.899

sd(acesData$meanAge, na.rm = TRUE)

## [1] 2.213
```

From our discussion of the *empirical rule*, we still might imagine 68% of the data should live between 21.7 years of age +/–2.2 years. Does this hold true? Subtracting and adding gives us a one standard deviation range from 19.5 to 23.9. Based on our quartile data, that looks *about* the right distance to be a bit larger than the middle 50%.

However, from the preceding example with penguin data, we pause for a moment to take a quick look at the histogram. As you see, Figure 6-24 does not quite look normal. Thus, we suspect the empirical rule may not quite fit these data. However, it does look close to normal, so we might expect something close to the empirical rule:

```
ggplot(data = acesData,
       mapping = aes(x = meanAge))+
  geom_histogram()

## 'stat_bin()' using 'bins = 30'. Pick better value with 'binwidth'.

## Warning: Removed 26 rows containing non-finite values (stat_bin).
```

As before, to check more closely, a quick count can help. The `acesData` has some missing age data, so remember to use the logical check of `is.na()` on the `meanAge` data. That check would find the missing values, so it must be negated with the exclamation mark. These data certainly look closer to normal on the histogram vs. the penguin data, and the end result of 0.56 is closer to the empirical rule:

```
count <- acesData[meanAge %between% c(19.5, 23.9) &
                    !is.na(meanAge),
                  .N]
count

## [1] 3679

count/nrow(acesData[!is.na(meanAge)])

## [1] 0.5597
```

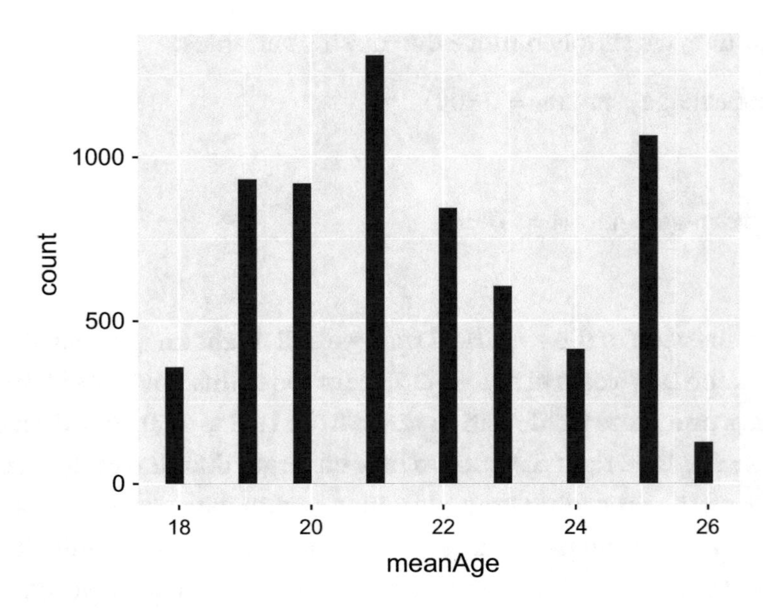

***Figure 6-24.** meanAge histogram*

6.6 Summary

This chapter introduced exploratory data analysis tools you can use to understand any data set. In particular, you learned how to quickly graph data using `ggplot()` into histograms, dot plots, and boxplots. You applied `median()`, `mean()`, `summary()`, and `sd()` to create summaries and develop understanding of a data set in terms of central tendency, position, and turbulence. Along the way, you compared data sets to two "perfect" models, one "normal" and one "uniform." Put together, you now possess the knowledge, skills, and abilities to compare and contrast any two data sets and understand similarities and differences. You put in a lot of work learning these ideas and can refer back to Table 6-1 as you practice your skills in the exercises.

***Table 6-1.** Chapter Summary*

Idea	What It Means
`hist()`	Base R plot for histogram.
`setdiff()`	Compares set x to set y and returns the items in x that are not in y.
`dotchart()`	Base R dot plot.

(continued)

Table 6-1. *(continued)*

Idea	What It Means
`ggplot()`	The grammar of graphics package main function, takes a data and mapping argument.
`aes()`	Takes `x =`, `y =`, `fill =` arguments from data columns.
`geom_histogram()`	Turns a ggplot into a histogram.
`geom_dotplot()`	Turns a ggplot into a dot plot.
`geom_jitter()`	Turns a ggplot into a set of points.
`geom boxplot()`	Turns a ggplot into a boxplot.
Normal data	The mean and median are the most common data points.
Uniform data	Rectangular data where every point is just as likely as any other point.
`mean()`	Calculates the arithmetic mean.
`median()`	Returns the data point in the exact middle of the data.
`quantile()`	Calculates the relative position of data elements.
Outliers	Data points far away from the mean and the bulk of the data.
Standard deviation	A measure of data turbulence.
`sd()`	Function call to compute standard deviation.
Empirical rule	Also called the 68-95-99.7 rule.
`summary()`	Numerical version of boxplot.

6.7 Practice for Mastery

Check your progress and grow through practice by working through some exercises. Comprehension checks ask critical thinking questions that may be best answered with a written or verbal response. Part of the art of statistics is successfully communicating results to your stakeholders or audience. Sometimes that audience is highly technical and other times very much not technical. Exercises are more direct applications of the concepts explored in the chapter.

Comprehension Checks

1. In your own words, what is the difference between the mean and the median? What is the process behind each? If you wanted a single number to describe the income of the "average" citizen of a country, which might be better? Why?

2. While researching age differences between Country A and Country B, you see the two countries share a median age of 36 and a mean age of 40. However, Country A has a standard deviation of 4.5 years, and Country B has a standard deviation of 10.2 years. You know age generally follows a mostly normal distribution (and the median and mean being close supports this). What might you expect about the overall histograms of these two countries? Which country is more likely to have a citizen who is 104 years?

Exercises

1. The best learning happens by doing. Copy the code from each example in this chapter that produces a graph. Then, switch the `x =` to a different column in the same data set (e.g., replace `x = flipper_length_mm` with `x = body_mass_g`). What have you learned about some of the other columns of data? Are any of them normal? Uniform?

2. Now that you did the column switch for graphs in the prior exercise, go ahead and use `summary()` and `sd()` on each of the new columns you explored. Compare and contrast the new means and standard deviations with those you have already seen. Compare and contrast the graphs. Are you starting to get a feel for the visual differences between bigger and smaller standard deviations?

3. Remember sample standard deviation – `sd()` – follows the formula $s = \sqrt{\frac{\Sigma(x_i - \overline{x})^2}{n-1}}$. How might the standard deviation be computed in the traditional fashion by hand? Well, at 344 rows, a

completely by hand process might be too much to ask. Instead, work through the following code and see what standard deviation is really doing:

```
#first you get a data.table with only lengths
standardDeviationpenguin <- penguinsData[,.(flipper_length_mm)]

#you need to know the mean length
standardDeviationpenguin[, Mean := mean(flipper_length_mm, na.rm = TRUE)]

#so far we have one entry per penguin flipper length and the overall mean.
head(standardDeviationpenguin)

##    flipper_length_mm  Mean
## 1:               181 200.9
## 2:               186 200.9
## 3:               195 200.9
## 4:                NA 200.9
## 5:               193 200.9
## 6:               190 200.9

#this is the residual difference between each length and the mean
standardDeviationpenguin[ , x_iMinusMean := (flipper_length_mm - Mean)]

#this lets us see three penguins (these happen to be from each species by
visual inspection)
standardDeviationpenguin[c(1, 153, 277)]

##    flipper_length_mm  Mean x_iMinusMean
## 1:               181 200.9      -19.915
## 2:               211 200.9       10.085
## 3:               192 200.9       -8.915
```

Notice that, so far, what we see in `x_iMinusMean` is the *difference* or **residual** of each flipper length from the average length. The -19.9 tells us the first flipper is quite a bit smaller than average. The second species' 10.1 tells us this flipper is longer than average.

Normal data have small standard deviations. That tells us the residual differences are always small in those data sets. Uniform data have larger standard deviations that tells us the residual differences are not always small.

Now, as it stands, if you summed up the x_iMinusMean values, the negative numbers would in some ways get "canceled out" by the positive values (and vice versa). That is not quite fair because notice the first penguin is the greatest overall distance from the mean of 201.

So, in mathematics, to get rid of negative numbers, we square them:

```
standardDeviationpenguin[ , x_iMinusMeanSquared := (x_iMinusMean^2)]

#this lets us see three penguins (these happen to be from each species by
visual inspection)
standardDeviationpenguin[c(1, 153, 277)]
```

```
##    flipper_length_mm  Mean x_iMinusMean x_iMinusMeanSquared
## 1:               181 200.9      -19.915              396.62
## 2:               211 200.9       10.085              101.70
## 3:               192 200.9       -8.915               79.48
```

Looking back at the *mean* formula, recall $\overline{x} = \frac{\sum x_i}{n}$. Notice the technique to find an arithmetic average is to sum up all the values and divide by the number of items.

Now that our column x_iMinusMeanSquared can be summed without cancelling anything out; that is precisely what we do for variance! Variance is just the arithmetic mean of the residual difference squares. It tells us the usual squared residual:

```
variance <- sum(standardDeviationpenguin$x_iMinusMeanSquared, na.rm = TRUE)
/ (nrow(standardDeviationpenguin) - 1)
```

In general, squared values do not make sense to humans, so we take the square root to get back to flipper lengths instead of squared flipper lengths. Thus, the square root of variance is standard deviation. You can check your work by using the sd() function:

```
sqrt(variance)

## [1] 14.02

sd(standardDeviationpenguin$flipper_length_mm, na.rm = TRUE)

## [1] 14.06
```

Having seen this code, go ahead and do this again yourself on the mtcarsData for the mpg column. Remember to check your work with sd()!

CHAPTER 7

Understanding Probability and Distributions

Now that you can describe the sample data sets, it is time to move toward analyzing your data. Progressing from descriptive to *inferential* statistics takes some thought and time. To recap, you have already learned a lot about using R. Then, you learned how to manipulate data and even describe the data through various summaries (numerical and graphical). **Inferential** statistics is the science and art of using samples to understand and predict the population behavior. To successfully make these inferences, you need some background mathematics.

Analyzing data builds from a mathematical foundation. Part of that foundation is fraction operations (e.g., addition, subtraction, multiplication, and division). While this chapter supposes you already understand fractions, there is no reason to go into supremely arcane fractions. The usual sorts of whole-number integers you might meet in an algebra course will do. If fractions are a little scary, take heart! We will, as always, use R to do the heavy lifting. Of course, the more you understand on your own, the more sense some of the computations will make. Now is a great time to refresh your knowledge of the algebra of fractions.

So if we suppose you already are practiced in fractions, what is this chapter? To continue increasing your growing experience of thinking like a statistician, you want to leverage fractions to understand probability. With an understanding of probability, you will be able to understand distributions. In fact, you already met two distributions. We called one *normal* and the other *uniform*.

Thus, while this chapter can feel a bit strange as we start with something completely different from our familiar data sets, your goal remains firmly fixed on increasing your experience and understanding of mathematics and R to make statistical sense of data.

M. Wiley and J. F. Wiley, *Beginning R 4*, https://doi.org/10.1007/978-1-4842-6053-1_7

By the end of this chapter, you should be able to

- Understand the basic features of probability.
- Apply probability to understanding the likelihood of various events.
- Remember certain common distributions.
- Evaluate various data samples using known distributions.

7.1 R Setup

As usual, to continue practicing creating and using projects, we start a new project for this chapter.

If necessary, review the steps in Chapter 1 to create a new project. After starting RStudio, on the upper-left menu ribbon, select *File* and click *New Project*. Choose *New Directory* ➤ *New Project*, with *Directory name* `ThisChapterTitle`, and select *Create Project*. To create an R script file, on the top ribbon, right under the word *File*, click the small icon with a plus sign on top of a blank bit of paper, and select the *R Script* menu option. Click the floppy disk–shaped *save* icon, name this file `PracticingToLearn_XX.R` (where XX is the number of this chapter), and click *Save*.

In the lower-right pane, you should see your project's two files, and right under the *Files* tab, click the button titled *New Folder*. In the *New Folder* pop-up, type `data` and select *OK*. In the lower-right pane, click your new *data* folder. Repeat the folder creation process, making a new folder titled `ch07`.

Remember all packages used in this chapter were already installed on your local computing machine in Chapter 2. There is no need to re-install. However, this is a new project, and we are running this set of code for the first time. Therefore, you need to run the following `library()` calls:

```
library(data.table)
```

```
## data.table 1.13.0 using 6 threads (see ?getDTthreads). Latest news:
r-datatable.com
```

```
library(ggplot2)
```

```
## Need help getting started? Try the R Graphics Cookbook:
## https://r-graphics.org
```

```
library(palmerpenguins)

library(JWileymisc)
library(extraoperators)
```

In addition, go ahead and set our three usual data sets into our familiar variables and `data.table` format. Something to think about, in terms of real-life use of statistics and R, are your "usual" data sets. Having a nice structure and process for getting those data into action (like we do in the following lines of code) can make a daily workflow much easier to start:

```
acesData <- as.data.table(aces_daily)
penguinsData <- as.data.table(penguins)
mtcarsData <- as.data.table(mtcars, keep.rownames = TRUE)
```

You are now free to learn more about statistics!

7.2 Probability

Imagine tossing a fair coin into the air, catching it, and seeing whether it landed on heads or tails. You already know you have an equal chance between the two options. Real-life events that end in a specific, measurable state are said to have **outcomes**. The study of **probability** has at its heart ratios of *specific* outcomes and *total possible* outcomes. In the case of flipping a coin, the total possible outcomes are two (heads or tails). The overall probability of coin flipping can thus be described by two ratios, $\frac{1}{2}+\frac{1}{2}=1$

Take a moment to think about this idea of $\frac{Specific Outcome}{Total\ Possible Outcomes}$. If we want to know the chance of flipping a coin to heads, the specific outcome is heads. There is only one way that can happen. The total possible outcomes are heads or tails. There are therefore two possible outcomes. The probability of flipping our coin to heads is mathematically described by a ratio of $\frac{1}{2}$. That ratio can be written as a decimal, 0.50. Of course, that decimal can be written as a percent too – 50%. All these are equivalent ways of saying the same thing.

Sometimes, in mathematics, we express ideas in a formula (e.g., the arithmetic mean formula from Chapter 6). One way to express this formula might be $P(H)=\frac{1}{2}$ which could be read in English as "the **P**robability of flipping a coin to **H**eads is one-half."

If one sums up each specific outcome probability, the total should sum to 1. In the world of coin flipping, we might write $P(H) + P(T) = 1$.

This brings us to some of the language of probability. For a single flip of a coin, you cannot get both heads and tails at the same time. We say these outcomes are **mutually exclusive** because they cannot both happen. Another way to say this is to say heads and tails are *complements* of each other. Recall, in Chapter 6, we took a brief dive into some common formulae. Then, as now, our goal with some of these is less about using them by hand ourselves and more about making sure you gain some familiarity with the language used in research. When talking about two different outcomes, mathematics often notates those in uppercase, Latin letters (e.g., A and B). However, because complements have this idea of being the opposites of each other, they are often notated by either $\overline{A}$ (pronounced “A bar”) or A' (pronounced “A prime”). More formally, you might see the following formula:

$$P(A) + P(\overline{A}) = 1$$

Much like the word “bat” has two different definitions depending on context (e.g., wooden stick or flying mammal), so too $\overline{x}$ is different from $\overline{A}$ (which is why A' is sometimes seen). The shift from the start to the end of the alphabet – as well as the case of the letters – gives context clues as to what is meant.

On a slightly different note, any *two* flips of a coin are **independent** because the outcome of the first flip has no influence on the outcome of the second flip. As with complements, there is a formula for independent events. In the case of flipping a coin, we might get heads the *first* time and heads the *second* time. Now, admittedly the mathematical notation here dives back into the need for context. If we want to describe the probability of heads and heads, we write $P(H\&H)$. As before, in mathematics, we often write formulae in a *general* way and then plug in specifics. Thus, our old friends A and B return for event A and event B (which in our example are heads and heads). This gives us the following formula for independent probability:

$$P(A \& B) = P(A)P(B)$$

If this formula holds true, the probabilities are indeed independent. On the other hand, if we know the probabilities are independent, we are for sure free to use this formula.

The reason a coin flip is independent is because we are **replacing** our options each time with the full collection of possible outcomes (e.g., heads and tails are both possible both times). On the second coin flip, your possible outcomes are exactly the same (e.g., have been *replaced* back to pristine starting condition). On the other hand, some types of probabilities are **not replacing**. For example, if you are eating chocolates from a tray of mixed candies, you will have different choices the next time you choose a treat.

Notice mutually exclusive and independent do not have the same mathematical definition; these are different ideas. Independent events simply have no influence on each other. Contrastingly, mutually exclusive events cannot both happen. With this language on our minds, let us explore some examples together.

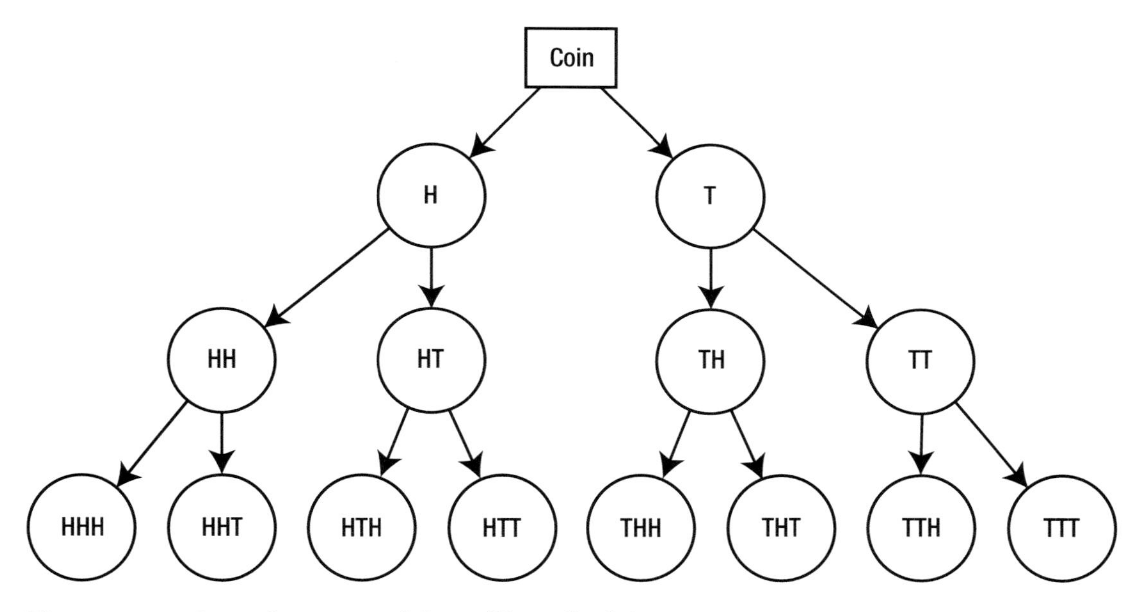

Figure 7-1. *Sample space of three flips of a fair coin*

Example: Independent

Returning to the idea of flipping a fair coin more than once, let us consider what happens when we flip a coin *three* times. In Figure 7-1, you see a *tree diagram* (made with the `DiagrammeR` package [10]). After the first flip, on the first row of coin outcomes, you see the two possible cases, heads or tails. After the second flip, on the second row of coin outcomes, you see the four possible cases. Lastly, after the third flip, you see the eight possible cases.

Each flip of the coin is *independent*. This makes sense on a logical level – the results of the first coin flip **do not** influence the second coin flip. This is actually really important. Humans (in general) are not naturally good at probability (us included). Instead, we must train ourselves to understand the mathematics behind chance. Take a look at the far-left branch after the second flip where the person standing there has HH. There is the same 0.5 to 0.5 chance of getting either heads or tails on the third flip. That coin is not "due" to get tails. Have you ever heard someone say, "We are due for a hurricane"? Probability does not work that way. Maybe you have heard "bad news comes in threes" – also not statistically true.

So what do we learn from independent events? Well, consider the probability of getting P(HTH) there on the third row. The specific outcome we want is HTH which is the third option from the left on the third row. Remember our general probability ratio? Specific outcomes divided by total possible outcomes gives us $\frac{1}{8}$ which is about 12%:

```
1/8

## [1] 0.125
```

However, to know that there are eight total possible outcomes requires us to build out this tree diagram. It would not take too many coin flips for that to get big fast!

Instead, we can leverage the fact that the coin flips are independent and use our formula $P(AandB) = P(A)P(B)$ for independent events. In particular, we *extend* that formula to cope with three flips – it uses the same logic:

$$P(H\,and\,T\,and\,H) = P(H)P(T)P(H) = \frac{1}{2}\frac{1}{2}\frac{1}{2} = \frac{1}{2*2*2} = \frac{1}{8}$$

```
(1/2)*(1/2)*(1/2)

## [1] 0.125
```

Using the independent outcomes formula allows us to calculate many independent events in a row quickly (without drawing out a diagram first).

So why study probability with statistics? Because it helps us to understand what we might expect in general. Suppose we told you we had given eight readers a coin and set them to flipping it three times each. Would you be shocked if we told you "One of our readers JUST flipped TTT!"? Probably not now – it is not too shocking since we had eight people throwing coins.

This is all to say that the imaginary *population* of coin flips in Figure 7-1 can be *sampled* by flipping coins. And if you know your population is big enough to include eight coin flippers, then TTT is not a hugely unexpected member of the sample. On the other hand, if you only had one person flipping a coin, you might not expect someone to flip TTT.

Example: Complement

In fact, let us take a look at what we might expect. In the prior example, you just calculated the chance of flipping P(TTT) = 1/8. What is the chance that someone *does not* flip TTT their first three tries? This is an example of **complement**. Taking the complement formula from before, you can edit it to show this exact scenario: $P(TTT)+P\left(\overline{TTT}\right)=1$. Remember $P\left(\overline{TTT}\right)$ is the complement of flipping TTT – in other words, it is the chance of **not** flipping all tails. A bit of algebra tells us

$$P(TTT)+P\left(\overline{TTT}\right)=1$$
$$P\left(\overline{TTT}\right)=1-P(TTT)$$
$$P\left(\overline{TTT}\right)=1-\frac{1}{8}$$
$$P\left(\overline{TTT}\right)=\frac{7}{8}$$

Just in case your fraction algebra is a bit rusty, R can do that computation for you just as well:

```
1 - (1/8)

## [1] 0.875

7/8 == 1 - (1/8)

## [1] TRUE
```

In any case, there is an 88% chance that someone would not flip all tails!

Just like the independent formula, the complement formula helps us more easily understand (without drawing a full diagram) what we might expect.

Again, think back to populations and samples. Suppose you entered our booth at a locale faire. We offer to let you win a large toy animal if you can flip HHH. You in fact flip TTT. You are suspicious and send in your friend to investigate. Your friend *also* flips TTT. Now suspecting something, you both call in a third friend to enter our booth, and, once again, they also get TTT. In this case, you expected the population (and indeed we sold you tickets) assuming it looked like Figure 7-1. Now, you have a random sample that has TTT, TTT, and TTT. Does it seem likely that your sample came from a world where both heads and tails were equally likely on that coin?

Or do you now suspect a weighted coin?

Based on the 88% chance of not getting TTT, you may well suspect something is wrong with that coin! More importantly, you just used probability and a sample to make a claim about a population. In other words, you just did statistics!

Probability Final Thoughts

As you see, probability is always a ratio of specific outcomes over all possible outcomes. Thus, probabilities are always going to be (as decimals) numbers between 0 and 1 (which we sometimes write as a percent). Because physically drawing out an entire tree diagram or some such is not always practical to determine the total possible outcomes, you learned some formulae to help compute probability. There are entire books on the study of probability full of ever more clever ways to count up the total possible outcomes of ever more complex populations. For us, here and now, what you learned should be enough to model simpler populations and learn more about statistics. Key to learning a skill set is getting comfortable with lifelong learning of additional techniques and new ways of thinking about topics.

That said, for many cases, you now have a language and tool kit to understand what might be expected from a population. If there is a way to set that population's total possible outcomes into probability ratios, those will sum up to 1. This gives you the ability to model populations (which is vital to comparing any sample to what we might expect that population to be). Much like the example of the faire booth, making a mental model of a population and then comparing the sample data to that population model can help us understand how accurate our population model is.

To do that, you must learn about distributions, which are simply ways of merging our histograms from Chapter 6 with your new earned knowledge of probability.

7.3 Normal Distribution

Functions in R take inputs and give outputs. In mathematics, a **function** takes one or more inputs (often called x) and maps them to an output (often called y). A probability **distribution** is a function that maps a value to a particular probability of occurring. Most probability distributions we deal with, such as the normal distribution, are not a single distribution, but a family of distributions. This is because the specific normal distribution we get depends on some parameters.

In particular, the normal distribution has *two* parameters:

- Mean (μ), also sometimes called the "location" because it controls the location from the center of the distribution.

-Standard deviation (σ), also sometimes called the "scale" because it controls the scale or spread of the distribution.

Formally, the normal distribution can be written

$$N(\mu,\sigma)$$

When we talk about a normal distribution, we may think of the standard normal distribution, which has mean zero and unit (or 1) variance/standard deviation:

$$N(0,1)$$

However, there are many normal distributions. Figure 7-2 shows a few different normal distributions, some with the same mean and different standard deviations, some with different means. What this shows is that to figure out the probability, it is not enough to specify that you are assuming a normal distribution; you must also specify which specific normal distribution by specifying its parameters: the arithmetic mean and standard deviation.

Can you spot the normal shape in Figure 7-2 that is the standard normal distribution? Hint: The highest part of the bell shape needs to be over zero. Recall the *empirical rule*; ±1 (the standard deviation of the standard normal distribution) should capture 68% of the area under that curve.

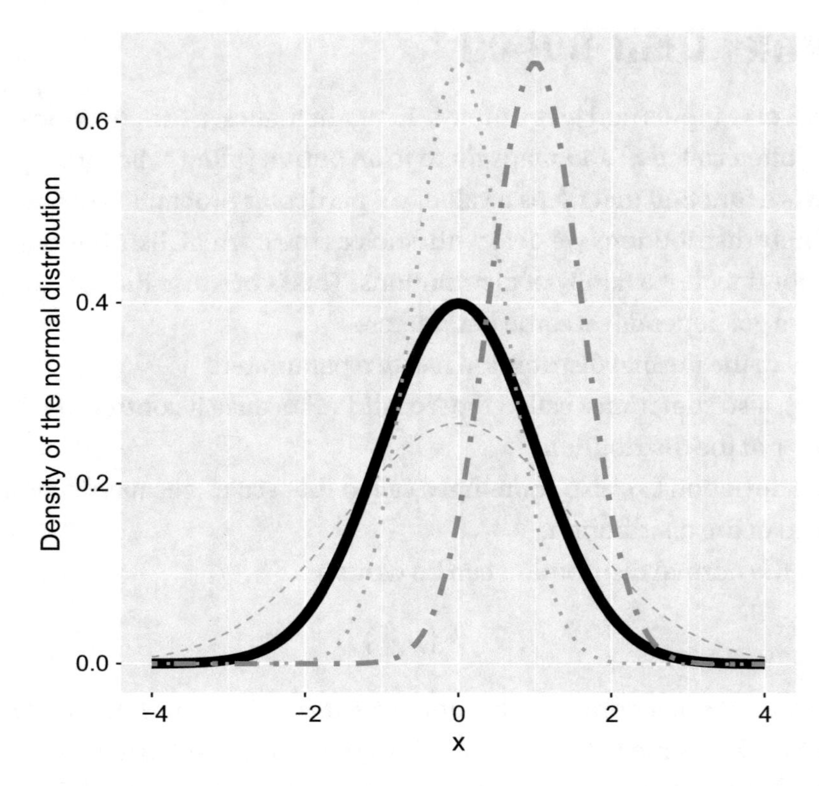

Figure 7-2. *Normal distributions with different means and standard deviations*

The answer is the solid, widest line.

In fact, you already saw another normal distribution, Figure 7-3, in Chapter 6. Back then, we called this a "perfect" example (and it was not bad). While the standard normal distribution is $N(0, 1)$ for mean and standard deviation, our old perfect example is $N(2.5, 0.75)$. Back when we only looked at histograms, we said to not pay much attention to the y-axis. However, take a look at the vertical y-axis now for these normal distributions.

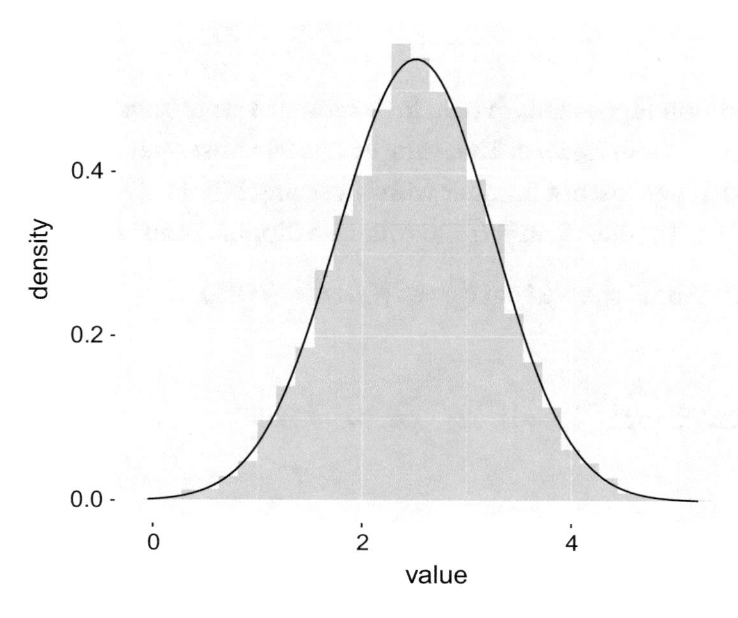

Figure 7-3. *Nearly perfect normal histogram with normal distribution superimposed*

Remember distributions are being introduced *after* probability, and there is always a method behind the chaos in mathematics. What the *density* numbers of the y-axis are telling you ties into probability. The definition of a probability **distribution** is a function that maps a value to *a particular probability of occurring*. Recall probabilities must live between 0 and 1.

The genius of distributions is the entire area under the curve is equal to 1. Thus, the relationship between our data and the probability of sampling data has been neatly connected in a single graph. This is actually a terribly clever idea (and in some ways was the big idea that turned some random mathematics into statistics). Finding just the right functions for distributions is all about taking the input of a mean and a standard deviation and turning that into an output that shows the probability. If we then graph the distribution function, we get the curves you see. The total probability has to be 1 (ergo 100% chance something happens).

Histograms by default have a y-axis of raw "headcount" rather than density/probability/weight. Thus, a histogram is sort of a proto-distribution in the making. It is not yet mature. You may recall back in our discussion of histograms we asked you to pretend to throw a dart at the histogram. Well, now there is a better way – mapping the data behind the histogram onto a distribution. To see how this works, join us in considering this first example.

Example

Consider the penguin flipper length data for a moment and recall the mean and standard deviation of that *sample*. That sample has 344 observations (although there are a couple NAs). If you are not familiar with surveying humans, 344 is actually a fairly decent sample size. The headcount (or in this case flipper count) matters:

```
mean(penguinsData$flipper_length_mm, na.rm = TRUE)
```

```
## [1] 200.9
```

```
sd(penguinsData$flipper_length_mm, na.rm = TRUE)
```

```
## [1] 14.06
```

Now examine Figure 7-4 for a moment. On the left side is the density y-axis of the normal curve. That normal curve is based on the *sample* arithmetic mean and the *sample* standard deviation. In other words, it is $N(201, 14)$.

However, when we draw a normal curve like this, what we are really doing is asking the question "Does it look like this sample, shown in the histogram, could have come from a population that looks somewhat like that normal curve?"

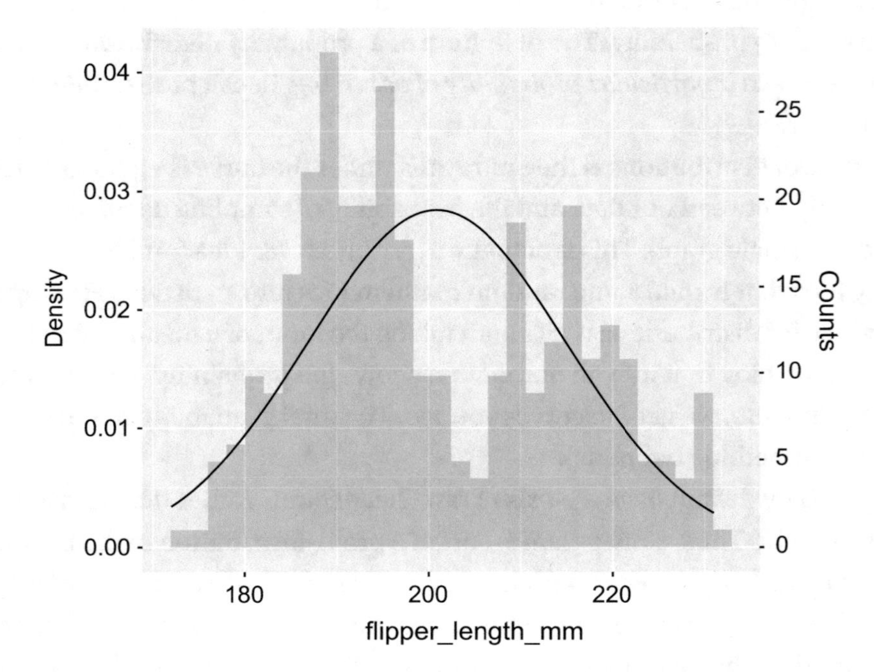

Figure 7-4. *Penguin flipper histogram with normal curve*

Here, it rather seems the answer may well be "no." The histogram and normal curve do not match up well. There is a rather large empty space exactly where we would most expect penguin flipper lengths. Could a random sample have accidentally gotten us here? Sure. Seems unlikely though, just as unlikely as if you and your two friends managed to flip only TTT from a "fair" coin at a faire booth.

Keep in mind that the penguin data actually has three different species of penguin. If you look at just the Adelie penguins, a different story emerges. Just like in Figure 7-2, there are families of normal curves. Notice Adelie penguins seem to have the shortest mean flipper length:

```
penguinsData[, mean(flipper_length_mm, na.rm = TRUE),
             by = species]
```

```
##      species    V1
## 1:    Adelie 190.0
## 2:    Gentoo 217.2
## 3: Chinstrap 195.8
```

```
penguinsData[, sd(flipper_length_mm, na.rm = TRUE), by = species]
```

```
##      species    V1
## 1:    Adelie 6.539
## 2:    Gentoo 6.485
## 3: Chinstrap 7.132
```

What if we edited our histogram and normal curve to only consider Adelie data? As you see in Figure 7-5, these data now look much more normal.

This example has been deliberately light on code. Your focus is meant to be on the images and building your intuition and understanding of normal distribution. While we continue that approach in the next example, it will soon be time to bring R back into the mix.

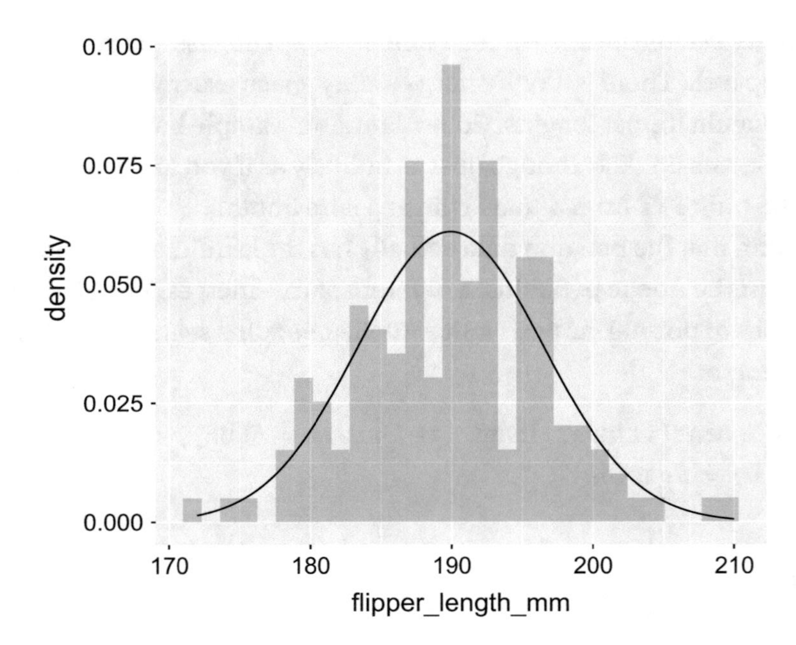

Figure 7-5. *Adelie flipper histogram with normal curve*

Example

We have mentioned more than once that a visual inspection of a histogram might tell us if the sample looks normal. To understand that rationale in more depth, we have to back up a bit and remember some facts about statistics.

The goal of statistics is to identify a population, develop some research questions about that population, select a sampling method that has a high likelihood of successfully capturing an unbiased sample, describe that sample in some depth, and then use that sample to understand the population.

The value of a distribution, such as a normal distribution, is that many measures fit a normal distribution. Thus, we can model the population on a normal distribution. However, if our sample does not appear normal, that may indicate one of two challenges to this philosophical approach. Either our sample was accidentally biased or our population is not in fact normal.

Recall in Figure 7-3 we used data simulated to be normal. In Figure 7-3, we used a histogram to plot the real data and then *also* superimposed a normal curve based on the sample mean and standard deviation. While it was a little rough around the edges (literally), this was clearly normal data.

Using the same data, now please consider Figure 7-6. The first, upper graph is two different graphs, one almost on top of the other. The solid shape is smoothed density plot of our almost perfect normal data. It is like the histogram other with the edges smoothed out a bit (this helps us compare better). The dashed line is, once again, a normal curve based on the sample's mean and standard deviation. As expected, this is a close fit.

Take a look at the x-axis. Rather than give a lot of x-axis values (like the histogram chart does), this graph focuses our attention on the five-figure summary (showing min, first quartile, mean, third quartile, and max). Notice the x-axis itself is broken up to show a sort of flattened box and whisker diagram. You can see where each quartile starts and ends visually.

So far, while Figure 7-6 does give a lot of information in a very small space, nothing is particularly better than our old method of histogram inspection.

It is the second, lower graph that makes this figure so useful. Not to get overly technical, this is based on what is called a quantile-quantile plot (more often a "Q-Q plot"). If our sample data are close to the expected distribution, then those dots should end up mostly on the line. In this case, they do. Thus, because we were testing against a normal distribution, we can be fairly confident this is a normal distribution.

In fact, notice the scale on the y-axis for this group of points. These are called *deviates* which is a mathematical word for *difference*. Thus, you can see from the scale of the deviates that the difference between any data point and the normal curve is actually not all that much. These numbers are all quite small decimals.

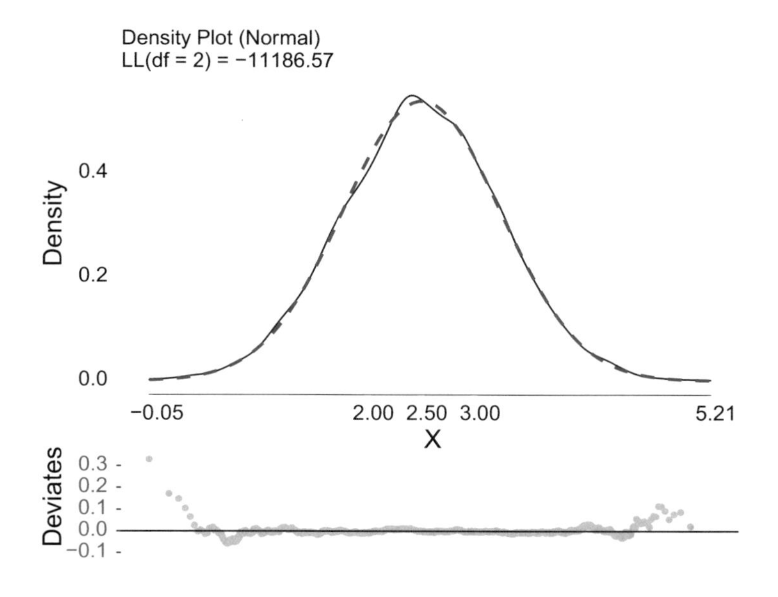

Figure 7-6. *Nearly perfect normal data in testDistribution() plot*

Contrast this with the nearly perfect uniform distribution we also introduced in Chapter 6 as shown in Figure 7-7. Again, in the top you can see the density function as a solid line vs. the dashed normal distribution based on the same mean and standard deviation. Because this data is uniform, you see the uniform quartiles in the x-axis.

It is the lower graph showing the deviation or difference where it becomes clear there are some significant differences from normal. Not only are there major areas where many points are off that line. Look again at the scale of the y-axis. These are not decimals – these are whole numbers. Taken together, this is a clear signal that uniform data are not normal (which of course we already knew).

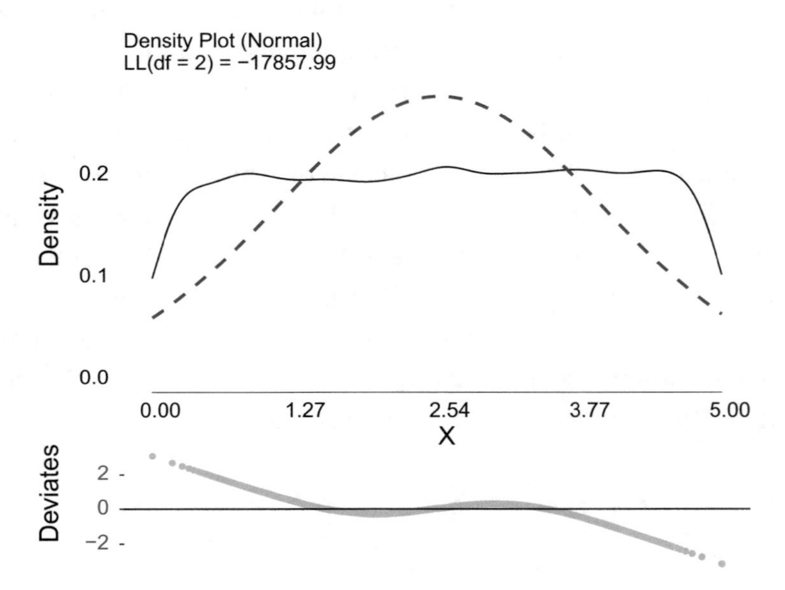

***Figure 7-7.** Nearly perfect uniform data in testDistribution() plot*

So now that we have seen this approach to detecting normal from a philosophical perspective, how can you use this in your own learning?

Example

The function we used is from the `JWileymisc` package and is called `testDistribution`. While it can take several arguments, for now we need just two. The first argument takes a data set, and the second argument takes a text string telling the function which distribution you expect.

You have already seen the normal distribution is in fact a family of functions. Not all distributions are normal. You have already also seen the *uniform* distribution. There are many more distributions beyond just these two. Thus, this second argument can take several options (although there are more distributions than this one function has yet been built to handle).

The testDistribution() function is actually fairly advanced, and some output is beyond our present needs. Thus, we are going to take our function results and assign them to variables. That way, we access the info currently required. All the same, as mentioned, you see the first argument set to data and the second formal to distr = "normal". Once this output is assigned to a variable (we chose testF for *test*Distribution *F*ull penguin data set), the plot() function creates Figure 7-8. As expected, the penguin data are not quite normal. The deviates plot has most data points being quite off the line:

```
testF <- testDistribution(penguinsData$flipper_length_mm,
                 distr = "normal")

plot(testF)
```

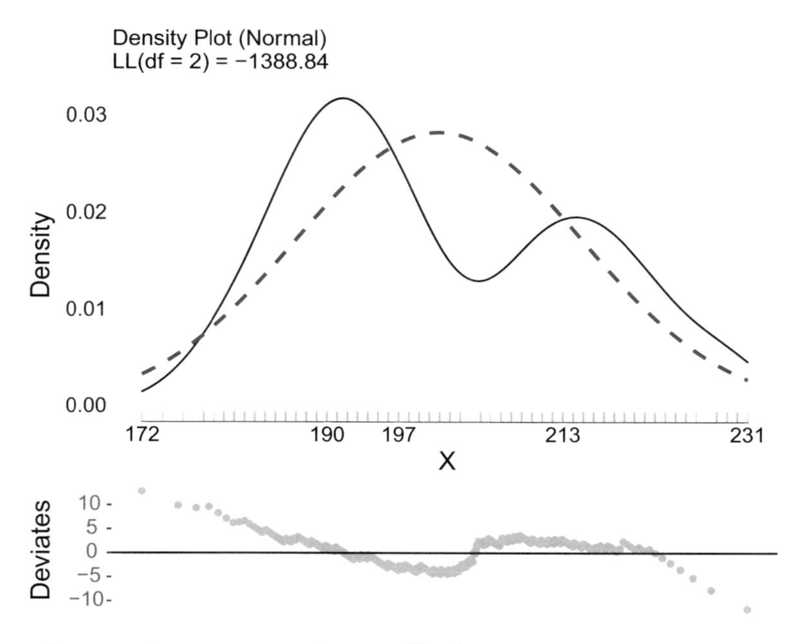

Figure 7-8. *All penguins testDistribution() plot*

Looking at the density plot, you can see the solid line of the penguin flipper density has two crests or *maxima*. In addition, on the x-axis, because this data set has a sensible number of points (our "perfect" data has 10,000 points), you can see there is a *rug* graph. In a rug graph, each line (seen on the x-axis) represents an individual observation. Now, in some cases, more than one observation are overlapping each other; this particular version does not show that density well (although the peaks of the density plot signal us some lines must have greater frequency). Graphs that are only rug graphs might use color scales to show frequency. The value for us looking at this particular rug graph is in visually seeing the "raw" data. Note particularly around the 172 mark on the x-axis, one can see there is a fairly wide spacing between data points. The spacing gets narrower under the peaks.

So far, you have seen examples of simulated normal and uniform data, as well as one real-world example that was not normal. What about some real-world examples that are normal? Remember the more examples you have to compare and contrast, the better you are able to use this function on your own real-world data.

Example

While the penguin data may not have looked normal when all three species are collapsed onto the same histogram, two of the species have samples that actually are normal as seen in Figure 7-9.

It is worthwhile discussing the `ggplot()` code that creates Figure 7-9 in depth. As usual for `ggplot()`, we first set the data argument, in this case `data = penguinsData`. Next, we map our aesthetics for the entire plot to be the flipper lengths on the x-axis. As before, `geom_histogram()` will give histograms. However, we use a new argument for the histogram. For the first time, we set the y-axis (vertical) aesthetic. In this case, we use `y = ..density..` to ensure the y-axis (which defaults to a count in histogram) instead gives us the probability or the *relative frequency*. The last line of code, `facet_grid()`, breaks the graphs into *col*umn*s* by the *var*iable*s* given. In this case, we use `cols = vars(species)`.

Taken altogether, one can see these all look fairly normal. Thus, we might suspect these samples came from normal populations:

```
ggplot(data = penguinsData,
       mapping = aes(x = flipper_length_mm)) +
  geom_histogram(aes(y = ..density..)) +
  facet_grid(cols    =    vars(species))
```

```
## 'stat_bin()' using 'bins = 30'. Pick better value with 'binwidth'.

## Warning: Removed 2 rows containing non-finite values (stat_bin).
```

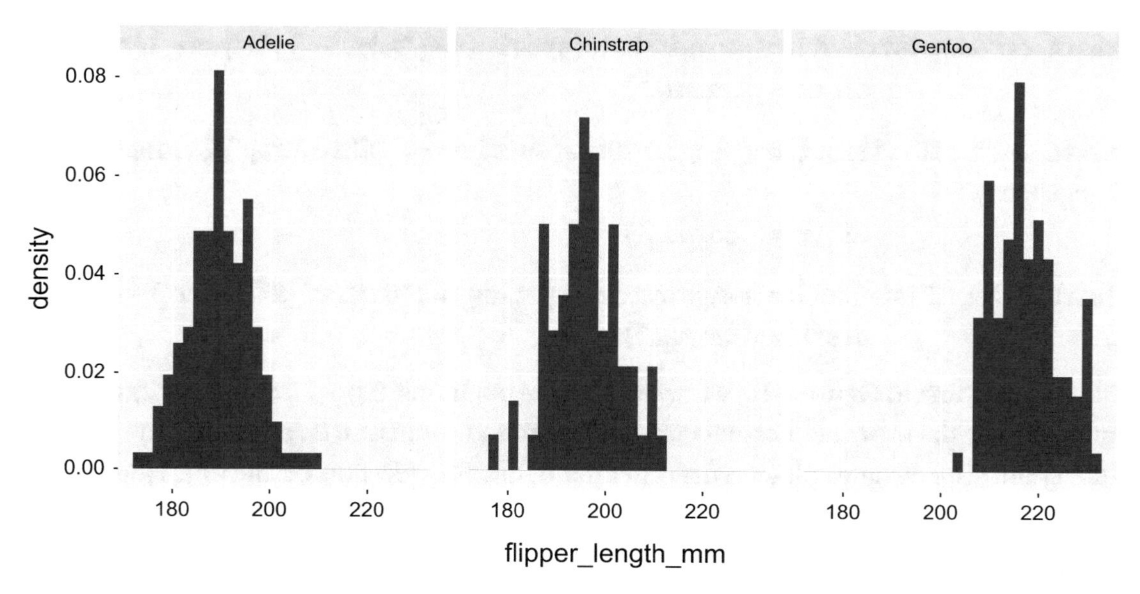

Figure 7-9. *Exploded species flipper histograms (density)*

It is worth exploring this in more depth. Remember our eventual goal is to use sample data (such as penguin samples) to understand features of the penguin population. If we find evidence that our sample is not behaving normally, then we need to decide if either our sample was biased or we need to change our hypothesized population distribution from normal to something else. Because there are many distributions possible, most introductory statistics books focus on normal distribution. So, on the one hand, if we cannot find evidence our penguins are normally distributed, most of the techniques described later in this book are not much use. On the other hand, if they are not normal, then at least we know one distribution model to avoid!

Of course, one of the virtues of using a programming language like R is that it eventually turns out a lot of the techniques between normal and other distributions are largely the same. This is helpful because it allows an introductory text like this to focus on only the most common distribution - normal - all while giving you almost all the same skills (and indeed function names) you want for other distributions.

The only difference in our approach to explore the penguin species one at a time is to control for species using the ith row selection feature of `data.table`. Using our earlier naming convention, the letter after `test` in our variable assignment is the species names:

```
testA <- testDistribution(penguinsData[species == "Adelie"]$flipper_length_mm,
                 distr = "normal")

testC <- testDistribution(penguinsData[species == "Chinstrap"]$flipper_
length_mm,
                 distr = "normal")

testG <- testDistribution(penguinsData[species == "Gentoo"]$flipper_length_mm,
                 distr = "normal")
```

Looking first at Figure 7-10, we see the Adelie penguins' flipper length *density* matches quite closely the expected normal density. This much could also be seen from the histogram. The deviates plot, on the other hand, has very few dots on the line. However, in general, they are contained inside a band around the line of +/- 1. That symmetry is often okay for a mostly normal distribution. In particular, notice the dots have a pattern of being slanted bands, where each band has at least one dot that touches the solid line.

```
plot(testA)
```

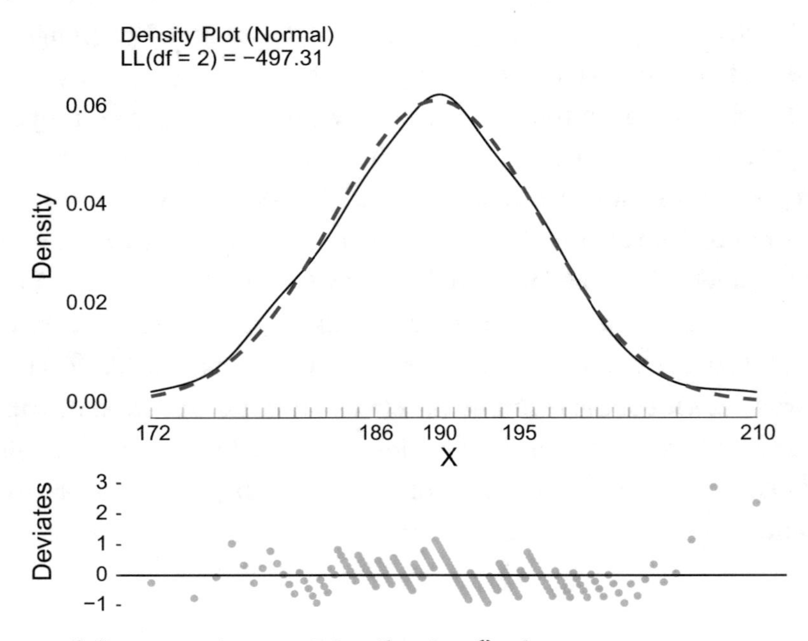

Figure 7-10. *Adelie penguins testDistribution() plot*

Looking next at Chinstrap penguins in Figure 7-11, we see a quite similar story. In particular, again notice in the middle of the deviates plot, there are many points which do indeed touch the normal line. While this was also present in Figure 7-10, we note here the far ends of the density plot (the solid line) are *above* the normal curve (the dashed line). This is a sign of more-than-normal observations at the extreme ends of the range of flipper lengths. If the density curve gets too tall on the ends, you have a uniform distribution. On the other hand, smaller sample sizes can also lead to this (selecting even a single instance of an extreme flipper length - short or long - can cause bias). It is one reason why larger sample sizes are always more ideal (in addition to larger sample sizes having a better chance of capturing more features of the population).

```
plot(testC)
```

The last penguin graph we look at is Figure 7-12 for Gentoo. Right away we see this is not looking very normal per the density plot. The histogram also was looking the least normal in Figure 7-9. Moving to the deviates plot, here are some key differences to see. In Figure 7-12, there are several dots that are on one side of the normal line all in a row. These are not evenly scattered on either side; they are all above (on the left side of that plot). Next, in the slanted bands near the middle of the deviates plot, you again see several bands that fall completely below the normal line. It matters that these bands do this together in a row; it would be different if one band was all below and the next band was all above the line. Contrastingly, in Figure 7-11, notice almost every slanted band touches the normal line, and any time one band does not, the next time a band does. This is a sign that Chinstrap penguin flipper lengths were normally distributed.

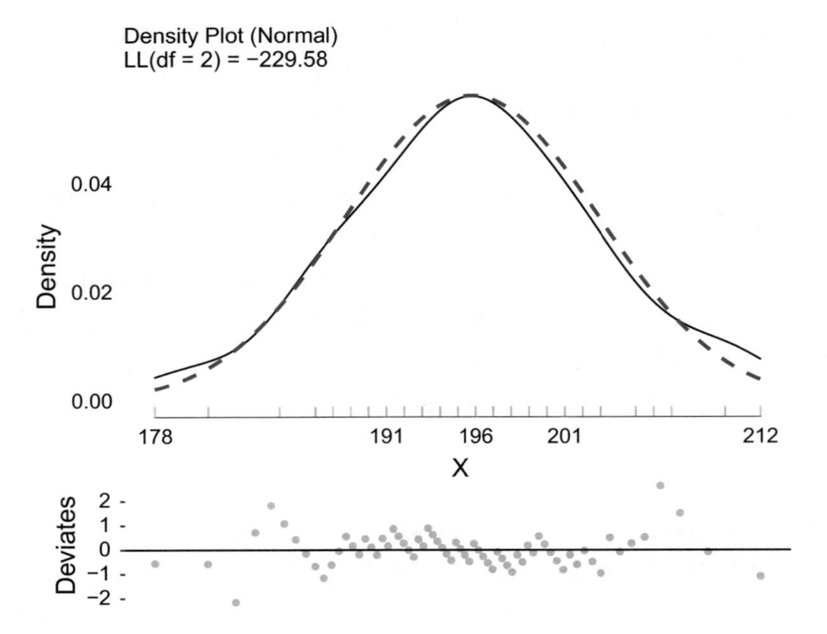

Figure 7-11. *Chinstrap penguins testDistribution() plot*

```
plot(testG)
```

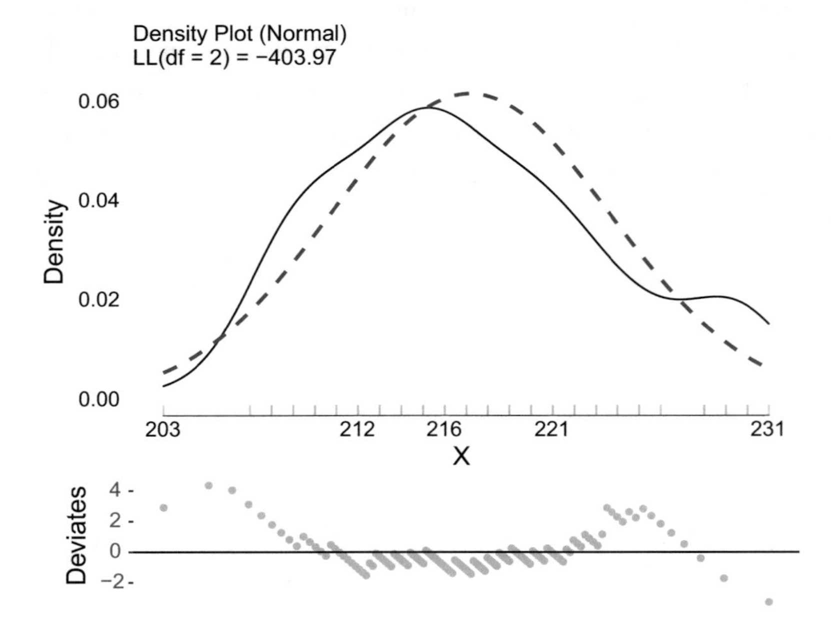

Figure 7-12. *Gentoo penguins testDistribution() plot*

As you have seen, by segmenting the penguin data into species, we discovered normality which had not been visible in the full data set. Might there be reasons the Gentoo penguin data set is not normal? One thing that comes to mind is the biological sex of the penguins; some species have a difference there. It seems the Gentoo in our data set are not necessarily as evenly segmented by sex; that may explain a shade of bias:

```
penguinsData[,
             .N,
             by = .(species, sex)]

##      species    sex  N
## 1:    Adelie   male 73
## 2:    Adelie female 73
## 3:    Adelie   <NA>  6
## 4:    Gentoo female 58
## 5:    Gentoo   male 61
## 6:    Gentoo   <NA>  5
## 7: Chinstrap female 34
## 8: Chinstrap   male 34
```

In any case, now you have seen some good ways to test for normalcy. While this book focuses on normal distributions, most modern statistical functions have several distributions that are easy enough to swap out. Thus, when checking a sample for likely distributions, there is a similar technique.

7.4 Distribution Probability

Provided sample data come from a normal population, there are some ways we can understand data in terms of the position of the elements of our data. Recall Figure 7-13 and the *empirical* or *68-95-99.7* rule.

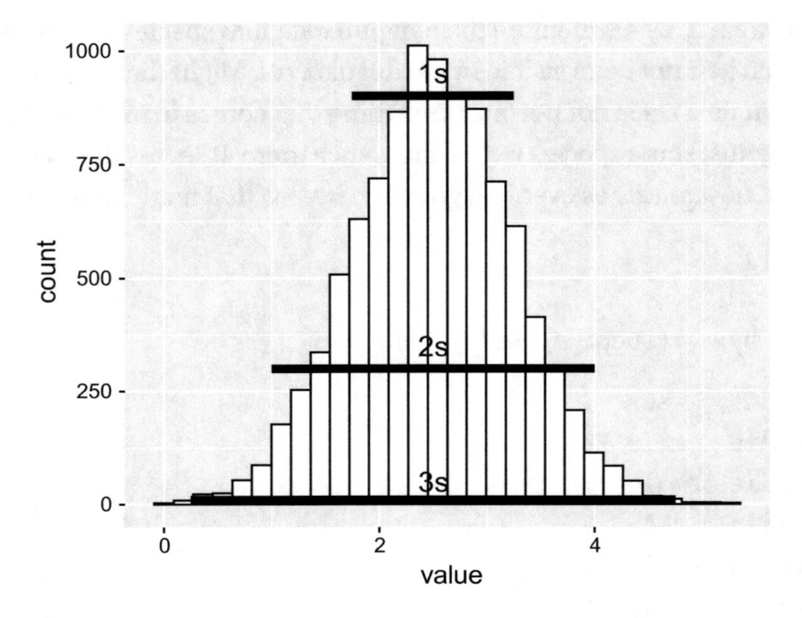

Figure 7-13. *Bars captured by the standard deviation ranges*

Each data point in a data set can be understood in terms of its standard score. In other words, does it live inside the first standard deviation band? The second? This standardized score is an example of unit conversion – a concept you may have met elsewhere to convert one unit to another (e.g., miles to kilometers or Fahrenheit to Celsius). In this case, the conversion is from whatever our data units are to standard units on the standard normal distribution $N(0, 1)$. This concept is called a **z-score,** and while we ideally use the population parameters, more often one uses the sample statistics as a proxy for the population parameters.

Z-score formula using population parameters is

$$z - score = \frac{x - \mu}{\sigma}$$

Z-score formula using sample statistics is

$$z - score = \frac{x - \overline{x}}{s}$$

Either way, you see the z-score gives one the difference between a data point x and the arithmetic mean, all divided by standard deviation. The resulting number, called a z-score, tells us in standard deviation units how far from the center (which is 0 of the standard normal distribution $N(0, 1)$ the data point is.

That may sound confusing, so let us take a closer look. Our perfect data set has a normal distribution $N(2.5, 0.75)$. In fact, that is a population measure. Keeping Figure 7-13 in mind, consider the number 2 from the sample data. Visually, from Figure 7-13, you see it clearly lives inside the first standard deviation bar. Additionally, it is clearly to the left of the center of the data (which has mean of 2.5 as you recall).

Z-score formula using population parameters for "perfect" data:

$$z - score = \frac{x - \mu}{\sigma}$$

$$z - score = \frac{2 - 2.5}{0.75}$$

$$z - score = \frac{-0.5}{0.75}$$

$$z - score = -0.66$$

The same mathematics in R yield the same answer:

```
perfectZscore <- (2-2.5)/0.75
perfectZscore
```

```
## [1] -0.6667
```

This z-score is *negative* which tells us the number 2 in the sample data is to the *left* of the mean. It is a number *smaller* than 1 in absolute value, so that tells us it lives in that first standard deviation band.

The same process, only this time on the data point x = 4.5, gives a different answer. Looking at Figure 7-13, you can see 4.5 is on the far-right side of the graph and just past the edge of the 2s bar. It lives to the right, so we expect a positive z-score. It is past the 2s bar, so we expect a number greater than 2 in absolute value.

Checking our mental logic in R shows we are correct:

```
perfectZscore <- (4.5-2.5)/0.75
perfectZscore
```

```
## [1] 2.667
```

What does this get us? Well, keep in mind the empirical rule. For normal distributions, we expect items outside the second standard deviation to be rare. Much like it would be strange to have you and your friends all flip TTT in a row for a fair coin flip in a faire booth, it would be strange to randomly sample the number 4.5 too often from a normal standard distribution of $N(2.5, 0.75)$.

How strange? The empirical rule states 95% of the data are inside the second standard deviation bar. Using the probability idea of *complement*, we know P(not inside 2s bar) = 1 – P(inside 2s bar) = 1 – 0.95 = 0.05. However, that 5% is split evenly on the both left outside part and the right outside part. So we can cut that number in half. This tells us the *left*, lower side of 4.5 has approximately 97.5% of the data points, while the *right*, greater side of 4.5 has approximately 2.5% of our data points. Note this probability estimate is based on the empirical rule, which (as we saw before) was a rough estimate. Also note this probability estimate supposed our data point was just larger than the two standard deviation line. Instead, as you can see in Figure 7-13, 4.5 is further to the right than where that 2s line ends. Therefore, we should imagine the probability of picking a point less than 4.5 is even larger than 97.5%.

How can we get a more accurate estimate? Naturally, you can use R to help with these sorts of calculations. Because what we just calculated was a probability of a **norm**al distribution, the function is called `pnorm()` and is pronounced "P, Norm" (roughly). It takes four inputs of the data point, the mean, the standard deviation, and a Boolean value that defaults to `lower.tail = TRUE`. The lower.tail is the same as the "probability of picking a point *less than*." Recall our estimate for that probability was more than 97.5%. Take a look at the following code and see how good our empirical rule estimate was:

```
pnorm(4.5,
      mean = 2.5,
      sd = 0.75,
      lower.tail = TRUE)

## [1] 0.9962
```

Close, yet not perfect. This is to be expected since our point, 4.5, had a z-score around 2.67 which is well to the right of the end of the 2s zone (which ended at two standard deviations to the right of the mean).

The value of the empirical rule (in these modern times of fast computing) predominately lies in the ability to estimate closely enough that you catch yourself making code mistakes. From a mathematical perspective, historically, only calculus could have computed the probability of 0.9962 so accurately. More recently (yet still in the past), one might have used "standard normal distribution table" to look up z-scores and estimate the probability. Today, we use `pnorm()`. The tables we used back when we learned statistics are now quaint (and honestly life is better using `R`).

Example

Mindful that the pnorm() function (just like the empirical rule) can only give accurate estimates when the underlying population distribution is normal, it is important to always check that normality assumption. From Figure 7-11 in a prior example, we know in the penguin data set, Chinstrap penguins' flippers can be supposed to be normal.

In order to more readily work on the Chinstrap penguin data, we assign that to a variable named chinstrapData. Additionally, we assign mean() and sd() to variables as well:

```
chinstrapData <- penguinsData[species == "Chinstrap"]
chinstrapMean <- mean(chinstrapData$flipper_length_mm)
chinstrapSD <- sd(chinstrapData$flipper_length_mm)
```

While you may have noted the z-score was not required to compute probability, it was useful to understand that probability. Rather than compute z-scores by hand, we can create a new column in chinstrapData using the scale() function. The scale() function is clever enough in a data table to compute the mean and standard deviation all on its own. As a confirmation check, we manually compute the z-score using $z - score = \frac{x - \overline{x}}{s}$ for the first row:

```
chinstrapData[, zScoreFlipper := scale(flipper_length_mm)]
head(chinstrapData)
```

```
##      species island bill_length_mm bill_depth_mm flipper_length_mm
## 1: Chinstrap  Dream           46.5          17.9               192
## 2: Chinstrap  Dream           50.0          19.5               196
## 3: Chinstrap  Dream           51.3          19.2               193
## 4: Chinstrap  Dream           45.4          18.7               188
## 5: Chinstrap  Dream           52.7          19.8               197
## 6: Chinstrap  Dream           45.2          17.8               198
##    body_mass_g    sex year zScoreFlipper
## 1:        3500 female 2007      -0.53612
## 2:        3900   male 2007       0.02474
## 3:        3650   male 2007      -0.39590
## 4:        3525 female 2007      -1.09698
## 5:        3725   male 2007       0.16496
```

```
## 6:         3950 female 2007         0.30517
```

```
(192-chinstrapMean)/chinstrapSD
```

```
## [1] -0.5361
```

Now having z-scores, let us think through some examples of using these to better understand Chinstrap penguins. Recall from Figure 7-11 the first quartile ends at $x = 191$. From your study of quartiles, you know that 25% of the data ought to be to the left of 191. Now, mindful that quartiles do *not* require a normal population, let us see if the `pnorm()` (which does require normal population) matches that estimate for `lower.tail = TRUE`:

```
pnorm(191, chinstrapMean, chinstrapSD, lower.tail = TRUE)
```

```
## [1] 0.2494
```

In fact it does! Now, that is not too shocking; the data did look normal per Figure 7-11. Additionally, the deviates plot you may recall is based on a Q-Q plot (where Q-Q stands for quantile-quantile). The way that the deviates plot works is via a comparison of Chinstrap flipper length quantiles to normal quantiles; provided the dots often fall on the line, the quantiles are matching.

What about the chance of getting a Chinstrap penguin with a flipper length longer than 191? Well, our quartile data suggests that ought to be 75%. Alternately, we can either use the probability idea of *complement* on the `pnorm()` just calculated or set `lower.tail = FALSE`. No matter what approach used, the answer is the same:

```
1 - pnorm(191, chinstrapMean, chinstrapSD, lower.tail = TRUE)
```

```
## [1] 0.7506
```

```
pnorm(191, chinstrapMean, chinstrapSD, lower.tail = FALSE)
```

```
## [1] 0.7506
```

Just as the population distribution of flips of a fair coin could help you understand the chances of you and two friends sampling TTT three times in a row, similarly, understanding the (normal) population distribution allows you to understand the likelihood of randomly finding many Chinstrap penguins with flipper lengths longer than 210. Could it happen? Yes. However, if it did, we may well become suspicious someone has been feeding those penguins!

```
pnorm(210, chinstrapMean, chinstrapSD, lower.tail = FALSE)
```

```
## [1] 0.02342
```

Example

Based on the sample data, what is the probability of randomly finding an Adelie penguin with flipper length greater than 202 mm?

Having already confirmed these sample data are not excluding normal parameters, per Figure 7-10, we have some evidence to proceed with modeling the population via normal distribution. Based on mean() and sd(), the Adelie population's normal distribution is estimated to be roughly $N(190, 6.5)$:

```
adelieData <- penguinsData[species == "Adelie"]
adelieMean <- mean(adelieData$flipper_length_mm, na.rm = TRUE)
adelieSD <- sd(adelieData$flipper_length_mm, na.rm = TRUE)
```

Using the sample statistics as rough estimates of the assumed normal population parameters, we use pnorm() to estimate a right-tail probability:

```
pnorm(202, # x > 202mm
      mean = adelieMean, #sample mean
      sd = adelieSD, # sample sd
      lower.tail = FALSE) #GREATER than is upper/right tail
```

```
## [1] 0.03273
```

It is not very likely to randomly see such large Adelie penguins!

To wrap up this example, we pose a thought puzzle. Look closely at the logic process in this example. We start off curious about seeing a long-flippered Adelie penguin. Inspection of the *sample* data seems to indicate the Adelie *population* could be normal (in other words, we found no evidence against our sample coming from a normal population in Figure 7-10). From there, we use the *sample statistics* of mean $\bar{x}$ and standard deviation s to estimate the (hypothesized to be) normal population distribution's parameters of μ and σ. **If** all our conjectures are correct, then our pnorm() probability estimate should be accurate.

That is a lot of assumptions though, is it not? Are we sure the Adelie population is normal? Technically no (at least not without using our knowledge of biology which is outside of this small sample of 152 or so penguins). Even if the Adelie population is normal, are we sure our sample statistics $\overline{x}$ and s are accurate or precise measures of the population parameters μ and σ? Again no (and again not without referencing some outside study of Adelie penguins); after all, the sample could be biased.

Is there any way we can be sure to have a normal population? What about getting a better sense of how accurate the *statistics* are at estimating the *parameters*? These are important questions that are answered in future sections and chapters.

7.5 Central Limit Theorem

You now have practice spotting samples of likely normal data. Now the challenge is taking the information on probability and distributions you learned and bringing that knowledge together to better understand just how useful samples are. Additionally, we seem to have made an odd choice to focus on only the normal distribution – why that one distribution when we have asserted more than once there are many? And, as just mentioned, can we ever be sure an underlying population is normal?

To answer these questions, let us first take a look at something rather strange. Consider two *random* samples from a standard normal distribution with $N(2.5, 0.75)$ (recall the first number is the mean and the second is the standard deviation). Just to make sure your numbers agree with ours, we again use the `set.seed(1234)` function. The functions in R that generate random data often start with a `r` for *r*andom. The specific function for *norm*al data is `rnorm` and is pronounced "R, Norm" (roughly). If we assign those two sets of random numbers pulled from such a distribution into two variables named `sampleA` and `sampleB`, we can see from the `summary()`) function our data have rather different *ranges*. That is to say, the minimum and the maximum are quite different from each other. This is to be expected in a random sample. However, take a look at the arithmetic mean. Those are actually quite close together:

```
set.seed(1234)
sampleA <- rnorm(n = 5, mean = 2.5, sd = 0.75)
sampleB <- rnorm(n = 5, mean = 2.5, sd = 0.75)

summary(sampleA)
```

```
##    Min. 1st Qu.  Median    Mean 3rd Qu.    Max.
## 0.741   1.595   2.708   2.236   2.822   3.313
summary(sampleB)
```

```
##    Min. 1st Qu.  Median    Mean 3rd Qu.    Max.
##   1.83    2.07    2.08    2.19    2.09    2.88
```

Why are the means so close to each other? The answer lies in the unbiased nature of a random sample. It would be strange if a random sample only pulled from the far-left side of the graph or the far-right side of the graph. If it did, that would be a sign of bias. Go back to thinking of the typical normal distribution shape as seen in Figure 7-6. In fact, our "perfect" normal data set we have been using for a demo was created with `rnorm()`. The highest points on that density plot are in the middle. It would be strange indeed to pull too many data points outside that middle. Thus, while any single random pull might give us a strange number, we would expect the mean of a sample pulled from the population to be fairly close to the actual population mean. When "averaged" together, the *consensus* of the sample data statistics gets quite close to the population data parameters (recall those words from our earlier definitions).

This feature of the sample mean *statistic* being close to the population means *parameter* σ is not unique to the normal distribution. Recycling the code we just used, we change `rnorm()` to `runif()`. This is pronounced "R, Unif" (roughly) for *r*andom *unif*-orm. Keeping in mind our "perfect" uniform data as seen in Figure 7-7, we can see those are not at all normal data. And we see the mean of the entire population is 2.5 from the quartiles on the x-axis of the density plot.

Looking now at samples `sampleC` and `sampleD`, we see the maximum values are not close to each other – nor are the first quartiles. Nevertheless, the sample arithmetic means are quite close to each other and close to the population mean of 2.5:

```
set.seed(1234)
sampleC <- runif(n = 5, min = 0, max = 5)
sampleD <- runif(n = 5, min = 0, max = 5)
```

```
summary(sampleC)
```

```
##    Min. 1st Qu.   Median   Mean  3rd Qu.    Max.
## 0.569   3.046    3.111  2.830    3.117   4.305
```

```
summary(sampleD)
```

```
##     Min. 1st Qu.   Median    Mean  3rd Qu.    Max.
##  0.047    1.163    2.571  2.063    3.202   3.330
```

What you are seeing is for two different population distributions, even a small random sample will have a mean quite close to the population mean. This makes sense, after all. We saw it from the beginning with mean salary; the mean salary was less extreme than any single salary. *The means of the samples cluster about the mean of the population*. While the word *theorem* in popular culture tends to be used to describe a *conjecture*, the word takes on a different definition in mathematics and science. In mathematics, **theorem** is essentially a mathematical truth – a provable fact. Think of some famous science theories such as *theory of relativity* or *string theory*. For statistics, the **central limit theorem (CLT)** states that random samples drawn from the same population distribution will have arithmetic means or sample sums fitting the normal distribution. Furthermore, the specific normal distribution of the sample means will match the following formula (the bigger n is, the better the match):

Normal distribution of sample means drawn from an arbitrary population distribution of mean μ and standard deviation σ

$$N\left(\mu, \frac{\sigma}{\sqrt{n}}\right)$$

What does the central limit theorem, ahem, mean?

Firstly, keep in mind the shape of the normal distribution. Most data points are bunched up in the middle. So the central limit theorem is telling you that for random samples, taking the arithmetic mean $\bar{x}$ or average is, more likely than not, going to yield an answer close to the population mean μ. What is shocking about this theorem is that it does not matter what your population distribution is; the sample arithmetic means are going to cluster in a normal shape! It sounds tough to believe, so we will explore that in an example in just a moment to let you get your own hands on this truth.

However, if the central limit theorem is true, what that tells you is that just like we found normal penguin data by species hiding inside the overall penguin data, so too we can find normal sample mean data hiding in *any* distribution.

This tells you exactly why, if we had to pick just one distribution to teach you, we would pick the one distribution that is part of all samples we might take. If you find it fascinating (as we do) that such a fact could be true, you may well wish to consider exploring the mathematical theory behind statistics (remember, in mathematics, theorems are *provably* true). If, on the other hand, you simply want to exploit this "fact of nature" and use it to understand the world around you better (which we often do too), you are in the right place.

This tells us something more though. Remember, in our last z-score example, we mentioned that one of our underlying hypotheses was that the population was normal. Because of the central limit theorem, regardless of the population distribution, we know the population *mean* distribution **must** be normal. The z-score will need to be adjusted just a bit, just like the normal distribution as adjusted.

Z-score formula using population parameters for sampling distribution of the mean is

$$z = \frac{\overline{x} - \mu}{\frac{\sigma}{\sqrt{n}}}$$

This formula may seem a bit confusing; we will walk through it step by step in an example. The important thing to keep in mind is that, by using the central limit theorem, we ensure the population is a normal distribution. This removes one of our question marks about using random samples to estimate populations.

Example

So how might we go about showing random samples means $\overline{x}$ from some arbitrary population distribution (that may not itself be a normal distribution) can turn into normal data? This is not a mathematical proof book. All the same, you must be wanting some sort of validation to this extraordinary claim! A good demonstration lies in `R`'s ability to simulate data. You already have all the tools you need; we just need to put those together in a slightly new way.

You just met the `runif()` function. This time, instead of using this to build a *sample*, we set `n = 100000` to create a *population*. You are also familiar with `data.table`; however, you have not yet built your own from scratch. The function to create a new one is `data.table()`. As inputs, it takes a name you choose (this will be your column name), and then you can set that column equal to some data. In this case, we choose the column name to be `uniform`, and the data we set that equal to is a random sample of uniform data that is suspiciously close to the uniform data you have already seen that we called a "perfect uniform" data set. Recall that `1:191` in Chapter 5 counted out the numbers 1–191. Here, we build a second row named `sample` filled with a repetition of

`1:1000` (R will recycle the data since it is too short compared to the 100,000 rows of the first column). This is an example of Kth sampling, where every 1000 random uniform distribution numbers is stuck into a sample. This happens again and again and again 100,000 times. Take a look at the first few rows of the `perfectUniform` data as well as the last few rows:

```
set.seed(1234)
perfectUniform <- data.table(uniform = runif(n = 100000,
                                             min = 0,
                                             max = 5),
                             sample = 1:1000)

perfectUniform

##         uniform sample
##      1:  0.5685      1
##      2:  3.1115      2
##      3:  3.0464      3
##      4:  3.1169      4
##      5:  4.3046      5
## ---
##  99996:  1.8576    996
##  99997:  0.9441    997
##  99998:  2.3871    998
##  99999:  4.4129    999
## 100000:  4.0344   1000
```

Next, just to make sure we have a population of uniform data, go ahead and look at the histogram using `hist()` as shown in Figure 7-14:

```
hist(perfectUniform$uniform)
```

Keeping in mind our data are in a data table, what we need are the 1,000 sample means. In other words, we want all rows, so there is no need for any ith selection or order. We need a new column that is the `mean()` of the `uniform` column; this is the column operation in the jth position. Lastly, we need this `by = sample`. Assign this to `sampleMeans` and inspect the resulting data table that has one row per sample:

```
sampleMeans <- perfectUniform[,
              .(SampleMean = mean(uniform)),
              by = sample]
sampleMeans

##        sample SampleMean
##    1:       1      2.189
##    2:       2      2.275
##    3:       3      2.528
##    4:       4      2.296
##    5:       5      2.493
## ---
## 996:     996      2.288
```

```
##  997:  997 2.537
##  998:  998 2.453
##  999:  999 2.589
## 1000: 1000 2.570
```

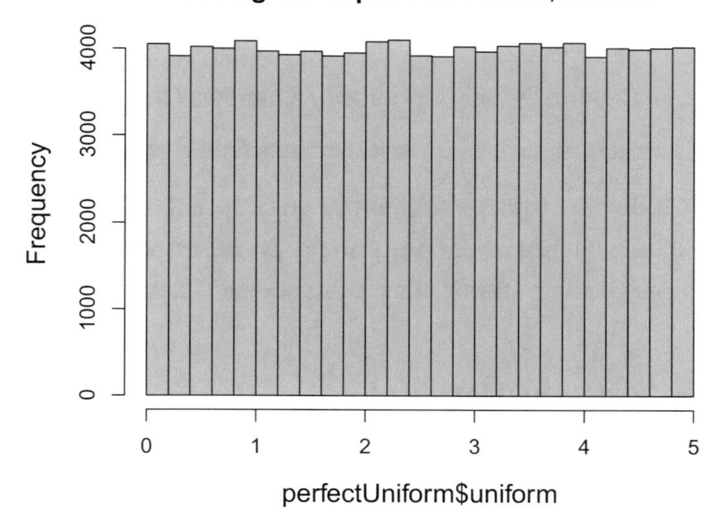

Figure 7-14. *Uniform distribution population N = 100,000*

Keep in mind our underlying population is a uniform distribution, and what we have are 1,000 samples pulled via Kth sampling (we could have used random – it was just easier to do Kth in this case). We once again use the histogram function `hist()`, and see Figure 7-15 looks rather normal:

```
hist(sampleMeans$SampleMean)
```

Fascinating, no?

This is what the central limit theorem assured us would be true – the distribution of the sample means is normal.

The clever "hack" one uses to leverage this is, if your population data are not normal and the statistical analysis you wish to use requires only normal data, you can coerce the data to behave normally by grouping individuals into samples and only making predictions on samples. This rather explains why, if we only have enough space in a beginning R textbook to explore one distribution with you, we best make it normal.

So how close were we to matching the central limit theorem prediction for this normal distribution?

Keep in mind the prediction was the following normal distribution:

$$N\left(\mu, \frac{\sigma}{\sqrt{n}}\right)$$

This prediction is on our original population, which was the uniform distribution `perfectUniform`. The second part, $\frac{\sigma}{\sqrt{n}}$, is sometimes called **standard error (SE)**. Taken together, we get the following information about our population's uniform distribution. This information, in turn, allows us to compute the *predicted* sample mean distribution. Notice the n in this case is the number of elements of each sample, not the number of samples:

```
mu <- mean(perfectUniform$uniform)
sigma <- sd(perfectUniform$uniform)
n <- 100000/1000

standardError <- sigma/sqrt(n)
standardError
```

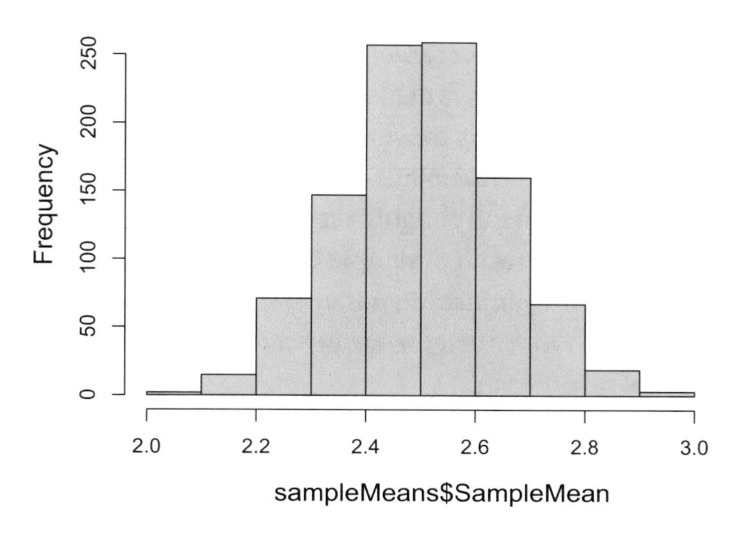

Figure 7-15. *1,000 sample means from a uniform distribution*

```
## [1] 0.1444
```

How well do these match with our actual sample mean results? Take a look, and you decide!

```
mean(sampleMeans$SampleMean)
```

```
## [1] 2.502
```

```
sd(sampleMeans$SampleMean)
```

```
## [1] 0.1454
```

The means are almost a perfect match. The expected standard error of 0.14 does not quite match the calculated sample mean standard deviation of 0.144. That is perfectly good and expected. The central limit theorem says this will only be true as the number of elements in each sample gets large – in other words, as *n* gets large. In fact, officially, we did not select our sample quite correctly. We used Kth sampling, not random sampling.

Still, this was very close.

Example

What about z-scores? Other than the conversion from the individual population data to the mean population data – which is achieved by dividing σ by the *standard error* – computing a z-score will work exactly the same way. Admittedly there is some overhead work required first to make that conversion.

Recall Figure 7-8 showed the full penguin data set was not quite normal in the sample. While this could be the result of sample bias (in which case the sample statistics may make bad estimates of the population parameters), it may be that across species, penguins have non-normal flipper lengths. In any case, using the central limit theorem can get us into a normal distribution.

Unlike our "perfect" data set, there are only 344 penguins. This is not quite enough to get the large number of samples we need. Fortunately, in our "perfect" data set, we cheated and used Kth sampling. In other words, we sampled *without* replacement. This was okay, because our data were in fact already randomly generated via `runif()`.

On real-life data, when converting from individual data to the sample mean distribution, sample *with replacement*. In other words, we will let `R` use a perfect random sample method, and the same penguin might be selected more than once. Just like a fair coin toss keeps both heads and tails "in play" for every flip, so too every penguin is up for flipper sampling every round. This will allow us to keep our samples sizes to be large enough (traditionally large enough is `n >= 30`).

To achieve this, you will use `R` to create a data table that can hold 1000 sample means that will be sampled 30 penguins at a time *with replacement*. You first must create a data table with a placeholder value for the `sampleMean`. In this case, set that to zero for now (it will change once the samples are pulled). What you have in `sampleMeansF` are 1,000 rows that are each uniquely numbered in the `sampleID` column and each have the default normal mean of `sampleMean = 0`. What you do in the following bit of code is create a shell and a placeholder for the means you are about to calculate on the samples you are about to pull:

```
sampleMeansF <- data.table(sampleID = 1:1000,
                           sampleMean = 0)
```

To make sure your results match ours, go ahead and use `set.seed(1234)` here. This is not something done in real-life data:

```
set.seed(1234) #ensures your results match ours; not used in real life
```

This is where you see a bit of new code technique. Welcome to the `for()` loop! Your goal is to take a random sample with replacement, take the mean of that sample, and store it in the correct row of your `sampleMeansF` data table. However, you must do this 1,000 times. Rather than copy and paste your assignment code 1000 times, you will use a `for()` loop.

Loops are a way to let R know you wish the same chunk of code to go on repeat. To help see exactly what is happening in the loop, notice that we *comment* the code line by line to make each part more understandable. The general structure of your loop is as follows (and this snippet is not meant to be run):

```
for(i in startInteger:endInteger){ #notice this starting bracket

  #code inside here gets repeated from startInteger to endInteger.

} #notice this ending bracket!
```

This loop you do need to run. It starts at 1 and runs the inside code 1,000 times. Using the already familiar to you `sample()` function, samples of `size = 30` are taken with replacement from the penguin data set. These samples are assigned to the value `randomSample`. In fact, after you run this code, you will see the last `randomSample` in your global environment as a list of 30 integers.

Once the random sample is pulled, then the *iterator* variable `i` is used to select the correct `sampleID`. That is a row selection operation in the `sampleMeansF` data table; thus, it goes in that first, ith position. Next, the `mean()` for that sample needs to be stored in the `sampleMean` column. Note this replaces the `sampleMean = 0` placeholder with an actual sample mean. This is the end of the loop iteration, and this is repeated for all values of `i` from 1 to 1000:

```
for(i in 1:1000){#for loop cycles 1 to 1000 times, getting a random sample
each time.

  #each  iteration of the  loop, a NEW, with-replacement sample of flipper
  lengths
  randomSample <- sample(x = penguinsData$flipper_length_mm,
       size = 30, #there are 30 flipper lengths in each sample
       replace = TRUE) #with replacement!
```

```
  sampleMeansF[sampleID == i, #row selection to CURRENT sample ID
               sampleMean := mean(randomSample, #assign sample mean
                                  na.rm = TRUE)]

}
```

Once the preceding code is done running (and it may take a minute), you will see in your global environment some new variables. You already noticed the `randomSample` variable which has the last `sample()` still in it. There should also be an `i` variable that is at 1000.

You now have a sample of 1000 means. Remember each sample has 30 random penguin flipper lengths and your `sampleMeansF` data table has a unique ID for each sample as well as the `mean()` of that sample. With your conversion from individual penguins to sampled groups of penguins done, it makes sense to use `testDistribution` on your sample means to decide if moving from individual penguin data in Figure 7-8 to mean penguin data (in samples of 30) in Figure 7-16 has yielded a normal enough data set. Compare the two figures; what do you think?

```
testMF <- testDistribution(x = sampleMeansF$sampleMean,
                           distr = "normal")
plot(testMF)
```

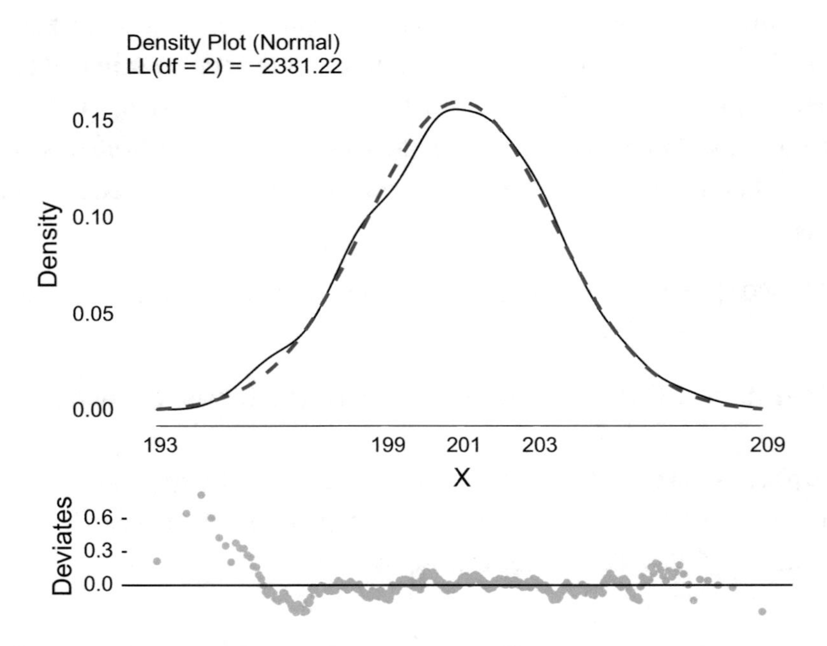

Figure 7-16. *Distribution of sample means in all penguins*

Figure 7-16 looks fairly normal, yes?

Next, you want to check how closely your `sampleMeansF` data fit the central limit theorem $N\left(\mu, \frac{\sigma}{\sqrt{n}}\right)$.

This time you do not have population-level data; so you must estimate the parameters using the statistics. Remember standard error is the standard deviation divided by the square root of the sample size:

```
xbar <- mean(penguinsData$flipper_length_mm, na.rm = TRUE)
stdDev <- sd(penguinsData$flipper_length_mm, na.rm = TRUE)
n <- 30

standardError <- stdDev/sqrt(n)
standardError

## [1] 2.567
```

If the sampling with replacement code worked, you expect it to match the central limit theorem values just calculated. Take a look at those computations vs. the actual `sampleMeansF` results:

```
xbar

## [1] 200.9

mean(sampleMeansF$sampleMean)

## [1] 201

standardError

## [1] 2.567

sd(sampleMeansF$sampleMean)

## [1] 2.491
```

That looks quite close to us; what do you think?

The data fit the theory fairly closely, the sample distribution looks normal enough, and thus z-scores can be used to consider probability of penguin samples.

Here is where one must be cautious about which variables go where. In the z-score formula for sampling distributions, we no longer have individual penguins. Instead, we have samples of penguins. Take a look at the z-score formula in English words and then compare and contrast the z-score for individual vs. sampling distributions.

Z-score in English words:

$$z = \frac{dataPoint - distributionMean}{distributionstandardDeviation}$$

Z-score for individual distributions:

$$z = \frac{x - \mu}{\sigma}$$

Z-score for sample distributions:

$$z = \frac{\overline{x} - \mu}{\frac{\sigma}{\sqrt{n}}}$$

As we have been doing (although it is not quite right), when you do not know parameter data, you can substitute statistics. Plugging into the z-score formula yields

Z-score for *penguin* sample distributions (values rounded):

$$z = \frac{\overline{x} - 201}{2.5}$$

Now you have your z-score set up, you can (finally) answer a probability question: *What is the chance of sampling 30 penguins and getting a mean of less than 195?*

As before, you use the same `pnorm()` function. Only this time, you set mean and standard deviation to the sampling distribution values:

```
pnorm(195,
      mean = mean(sampleMeansF$sampleMean),
      sd = sd(sampleMeansF$sampleMean),
      lower.tail = TRUE)

## [1] 0.008206
```

As you see, the chance of sampling 30 penguins which "average" to such a low mean is not very probable.

As you also see, the standard deviation of the sampling distribution is *smaller* than the underlying individual distribution deviation. Part of the fascinating usefulness of the central limit theorem lies in this truth that as sample size increases, the overall deviation (or standard error) decreases.

This has applications beyond forcing a penguin data set normal. Imagine instead of penguins you had company stocks. Randomly sampling from the universe of stocks (with replacement) means you may buy more stock from one company than another. And yet, even though your stock portfolio average might be expected to stay the same, your standard deviation is expected to shrink (by quite a lot). This of course reduces the chance that your stock portfolio will have wild growth (in other words, ends up on the far-right tail of the distribution). However, it also reduces the risk that your stock portfolio has wild loss (in other words, ends up on the far-left lower tail of the distribution).

Example

In our last example, let us consider our acesData. Thinking about the 191 participant ages, let us consider this question. Suppose we wanted to chat in a face-to-face focus group with 30 of our participants. What is the chance that the average age of our randomly selected focus group is more than 24? In other words, what is the chance that a randomly selected group of 30 study participants is mostly at the upper end of the age range?

Other than switching out data sets, we use the exact same process as before. In particular, much code will be recycled.

Using code from Chapter 6, we reduce our acesData down to only those users who entered an age and only one copy of each:

```
unduplicatedAcesAge <- acesData[! is.na(Age),
                                .(UserID, Age)]
unduplicatedAcesAge <- unique(unduplicatedAcesAge)
```

Next, using code from the prior example, we *bootstrap* 1000 samples from that data set (with replacement). Notice we are not only recycling our code; we mostly use the same variable names too. While this does overwrite those values in the global environment, that is not a risk in this case since we are done with the prior examples:

```
sampleMeansA <- data.table(sampleID = 1:1000,
                           sampleMean = 0)

set.seed(1234) #ensures your results match ours; not used in real life
```

```
for(i in 1:1000){#for loop cycles 1 to 1000 times, getting a random sample
each time.

  #each iteration of the loop, a NEW, with-replacement sample of ages
  randomSample <- sample(x = unduplicatedAcesAge$Age,
       size = 30, #there are 30 ages in each sample
       replace = TRUE) #with replacement!

  sampleMeansA[sampleID == i, #row selection to CURRENT sample ID
             sampleMean := mean(randomSample, #assign sample mean
                                 na.rm = TRUE)]

}
```

As always, it is important to check that the distribution is normal. While this distribution does have a small dent in the top Figure 7-17, you can see from the deviates plot that there is good coverage of the normal line. It is close enough we can risk it. As a sidenote, there is no magic rule about 1000 samples. You could just as easily bump that up to 10,000 (and it would more likely ensure a normal distribution). However, it would take more time to compute (you may already notice a short wait for 1,000):

```
testAces <- testDistribution(x = sampleMeansA$sampleMean,
                             distr = "normal")
plot(testAces)
```

Next, we check how closely our sampleMeansA data fit the central limit theorem $N\left(\mu, \frac{\sigma}{\sqrt{n}}\right)$.

This time you do not have population-level data; so you must estimate the parameters using the statistics. Remember standard error is the standard deviation divided by the square root of the sample size:

```
xbar <- mean(unduplicatedAcesAge$Age, na.rm = TRUE)
stdDev <- sd(unduplicatedAcesAge$Age, na.rm = TRUE)
n <- 30

standardError <- stdDev/sqrt(n)
standardError

## [1] 0.4086
```

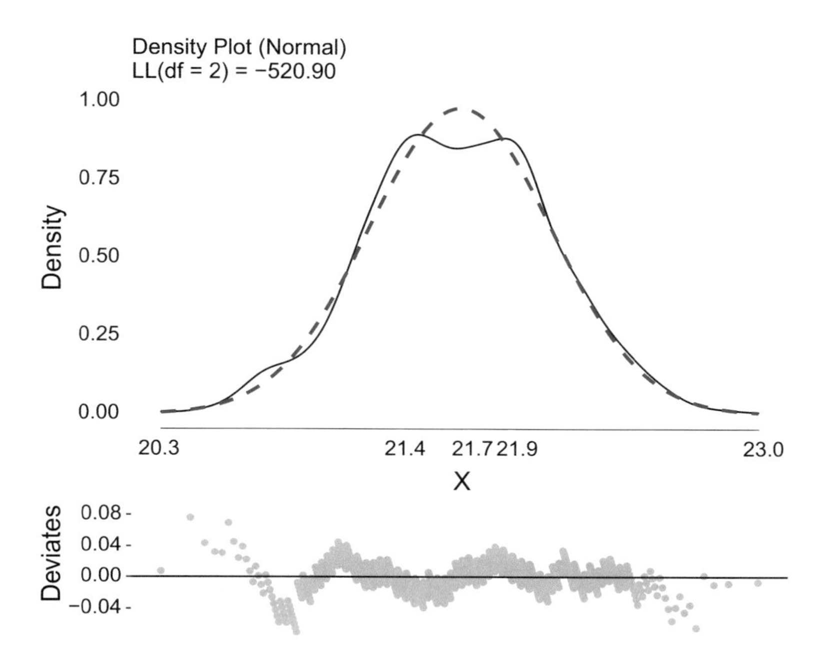

Figure 7-17. Distribution of sample means of ages in aces daily data

If the sampling with replacement code worked, you expect it to match the central limit theorem values just calculated. Take a look at those computations vs. the actual sampleMeansA results. Everything seems to be in order – these are all close matches:

```
xbar

## [1] 21.66

mean(sampleMeansA$sampleMean)

## [1] 21.67

standardError

## [1] 0.4086

sd(sampleMeansA$sampleMean)

## [1] 0.4076
```

The question we are trying to answer is "What is the chance that the average age of our randomly selected focus group is more than 24?"

As before, you use the same `pnorm()` function. Only this time, you set mean and standard deviation to the sampling distribution values. Looking back at Figure 7-17, you do not expect to see a sample average larger than 24. In fact, the x-axis stopped at 23.

Computing the `pnorm()` and remembering to set `lower.tail = FALSE` since we are curious about *more than* 24 yields a strange-looking answer:

```
pnorm(24,
      mean = mean(sampleMeansA$sampleMean),
      sd = sd(sampleMeansA$sampleMean),
      lower.tail = FALSE) #more than 24
```

```
## [1] 5.158e-09
```

`5.158e-09` is scientific notation. Look at the number after the e; this number tells us how many places to move the decimal. Since it is a *negative* 9, we will move the decimal to the *left*. In other words, the probability of getting a sample of 30 participants who together average 24 and up is 0.000000005158. Not likely.

Notice there are in fact 47 participants who are 24 and up in age. Thus, it is definitely *possible* to randomly pick 30 participants who average at or above 24. It is simply not likely. Again, thinking back to the faire coin, what is the chance of you and your friends all flipping TTT? Possible? Sure. Is it more likely the faire coin is not fair? Indeed.

```
unduplicatedAcesAge[Age >= 24, .N]
```

```
## [1] 47
```

With that, you have learned enough for now. It is time to practice.

7.6 Summary

Using sample data statistics to understand population parameters requires understanding if the statistics match (or do not match) what we suspect (or *hypothesis*) about the parameters. In other words, based on what we suspect $N(\mu, \sigma)$ is, does our sample data match that distribution? In turn, distributions themselves are clever applications of probability, where the total area under the distribution curve equals 1 (which translates to 100% chance of all possible outcomes).

Additionally, you used z-scores to understand the probability of selecting either individual data points or sample means lower than or greater than some cut-off. This allows you to see the chances of some extreme value occurring (which can be used in risk assessment/reduction). It can also, as hinted at via the faire coin toss, be used to check if, based on sample data, the population is likely to be what you believed/were told.

As you practice and hone your ability with these techniques, Table 7-1 should be a good reference of the key ideas you learned in this chapter.

Table 7-1. *Chapter Summary*

Idea	What It Means
Inferential statistics	The "analysis" side of statistics; makes a decision.
Outcomes	Measurable, end state of an event (e.g., flipping a coin to H).
Probability	Outcomes sum to 1; $\frac{SpecificOutcome}{TotalPossibleOutcomes}$.
Mutually exclusive	Two events are separate and cannot both happen.
Complement formula	$P(A)+P(\bar{A})=1$.
Independent	Events or outcomes that do not probabilistically influence each other.
With replacement	Each random draw starts with the same environment.
Probability distribution	A function mapping specific outcomes to probabilities; area under curve is 1.
Normal distribution	A specific probability distribution fully defined by `mean()` and `sd()`.
`testDistribution()`	A `JWileymisc` test of data to distribution.
`plot()`	Draw a density and deviates plot used to detect normal data.
`...density..`	A `geom_histogram() aes()` modifier to have y-axis density.
`rnorm()`	Random normal; requires sample size, mean, and standard deviation.
`runif()`	Random uniform; requires sample size, min., and max.
`data.table()`	Given titled column data, creates a from-scratch `data.table`.
`pnorm()`	Probability of a normal distribution; requires value, mean, sd(), and lower or upper tail.
`for()`	For loops allow code to be repeated many times.
`pnorm()`	

7.7 Practice for Mastery

Check your progress and grow through practice by working through some exercises. Comprehension checks ask critical thinking questions that may be best answered with a written or verbal response. Part of the art of statistics is successfully communicating results to your stakeholders or audience. Sometimes that audience is highly technical and other times very much not technical. Exercises are more direct applications of the concepts explored in the chapter.

Comprehension Checks

1. Suppose your friend is playing a betting game based on random chance (e.g., poker, lottery tickets, roulette). Your friend tells you: "I lost the last 200 times. I am going to double my bet value because I must be due for a win." Based on Figure 7-1 and the concept of independent probability, what do you tell your friend?
2. Suppose you lived in a city where a web page suggested the local population had an average (mean) income of $50,000 and that 68% of the population earned between $25,000 and $75,000. In other words, mean is 50K, and standard deviation is 25K. Suppose you sample 40 people at random and find the average salary is $34,000. Does it seem likely that the web page was correct about the population deviation? In other words, does the sample match what you might expect?

Exercises

1. Revisiting the prior question, where the population was claimed to have a mean of 50K and standard deviation of 25K, what is the probability of randomly selecting one person who has an income at or below $34,000? What is the probability of randomly selecting 40 people who average out at or below $34,000?

2. Using `testDistribution()` and `plot()` as shown in the examples, evaluate the columns used so far of `acesData` for normality.

3. Choose one of the examples that used a `for()` loop to pull random samples with replacement to create a sampling distribution of the means as an application of the central limit theorem. Change the number of loop iterations from 1,000 to 10,000. Time the two loops to completion and discuss the difference. Continue through the remainder of the example and compare and contrast the plot of the `testDistribution()`. Which version appears more normal? Are the distribution means different? How about the standard deviations?

4. Using a `for()` loop and random samples with replacement is part of a technique sometimes called a *bootstrap* or *bootstrapping*. Do some research (perhaps via a web search engine) on bootstrapping and statistics.

CHAPTER 8

Correlation and Regression

So far, you have worked with one variable at a time. From exploratory data analysis (EDA) to detecting normality, the focus has been on a single variable such as flipper lengths or ages. Your first steps in predictive analytics – normal probabilities and z-scores – were also for only a single variable. While tallies and counts of a single variable are quite useful, predictive analytics often seeks to understand the relationship between two or more dimensions of data. The eventual goal is to understand the interplay between many variables, or columns, of data. Many real-world questions get better answers with more complex models. Many real-world questions indeed require answers that consider several inputs. This chapter, however, starts the exploration with only two variables. This keeps things easier to visualize and understand. Later chapters build on the techniques and foundation you learn here. You will know you are on the right track when, by the end of this chapter, you are able to

- Understand independent and dependent variables.
- Analyze the relationship between two variables using correlation.
- Create a linear regression between two variables.

There are some big formulae in this chapter; you do not need to memorize these. The purpose of showing them is to both give you access to background information and set up the results of the formula (which is the part you will see and use in the examples). Background (or, in mathematical terminology, *theory*) is a tricky business. Too little and one may misapply techniques. Too much too early and one may lose interest. The formulae are something to glance through on a first pass; later, on a second (or third) read, there is value in digging into these ideas. One last note to set yourself up for success in this chapter: Consider reviewing the types of data (nominal, ordinal, interval, and ratio) from Chapter 5.

M. Wiley and J. F. Wiley, *Beginning R 4*, https://doi.org/10.1007/978-1-4842-6053-1_8

8.1 R Setup

As usual, to continue practicing creating and using projects, we start a new project for this chapter.

If necessary, review the steps in Chapter 1 to create a new project. After starting RStudio, on the upper-left menu ribbon, select *File* and click *New Project*. Choose *New Directory* ➤ *New Project*, with *Directory name* `ThisChapterTitle`, and select *Create Project*. To create an R script file, on the top ribbon, right under the word *File*, click the small icon with a plus sign on top of a blank bit of paper, and select the *R Script* menu option. Click the floppy disk–shaped *save* icon, name this file `PracticingToLearn_XX.R` (where XX is the number of this chapter), and click *Save*.

In the lower-right pane, you should see your project's two files, and right under the *Files* tab, click the button titled *New Folder*. In the *New Folder* pop-up, type `data` and select *OK*. In the lower-right pane, click your new *data* folder. Repeat the folder creation process, making a new folder titled `ch08`.

Remember all packages used in this chapter were already installed on your local computing machine in Chapter 2. There is no need to re-install. However, this is a new project, and we are running this set of code for the first time. Therefore, you need to run the following `library()` calls:

```
library(data.table)
```

```
## data.table 1.13.0 using 6 threads (see ?getDTthreads). Latest news:
r-datatable.com
```

```
library(ggplot2)
library(visreg)
library(palmerpenguins)
library(JWileymisc)
```

For this chapter, we work with just the evening survey data from one day of the ACES daily dataset, so each person contributes just one observation. For no reason in particular, we selected March 03, 2017:

```
acesData <- as.data.table(aces_daily)[SurveyDay == "2017-03-03" &
SurveyInteger == 3]
str(acesData)
```

```
## Classes 'data.table'  and 'data.frame':       189 obs. of 19 variables:
```

```
## $ UserID              : int  1 2 3 4 5 6 7 8 9 10 ...
## $ SurveyDay           : Date, format: "2017-03-03" ...
## $ SurveyInteger       : int  3 3 3 3 3 3 3 3 3 3 ...
## $ SurveyStartTimec11 : num  0.518 0.428 0.529 0.423 0.391 ...
## $ Female              : int  0 1 1 1 0 1 0 0 0 1 ...
## $ Age                 : num  21 23 21 24 25 22 21 21 22 25 ...
## $ BornAUS             : int  0 0 1 0 0 1 0 1 0 1 ...
## $ SES_1               : num  5 7 8 8 5 5 6 7 4 6 ...
## $ EDU                 : int  0 0 0 0 1 0 0 0 0 1 ...
## $ SOLs                : num  NA 33.1 29.63 3.41 100.47 ...
## $ WASONs              : num  NA 0 0 2 2 1 NA 1 1 1 ...
## $ STRESS              : num  2 1 4 4 0 0 0 4 0 1 ...
## $ SUPPORT             : num  5.93 3.94 8.84 4.45 2.1 ...
## $ PosAff              : num  1.55 3.17 3.51 1.7 2.22 ...
## $ NegAff              : num  1.39 1 1.69 1.38 1.08 ...
## $ COPEPrb             : num  2.05 1.93 3.21 2.12 1 ...
## $ COPEPrc             : num  2.31 1.69 3 1.7 1.1 ...
## $ COPEExp             : num  2.24 1.8 2.05 2.93 1 ...
## $ COPEDis             : num  1.94 2.13 2.53 1.21 1.86 ...
## - attr(*, ".internal.selfref")=<externalptr>
```

Of course, we also use our familiar penguin and car data sets:

```
penguinsData <- as.data.table(penguins)
mtcarsData <- as.data.table(mtcars, keep.rownames = TRUE)
```

You are now free to learn more about statistics!

8.2 Correlations

Often we are interested in how two variables are related to each other. One way to achieve this is via a **scatter plot** showing one variable on the horizontal x-axis and another variable on the vertical y-axis. Figure 8-1 shows a scatter plot with the horsepower (hp) and miles per gallon (mpg) of different cars from the mtcars dataset. Visually, we can see cars with higher miles per gallon also seem to generally have a lower horsepower.

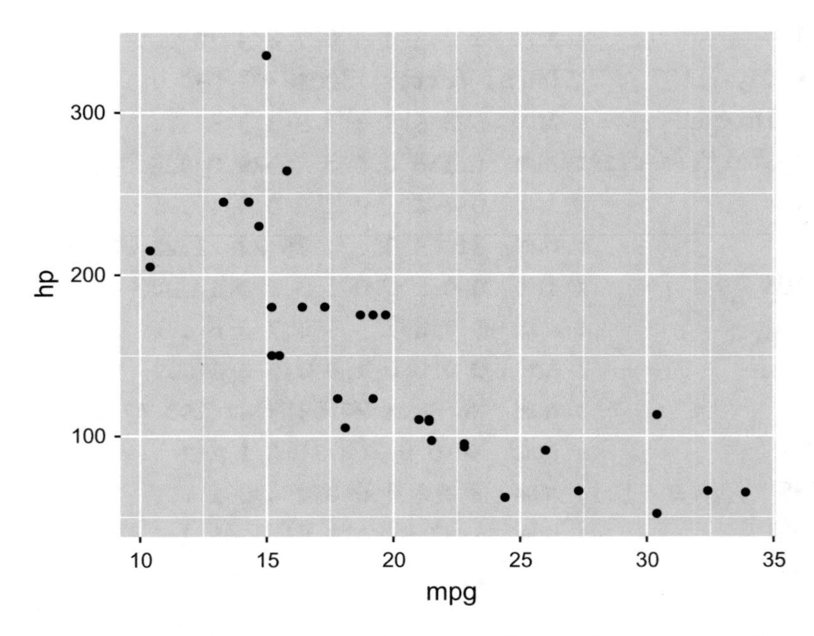

***Figure 8-1.** Scatter plot of miles per gallon (mpg) and horsepower (hp) in the mtcars dataset*

You may recall, from Chapter 1, a similar plot using mpg and wt. This time, we use the `ggplot()` function you now know. As you might suspect, we need to learn a bit more about graph building! Because you are now comparing two *dimensions* of data, the `mapping` needs to include both an x variable and a y variable. As always, that is set in the aesthetic area.

Additionally, you need a new visual geometry – in this case `geom_point()` – to provide a point plot of each (*x*, *y*) coordinate point in the data set:

```
ggplot(data = mtcarsData,
       mapping = aes(x = mpg,
                     y = hp)) +
  geom_point()
```

We can also see a scatter plot for stress and negative affect (mood) from the ACES daily dataset in Figure 8-2. This dataset is larger, and the variables are more skewed: there are many low values (people with low stress and low negative affect). We will look at these two examples in this chapter to see how these influence results:

```
ggplot(data = acesData,
       aes(x = STRESS,
```

```
         y = NegAff)) +
  geom_point()

## Warning: Removed 2 rows containing missing values (geom_point).
```

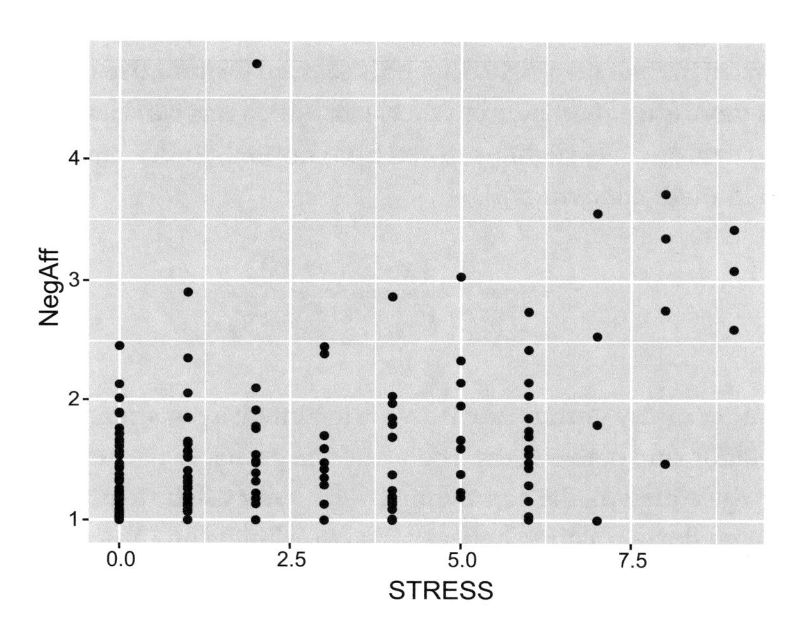

Figure 8-2. *Scatter plot of stress and negative affect (mood) in the ACES daily dataset*

Compare and contrast these two examples that explore the relationship between two variables or two dimensions. We spent some time developing ways to discuss the turbulence of data via variance and standard deviation and the central tendency of data via median and arithmetic mean. In a similar fashion, we need to develop a language to discuss the relationship between two different variables.

Correlation coefficients are a standardized way of quantifying or discussing the *linear* association or relationship between two variables. We look at some different correlation coefficients that are common in this chapter. Depending on the type of data we have, there are some different correlation formulae we might use.

Parametric

Perhaps the most common correlation method is Pearson's product moment correlation coefficient (PPMCC), *r*. This method is appropriate for paired data that are quantitative and either interval or ratio data (recall Chapter 5 discussion). Additionally, one supposes the data are normal in their own right. Indeed, the word **parametric** (the heading of this section) means there are parameters or criteria for which this correlation will (and will not) work. If the data are only slightly skewed from normal, then Pearson's may still be close enough (although not perfect):

$$r_{xy} = \frac{\sum_i^N (x-\bar{x})(y-\bar{Y})}{\sqrt{\sum_i^N (x-\bar{x})^2 \sum_i^N (y-\bar{y})^2}}$$

What this rather messy formula measures is the intensity or strength of the straight-line or linear relationship between the x data and the y data. In particular, due to the mathematics of the equation, the correlation coefficient *r* will be a number between -1 and 1. The stronger the correlation between the two dimensions, the closer to -1 or 1 the result will be. In this case, a strong correlation is one that looks almost like a straight line (also called **linear**). Positive correlation tells us the two values increase/decrease in sync together. Negative correlation tells us the two values move in opposite directions.

Example

While the formula is good to see (and understand as a longer-term goal), it is far too tedious to use by hand on large data sets. `R` implemented this formula as a correlation function, `cor()`, that takes two arguments – x and y values. Keeping in mind Figure 8-1, we expect a negative number. This is because as `mpg` increases, `hp` decreases. Thus, the two values move in opposite directions. Additionally, the dots in the scatter plot were mostly in a messy line. Thus, we also expect a number closer to -1 rather than just to the left of 0 on the number line.

Observing the function output, our intuition was correct:

```
cor(x = mtcarsData$mpg, y = mtcarsData$hp)
```

```
## [1] -0.7762
```

This is a strong, negative correlation.

Example

Figure 8-2 is different from the preceding example. Firstly, as STRESS increased on the x-axis, the NegAff on the y-axis started to show some higher points. Secondly, while that is somewhat true visually, you could certainly point to at least one point where NegAff was higher in the middle of the graph rather than at the far-right side. This is not very linear.

Thus, we would expect an answer that is closer to 0 in the middle. Still, over all, from left to right on the x-axis, the points mostly get taller on the y-axis. Thus, you might expect a positive answer.

You may have noticed when Figure 8-2 was created, there was a warning about dropping two missing values. In general, missing values can either be an entire missing (x, y) pair, or they can be two partial pairs – one missing x and the other missing y. In the case of missing data, there can be different ways to handle such things. The simplest way is to require all pairs to be complete, which is achieved by a third input of `use = "pair"`. We caution you there are risks associated with this process and that learning more from our statistical programming and data models [22] book could be useful.

That said, for many years, the default approach in statistics is to simply delete rows with any missing data. In the following code, we use `cor()` in the `data.table` method. It is a column operator and as such belongs in the `j`-th, column operation position:

```
acesData[,
         cor(x = STRESS,
             y = NegAff,
             use = "pair")]

## [1] 0.5029
```

As predicted, the relationship is not as strong as the prior example. Also as predicted, the relationship is positive.

Example

Let us consider one last example. Suppose we built a `data.table()` with *x* values ranging from 1 to 100 and *y* values being the cubes of the x data:

```
cubicData <- data.table(x = -100:100)
cubicData[, y :=  x^3]
```

Looking at the scatter plot in Figure 8-3, we see this is definitely not a straight-line relationship between x and y:

```
ggplot(data = cubicData,
       mapping = aes(x = x,
                     y = y)) +
  geom_point()
```

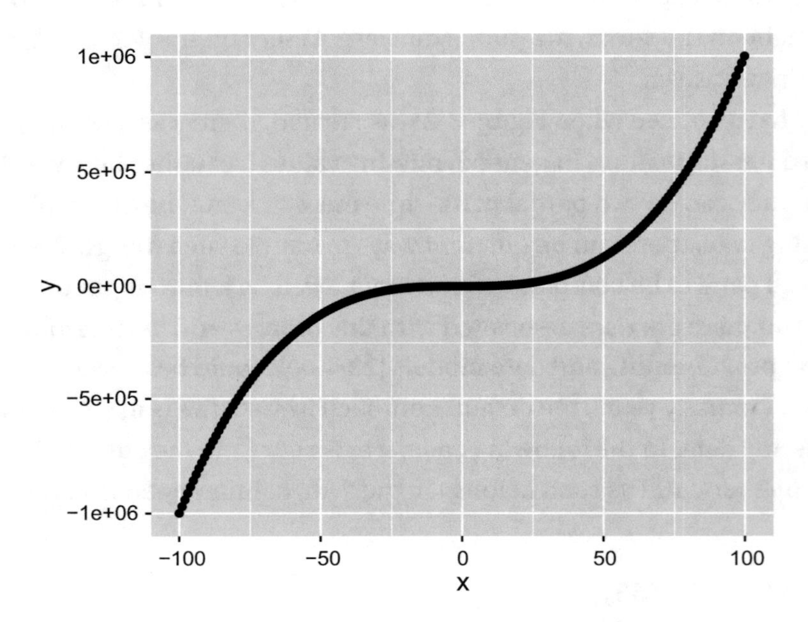

***Figure 8-3.** Scatter plot of stress and negative affect (mood) in the ACES daily dataset*

However, the correlation appears to be a strong, positive correlation – because it is:

```
cubicData[,
          cor(x = x,
              y = y)]
```

```
## [1] 0.9165
```

Thus, for us to interpret Pearson's correlation well, we need our assumptions to be met first. In addition to the data needing to be interval or ratio data, the scatter plot should appear mostly linear. Additionally, each variable (e.g., x or y) ought to be normal. In the case of our cubic data, these data are not jointly linear and not individually normal. Thus, while we can tell from the scatter plot (and indeed may know from common sense) that x and y have some sort of jointly upward trending relationship, Pearson's correlation is not the ideal way to detect this trend.

Non-parametric: Spearman

Non-parametric tests lack assumptions about the data.

In the prior section, there were several assumptions for Pearson's correlation, which measured *linear* relationships. In other words, a change (increase/decrease) in x value is imagined to predict a corresponding change in y value. In fact, Pearson's correlation imagines there is a constant ratio in the x-y relationship. In other words, no matter where we are on the x-axis, moving to the right by one unit ought to (roughly) result in the same change (up or down) on the y-axis.

Sometimes, while there is still a relationship between x and y, the connection is not so consistent. This can occur if x and y are ordinal or ranked data. In other words, depending on where we are on the x-axis, moving to the right by one unit ought to (roughly) result in different amounts of change (still in the same direction though) on the y-axis.

Spearman's rank-order correlation coefficient, ρ (this is the Greek letter pronounced "rho"), can be calculated as Pearson's correlation on the rank-ordered data, denoted by Rx and Ry for the rank of x and the rank of y. The only assumption on data for this test is that the data are ordinal (which is not much of an assumption):

$$\rho_{xy} = \frac{\sum_i^N \left(Rx - \bar{R}x\right)\left(Ry - \bar{R}Y\right)}{\sqrt{\sum_i^N \left(Rx - \bar{R}x\right)^2 \sum_i^N \left(Ry - \bar{R}y\right)^2}}$$

All the same, Spearman will still give us the usual correlation output that lives between -1 and 1.

Example

In R, we get Spearman's correlation directly using another `cor()` argument, `method = "spearman"`. In fact, technically, in the earlier section, we were using `method = "pearson"`. However, as that is the default, we did not have to write it out. In practice, it can be helpful to hard-code the method – especially when you need other people to read and understand your code and analysis choices.

In the case of Figure 8-1, notice that from 15 to 20 on the mpg x-axis, there seems to be a fairly large drop from 300 to just over 100 on the hp y-axis. Contrastingly, from 25 to 30 on the mpg x-axis, the hp y-axis is fairly flat. Thus, if we move to Spearman's correlation – which is focused on rank rather than a constant change ratio – we might expect a stronger signal:

```
cor(x = mtcarsData$mpg,
    y = mtcarsData$hp,
    method = "spearman")

## [1] -0.8947
```

We in fact get a stronger signal. Notice this is still a negative correlation – as mpg increases, hp decreases. It is simply closer to -1 in this case because while these data are quite linear, they do change in their relationship amount.

Example

In the same fashion, even if data are missing, we can use Spearman's method inside a data table. While by itself this only seems to save us some $ typing, it can be useful to use this in connection with other summary methods. Introducing it here makes lighter work later. Again, due to the missing data, the `use = "pair"` argument is included. If we did not include that argument, the output would be an `NA`:

```
acesData[, cor(x = STRESS,
               y = NegAff,
               use = "pair",
               method = "spearman")]

## [1] 0.4328
```

Here, the Pearson's and the Spearman's correlations were rather close together.

Non-parametric: Kendall

Like Spearman's, Kendall's is non-parametric – the only requirement is the data be quantitative, ordinal. To understand how Kendall's method works, let us consider a "toy" example as seen in Table 8-1. A **concordant** pair occurs when the two values in the pair rank an item the same. In particular, as long as the ranking moves in the same direction, a pair will be concordant. Imagine two reviewers are rating stuffed animal toys. If both reviewers rate a toy the same, that would be concordant. In fact, as long as both reviewers are moving in the same direction, that stays concordant (look at the following stuffed bear - both reviewers rate the bear as "lower" than a rabbit). However, it is when the reviewers move in different directions (look at the stuffed dog) that one gets a **discordant** pair.

Table 8-1. *Toy Example*

Toy	Reviewer A	Reviewer B	Pair Type
Stuffed Rabbit	1st	1st	Concordant
Stuffed Bear	2nd	3rd	Concordant
Stuffed Dog	3rd	2nd	Discordant

Putting this into mathematics to get Kendall's rank correlation coefficient, τ (the Greek letter "tau"), we use the number of concordant pairs of ranks, n_c; the number of discordant pairs, n_d; and the number of ways to choose two items from the total number of pairs, n_0, where n is the number of pairs:

$$n_0 = \frac{n(n-1)}{2}$$

This leads to τ being defined as

$$\tau_{xy} = \frac{n_c - n_d}{n_0}$$

If there are more concordant pairs, for example, (x_2, y_2) have the same rank, then n_c will be higher and n_d will be lower and τ will approach +1, being exactly +1 if all pairs have concordant ranks. Conversely, if all pairs are perfectly discordant, then τ will be -1.

If there are ties in the ranks, then R uses a slightly more complex formula, τ_B, that adjusts for the number of ties.

Going back to our toy example, we see there are $n_c = 2$ concordant pairs and $n_d = 1$ discordant pair. Similarly, with three toys total, we see that $n_0 = \frac{3(3-1)}{2} = \frac{6}{2} = 3$. Thus, $\tau_{xy} = \frac{2-1}{3} = \frac{1}{3}$.

To confirm this in R, we can use the data.frame() function to build a data set named toy:

```
toy <- data.frame(toy = c("rabbit", "bear", "dog"),
                  ReviewerA = c(1, 2, 3),
                  ReviewerB = c(1, 3, 2))
Toy
```

```
##      toy ReviewerA ReviewerB
## 1 rabbit         1         1
## 2   bear         2         3
## 3    dog         3         2
```

Of course, in R, there is no need to plug into formulae by hand. The cor() function gives Kendall's τ directly using method = "kendall". Notice $\frac{1}{3} = 0.\overline{3333}$:

```
cor(x = toy$ReviewerA,
    y = toy$ReviewerB,
    method = "kendall")
```

```
## [1] 0.3333
```

The computationally tedious part of Kendall's τ is sorting through the pairs and deciding which ones are discordant. In particular, this is tedious for examples like the stuffed bear – where the two pairs are both ranked lower than other things, yet are not both the same rank. Historically, for larger data sets, Spearman's ρ was easier to calculate. These days, as it makes no difference to R, Kendall's τ is perhaps preferred.

Example

Taking one last look at the mtcarsData via Kendall's, we again see a strong, negative correlation:

```
cor(x = mtcarsData$mpg,
    y = mtcarsData$hp,
    method = "kendall")
```

```
## [1] -0.7428
```

Example

Taking one last look at the acesData via Kendall's, we again see a weaker, positive correlation:

```
acesData[, cor(x = STRESS,
              y = NegAff,
              use = "pair",
              method = "kendall")]

## [1] 0.3342
```

Correlation Choices

Before we conclude this section, it helps to understand when to use the various correlations.

Pearson's product moment correlation coefficient (sometimes abbreviated PPMCC and often simply called correlation) is one of the older correlation methods. It is used to detect a linear relationship between two dimensions of data (e.g., two different columns of data). Because linear relationships are continuous, the underlying data also ought to be continuous. Thus, the assumptions on the data are that the relationship ought to be linear and the underlying data are interval or ratio data.

Spearman's rank correlation coefficient (often called Spearman's ρ) is a more recent invention. As is often the case in data science and statistical methods, data are messy. Thus, the methods used to understand data must adapt older structures for current applications. As a non-parametric correlation, it is detecting the relationship between two variables without preconditions on the data. There is the assumption of quantitative (one does compute it after all). Thus, the data must be at least ordinal. Computationally, Spearman's is faster to calculate than Kendall's. In practice, this is less of a concern today than a decade ago due to modern hardware.

Kendall's rank correlation coefficient (often called Kendall's τ) is also more recent. It is also non-parametric. Of course, the data also must work in a formula; thus, quantitative and ordinal are minimal requirements. While the precise interpretations of Spearman's and Kendall's are different, they both detect the presence of a positive or negative relationship between two variables. For this text, they are essentially the same (both have their uses in specific, real-world applications).

8.3 Simple Linear Regression

Introduction

We mentioned in "Correlations" that Pearson's correlation supposes a linear or straight-line relationship between the data points of the scatter plot. If the data are mostly correlated in a linear fashion, as perhaps we saw in Figure 8-1, it may make sense to build a straight line through those points. While we might draw several lines that fit inside the points, the goal is to find the *best* line. The best line will be as close as possible to every point. That is tougher to do than it might seem (although R will make it easy enough). Such a line is called a **regression** line.

Now, something we did not say in the "Correlations" section needs to be said now. In mathematics (and programming), functions take *inputs* and generate an *output*. Traditionally, the input is an x value and the output is a y value. In the world of statistics and regression lines, we are often in the business of performing a predictive analysis. Thus, the x-axis input is called a **predictor**. It is also sometimes called the **explanatory** variable or the **independent** variable. The y-axis output is called the **outcome** or **dependent** variable.

Part of the challenge of a regression line is that line rarely hits all the points on the plot (unless the correlation was a perfect -1 or 1). Again, looking at Figure 8-1, there is no single line that can touch every point. Thus, even the *best* line will "miss the mark" frequently. Now, keep in mind the usual equation of a straight line is $y = b_0 + b_1 * x$. However, many points on our scatter plot will be either a little above or a little below the line. Thus, a simple linear regression is based off the straight-line equation yet includes a small error value to account for that variance. The word error starts with an "e," so it is not too shocking to see that when using a Greek letter for error, E (pronounced ep-sil-on) is used. For the *i*th pair, the equation is

$$y_i = b_0 + b_1 * x_i + \varepsilon_i$$

where

- y is the outcome or dependent variable.
- x is the predictor/explanatory/independent variable.
- ε is the residual/error term.

- b_0 is the intercept, the expected (model-predicted) value of y when x = 0, written $E(Y|x = 0)$.
- b_1 is the slope of the line and captures how much y is expected to change for a one-unit change in x.
- The point is (x_i, y_i).

The model **parameters**, b_0 and b_1, the regression coefficients, are the same for all participants. The subscript i indicates that each person (or observation) has its own value of y and its own value of x, and because the model is not likely perfect, there will be some unexplained residual, ε_i (Greek letter epsilon). In particular, we call this ε_i because for each data point (x_i, y_i), the epsilon/error will vary. In other words, the model will be closer to some points than it will to others. So some points will have very little error, while others will have more.

If we want to talk about only what is predicted based on the regression coefficients (the model parameters) rather than the "true value" of y_i, we call the *predicted* value y_i . In an equation, we write it

$$y_i = b_0 + b_1 * x_i$$

This formula leaves off the residual error term, ε_i, because we are talking about the model, not the actual values of y_i which are the model plus the error.

This residual error is more formally written as

$$\eta = b_0 + b_1 * x_i$$

where the Greek letter η (eta) is used to indicate the linear predicted values from the model for all observations. The bold font on η is to indicate that it is a vector of all observations, not just for the ith person. We will discuss this more in the "Assumptions" section.

Visually, the intercept, b_0, and slope, b_1, coefficients/parameters look like those in Figure 8-4.

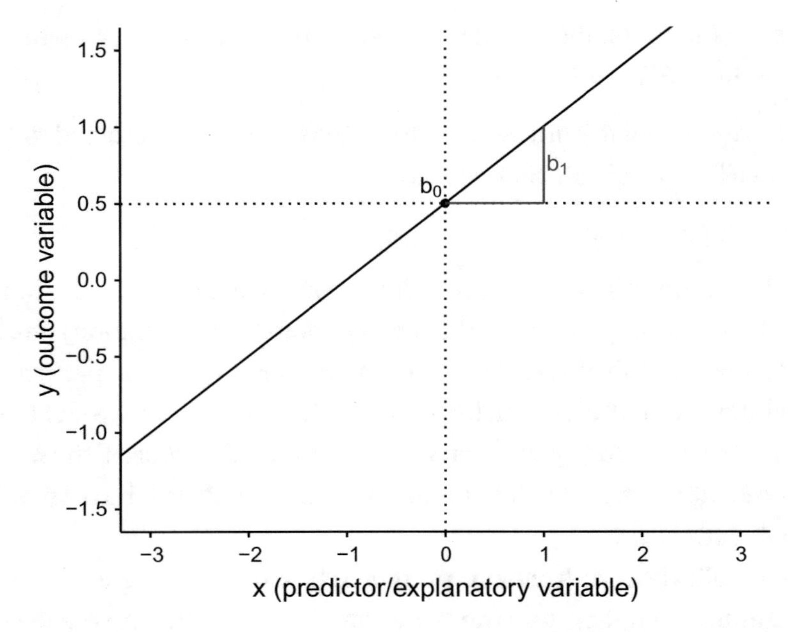

***Figure 8-4.** Straight line graphed from a simple, linear regression model*

Variations in the sign of the intercept indicate the height, sometimes called the **level**, of the line is shifted. Positive values of the intercept, that is, $b_0 > 0$, will shift the line up above the x-axis at $y = 0$. On the other hand, negative values of the intercept, that is, $b_0 < 0$, will shift the line down below the x-axis at $y = 0$, as shown in Figure 8-5. Just because the intercept changes does not mean the slope changes.

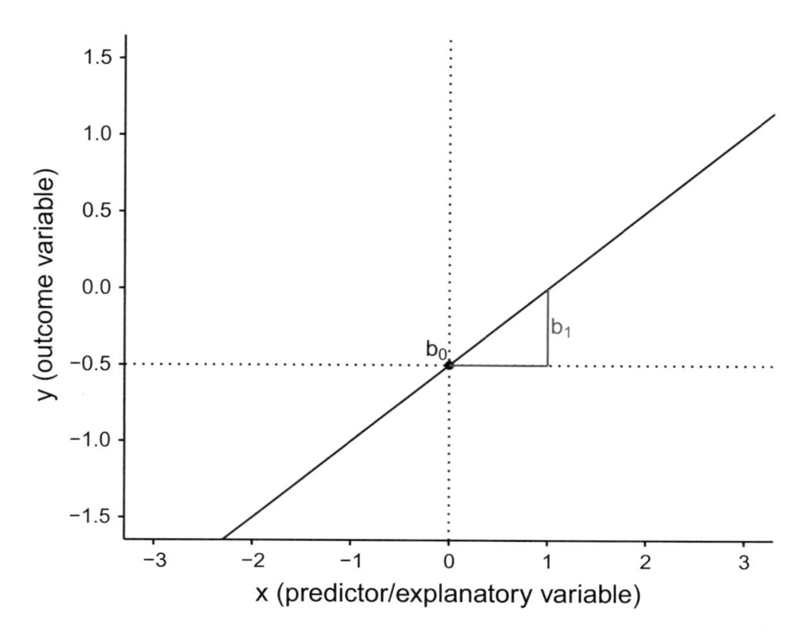

Figure 8-5. *Straight line graphed from a simple, linear regression model with another intercept value*

Changes to the slope indicate the steepness of the line is shifted. Larger positive values indicate a more positive (steeper) slope, such that as x increases, y also is expected to increase. Negative values further from 0 indicate a more negative (also maybe steeper declining) slope, such that as x increases, y is expected to decrease. The line in Figure 8-6 is somewhat close to the line we might eventually expect for Figure 8-1.

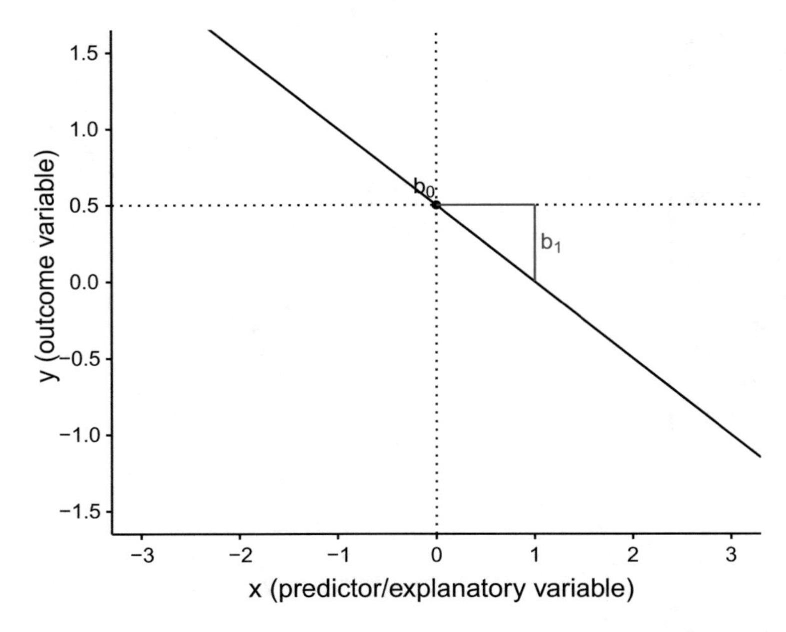

Figure 8-6. *Example image of a simple, linear regression model with a different slope*

In simple linear regression, we are not creating an arbitrary line; we are creating the line of best fit. Here, "best fit" means the line that minimizes the sum of squared residuals. Another way to say this is that the best fit line will be the one line that is the *closest* to *all* points. If you moved the line of *best* fit closer to some point, it would end up being further away from some other points (and would no longer be the best-fitting line). Again, as you work through this theory, remember `R` will do the heavy lifting. Our goal is you understand enough that the output from `R` makes sense and that you can explain your results to some suitable audience (e.g., your clients or boss or colleagues).

To better understand residuals, consider this visual look at the residuals in a simple linear regression. Figure 8-7 shows the scatter plot of the two `mtcars` variables, the horsepower, `hp`, and their miles per gallon, `mpg`. The solid black line is the regression line of best fit, and the blue lines between each observed value and the line are the residuals. The residuals capture the difference between what the model predicted and what was actually observed.

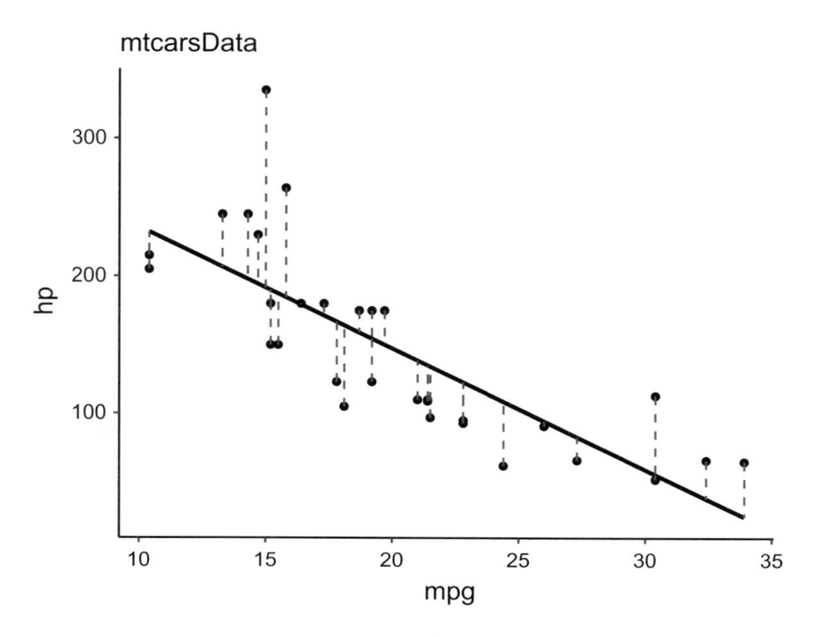

Figure 8-7. *Example showing data points (black points), linear regression line (black line), and residuals (blue, dashed lines)*

That leads to another way we can think about and write the equation for residuals – the difference between the observed and predicted outcome values:

$$\varepsilon_i = y_i - \hat{y}_i$$

When we fit a linear regression, equations are solved to give *estimates* of b_0 and b_1 that minimize the sum of squared residuals, and those parameter estimates (or regression coefficients) produce the line of best fit. Before we discuss how R can help you create the same results we got (and give the actual estimates for b_0 and b_1), we must discuss the assumptions.

Assumptions

We already discussed the basics about linear regression. As with many statistical models, linear regression requires some *assumptions* to be true in order for the results to be valid. The following information is deeper background on the theory. If you are designing your own studies, this will be important to work through and understand. That said, it can be worked through multiple times – each read through will give you more insight. However, if you are wanting to jump directly into coding linear models, focus on the assumption summary.

We have shown the standard form for a linear regression equation showing the intercept, b_0, and the slope, b_1, which together with some data on the predictor variable, x_i, are used to predict the outcome for a specific observation, y_i. The part the predictions cannot explain are the residuals, ε_i. You saw an example of those residuals in Figure 8-7 as the blue, vertical lines. They are the amount of error between the prediction of the line and the actual data points:

$$y_i = b_0 + b_1 * x_i + \varepsilon_i$$

Although the regression coefficients, b_0 and b_1 are the same for all observations, the values of the predictor variable, x_i, can be different for different observations. Thus, you can get a different predicted value for each observation, depending on the score of x_i.

We also saw earlier the predictions can be defined as η (pronounced "eta"):

$$\eta = b_0 + b_1 * x_i$$

Using this piece, we could simplify the previous equation as follows:

$$y = \eta + \varepsilon$$

Here we use bold fonts to indicate they hold true for every (x_i, y_i) pair; these pairs are sometimes called **vectors**. Rewriting our regression equation highlights the basic building blocks of linear regression. In words, we could say "our outcome, *y*, is equal to the sum of our model predictions, $\boldsymbol{\eta}$, and an unexplained (residual) part, $\boldsymbol{\varepsilon}$."

With all of these pieces, now we can talk about the four assumptions of a linear regression model.

Linearity

The **linearity** assumption in linear regression is that the association between the predictor, $\boldsymbol{x}$, and the outcome, $\boldsymbol{y}$, is a linear association. This is required because the regression coefficients we estimate, b_0 and b_1 define a straight line. If the true association between $\boldsymbol{x}$ and $\boldsymbol{y}$ was non-linear, we will not capture it well with a line.

As with many things in statistics, in the real world, most things are not perfectly linear or perfectly non-linear. In practice, you are often looking for whether the association is *about* linear. The question becomes whether using a line to model the association is sufficiently accurate that the line is *useful enough*.

One way to assess linearity is to use a scatter plot. Here are two examples in Figure 8-8. Although neither example is perfectly linear, Figure 8-8 Panel B is close enough that a linear regression line is useful or helpful for understanding the main association between the predictor and outcome:

```
## Registered S3 methods overwritten by 'car':
##   method                          from
##   influence.merMod                lme4
##   cooks.distance.influence.merMod lme4
##   dfbeta.influence.merMod         lme4
##   dfbetas.influence.merMod        lme4
```

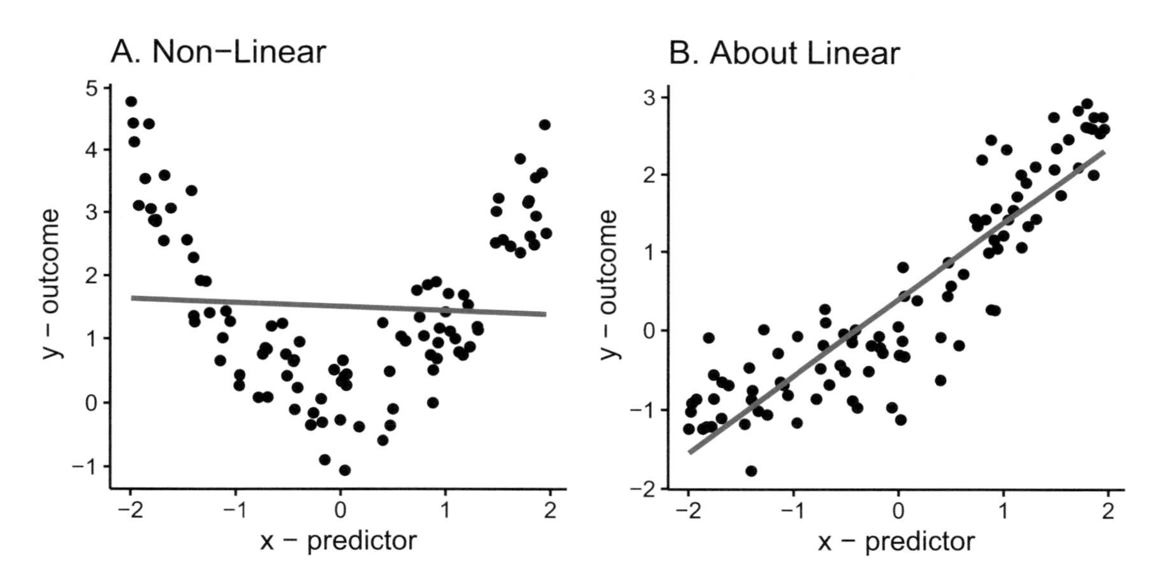

Figure 8-8. *Example showing a very non-linear association (Panel A) and an approximately linear association (Panel B) between a predictor, x, and outcome, y. Linear regression lines are shown in the blue lines*

Normality

Linear regression also assumes a **normal** distribution. But in linear regression, what does it really mean when we say normality is assumed? What we assume is that the outcome is conditionally normal, conditional on the model. In practice, this tends to be written one of two ways, which are equivalent:

$$y \sim N(\eta, \sigma_\varepsilon)$$

This is read, "Y is distributed as a normal distribution with mean equal to η and standard deviation σ_ε." The idea here is we do not require y itself to follow the same normal distribution. We assume that for any given predicted value, η, the observed values of y are distributed normally *around* that predicted value.

In practice, that is more difficult to test or check. If you look at it, we do assume y follows a normal distribution, but different normal distributions with the means being different for each value of η. So that we can check the assumption with just **one** normal distribution, we can make a small variation. If you create a new variable by subtracting the mean from a variable, the new variable will always have mean 0. We can use this to make our assumption checking easier. Remember our simplified high-level regression equation:

$$y = \eta + \varepsilon$$

If we rearrange by subtracting η from both sides, we get

$$y - \eta = \varepsilon$$

The residuals, ε, are the difference between the observed outcome and the predictor (those vertical, blue, dashed lines from Figure 8-7). Since the linear predictor is the assumed mean of the normal distribution for each value of y, if we subtract it, the result should be mean 0.

This allows us to write the normality assumption another way:

$$\varepsilon \sim N(0, \sigma_\varepsilon)$$

The only difference is whether we use the residuals and assume a mean of 0 (because residuals are always constructed to have mean 0) or the raw outcome variable, but assume mean equal to η. In effect, we are assuming that the outcome is distributed normally around the regression line, with standard deviation based on the standard deviation of the residuals.

However, the residuals are much easier to test, because now we are just assuming one specific normal distribution and we can use graphs and other methods to assess whether the residuals indeed roughly follow a normal distribution with mean 0 and standard deviation σ_ε or not.

In other words, the normal assumption for linearity is the residuals - the error between the prediction line and the actual point - should be a normal distribution with mean 0.

Homoscedasticity

Related to the normality assumption is the assumption of **homoscedasticity**. *Scedasticity* refers to the distribution of the error or residual terms; *homo* means the same. Thus, the assumption is that the error terms have the same spread or amount of variability.

The assumption of homoscedasticity means we assume for every value of η, the variance (spread) of the residuals around it is the same. In other words, those vertical, dashed, blue lines in Figure 8-7 ought to be balanced. This assumption is related to the same equation we saw for normality:

$$y \sim N(\eta, \sigma_\varepsilon)$$

There is a single standard deviation, σ_ε, for all observations. In practice, this means that the variability of observations around the predicted value is the same at all levels of predictions. When the variance is not the same across the range of η, then we say there is heteroscedasticity, different spread of the errors (residuals).

One way to assess homoscedasticity is to use a scatter plot. Here are two examples in Figure 8-9. In both cases, there is a linear and normal association. However, in Panel A, the variance increases at higher values (heteroscedasticity), whereas in Panel B, the variance is consistent (homoscedasticity). If we were to add vertical, dashed, blue lines to Figure 8-9, those lines would be balanced in Panel B, yet in Panel A they would not be balanced.

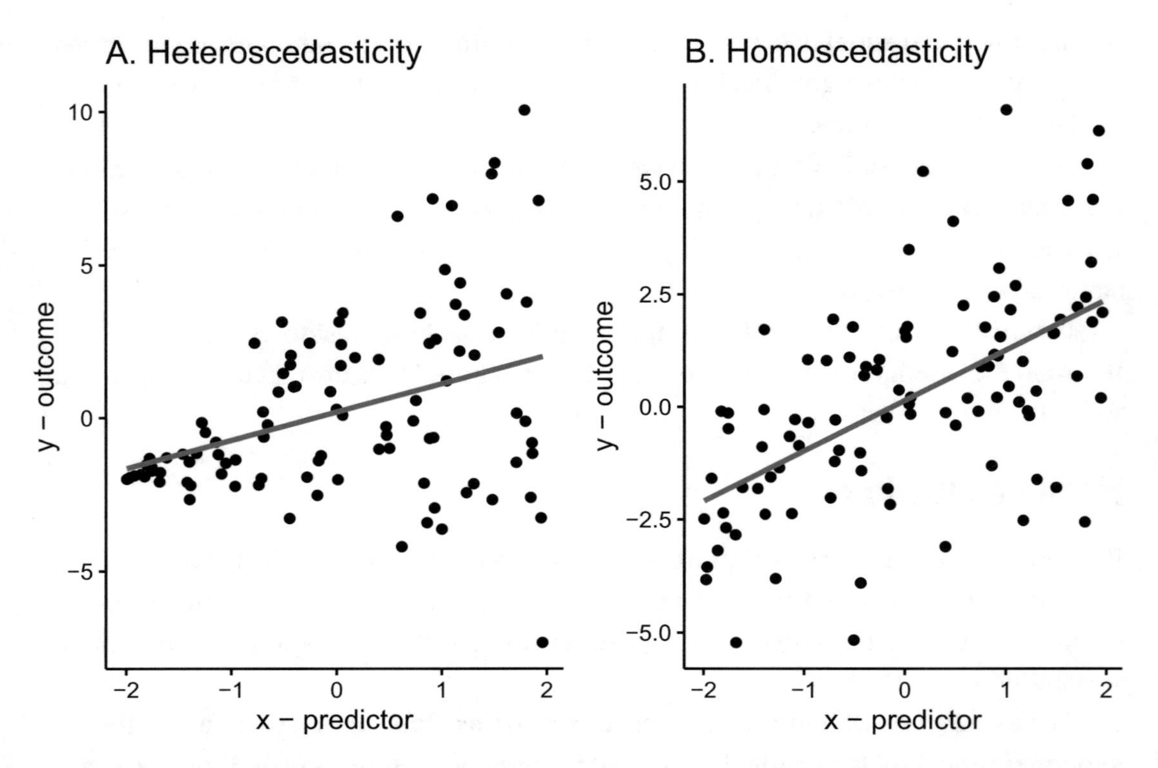

***Figure 8-9.** Example showing inconsistent residual variance, heteroscedasticity (Panel A), and consistent or equal residual variance, homoscedasticity (Panel B)*

One final note on homoscedasticity: When the residual variance is homogenous (e.g., the same), it will always look the same and more or less like Figure 8-9 Panel B. However, when there is heteroscedasticity, it can appear in many ways. The variance may increase, decrease, or fluctuate. Heteroscedasticity is not any single pattern. Rather, it is the absence of consistent, homogenous variance.

Independence

The last assumption for linear regression is that the observations are independent of each other, called **independence**. This assumption means that the values from one observation are not related to or driving values of another observation. In general, this is not something that we "test" statistically. It is something we assume based on the design of the study or the nature of the data collected.

For example, suppose you randomly sampled different people and asked them to answer some questions once. In this case, we would assume the observations are independent. One person's responses should not systematically be related or dependent on another person's responses.

Conversely, suppose we randomly sampled different families and then asked everyone from each family to answer a set of questions. In this case, we could not assume the observations are independent. People from the same family are more likely to answer similarly to each other. As an extreme example of this, suppose that one question was "What is your household income?" If we ask that to two to four people from the same household each time, even if we have a total of say 100 observations, if those observations come from only 35 different households, we will not have 100 independent measures of household income. Another common case of observations that would not be independent is when clusters of observations are sampled (e.g., sampling different classrooms and then all students from a classroom). Another example is in repeated measures or longitudinal studies; if you ask the same person to answer questions every year for several years. Observations within a person are not independent.

Although an important assumption, the **independence** assumption is not normally tested because either from the way the data are collected we believe the observations are independent or if you know they are structured (repeated measures in a person, multiple observations from the same family, etc.), you would need to use a different analysis that accounts for the non-independence.

Linear Regression Assumption Summary

The four assumptions for linear regression can be summarized as follows:

- *Linearity* between the x predictor and the y outcome variables. May be visually tested via scatter plot (see Figure 8-8).
- *Normality* of the residuals about mean 0. May be visually tested via the `testDistribution()` function.
- *Homoscedasticity* of the residuals about the regression line. May be visually tested via scatter plot (see Figure 8-9).
- *Independence* of each observation. May be philosophically tested via unbiased study design.

R^2: Variance Explained

Both regression and Pearson's correlation share a linearity assumption. Thus, regression lines satisfy the requirements to use this correlation, which has an additional use. Recall this correlation is written as r. Squaring this to r^2 yields the **coefficient of determination**. Because r lives between -1 and 1, r^2 lives between 0 and 1. Thus, r^2 can be thought of as a percent. The coefficient of determination tells us (for the regression line) the percent of the change (or variation) the outcome, y, variable has that comes from the input x variable.

One result of this is if r^2 is high, then we may have found a good predictor variable. When r^2 is not so high, we may want to search for a better predictor. Of course, in many cases, any single predictor will not yield a strong r^2. At that stage, we may want more than one predictor. Such a topic is saved for Chapter 11.

Keeping in mind our study of probability back in Chapter 7, we know the complement of r^2 is $1 - r^2$. This complement is the quantity of change in y that is not determined by input x. Again, if this is large, you are likely soon to be in the business of finding some additional predictors and in need of Chapter 11.

Linear Regression in R

You have gone some time learning some new thoughts without practicing them in `R`. It is time to fit this theory into practice. Most of the functions you have already met. The two new ones are `lm()` for *l*inear *m*odel and a `ggplot()` visual that creates a smooth line named `geom_smooth()`.

Along the way, in addition to calculating the intercept b_0 and the coefficient b_1, your data must also pass the four assumptions. Of course, it will also help to see a graph of your linear regression.

Example

Consider the `penguinsData` along with the `flipper_length_mm` you met before as well as a new variable, `body_mass_g`. Keeping in mind we know nothing of penguins, suppose it is easy to measure flipper lengths via some sort of clever method using binoculars and some key landmarks of known size. On the other hand, suppose that it is comparatively tougher to weigh a penguin (one supposes the penguin must be gently captured and then safely released). In this case, we might wish to use `flipper_length_mm` as the independent variable or predictor and have `body_mass_g` as the dependent variable or outcome.

This would require we find the linear regression line and equation. In other words, we must calculate the intercept b_0 and the coefficient b_1 for penguin flippers and weights. To do this, we must first confirm the data fit the four assumptions of linear regression.

The linearity assumption is verified via a visual inspection of the scatter plot in Figure 8-10. Upon inspection, it does seem the data look to move in a roughly linear fashion. This graph is created with a fairly standard `ggplot()`. The new geometry visual - `geom_point()` - is used:

```
ggplot(data = penguinsData,
       mapping = aes(x = flipper_length_mm,
                     y = body_mass_g)) +
  geom_point()

## Warning: Removed 2 rows containing missing values (geom_point).
```

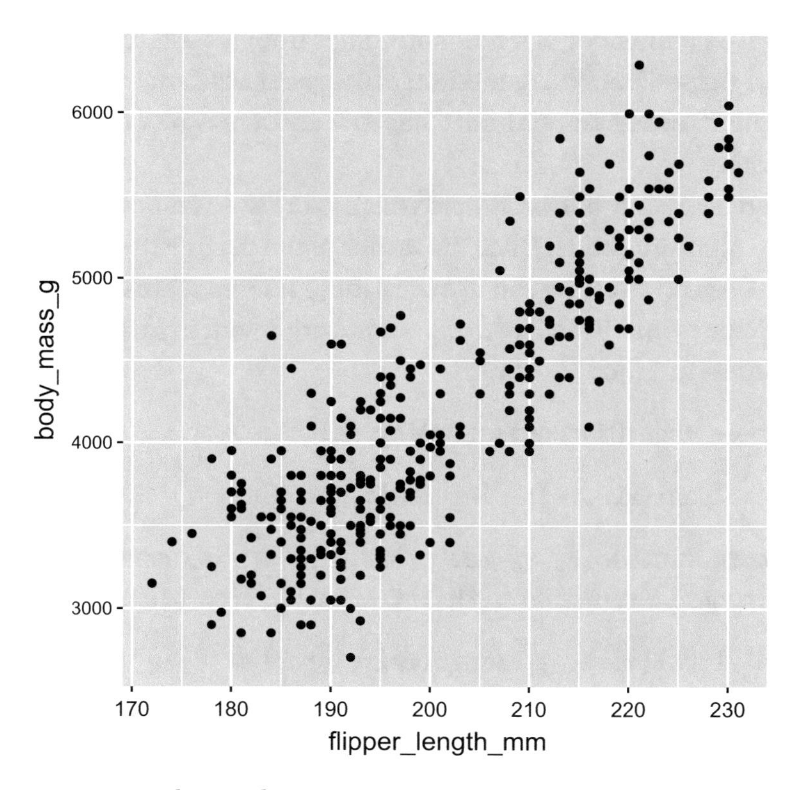

Figure 8-10. *Penguins data. Flipper lengths vs. body mass*

To verify the normality assumption on the residuals requires us to first compute the linear regression line. Without the line, we cannot calculate the residuals (recall Figure 8-7 showed the residuals as blue lines). Thus, you must learn how to compute a linear regression – and indeed do so – before ever knowing if your work will pay off!

Linear regressions are *l*inear *m*odels. Thus, the function to fit a linear model is `lm()`. Fitting a model often generates much data (the residuals are only one part of the model). Thus, rather than simply fit the model, we assign it to a variable. It is common to use the variable `m` for *m*odel. To distinguish from other models, we name this one `mPenguin`.

The function `lm()` takes two arguments. The first is a `formula =` argument. R has a standard formula process that uses ~ to break the dependent y variable from the independent x variable(s). As we are using flipper lengths to predict the body mass outcome, our formula will be `body_mass_g ~ flipper_length_mm`:

```
mPenguin <- lm(formula = body_mass_g ~ flipper_length_mm,
               data = penguinsData)
```

To confirm the normality assumption, we could use `testDistribution()` on the residuals. However, remember that function comes from the `JWileymisc` package. There is another helper function named `modelDiagnostics()` that will generate both normal assumption testing graphs from Chapter 7 and a new plot that will help us check homoscedasticity.

The function takes only one argument which is the linear model we assigned to `mPenguin` in the preceding code. Once the results of the diagnostics are assigned to the new variable `mPenguinTests`, we can use the `plot()` function. This is technically two graphs for the price of one. Thus, we use a second argument to `plot()` that sets the number of columns to 2 (`ncol = 2`):

```
mPenguinTests <- modelDiagnostics(mPenguin)

## Warning in .local(x, ...): singularity problem

## Warning in rq.fit.sfn(x, y, tau = tau, rhs = rhs, control = control,
...): tiny diagonals replaced with Inf when calling blkfct

## Warning in .local(x, ...): singularity problem

## Warning in rq.fit.sfn(x, y, tau = tau, rhs = rhs, control = control,
...): tiny diagonals replaced with Inf when calling blkfct
```

```
plot(mPenguinTests,
     ncol = 2)

## 'geom_smooth()' using formula 'y ~ x'
```

The resulting Figure 8-11 has two graphs side by side then in those two columns. The left side is the familiar plot from the testDistribution() function (only this is on the residuals rather than the raw data). As you see, it seems to fit the normal curve fairly well in the density plot. The deviates plot does admittedly leave something to be desired. However, notice the size of the deviate changes ranges from -0.1 to 0.4. This is not a very tall range; all these points are close to that normal line. Additionally, while the edges are somewhat above the normal line, the middle goes below. Thus, the dots do hit that line several times. Is it perfectly normal? No. Is it close enough to use the model? Well, it is close enough to say the model passes the normalcy assumption on residuals.

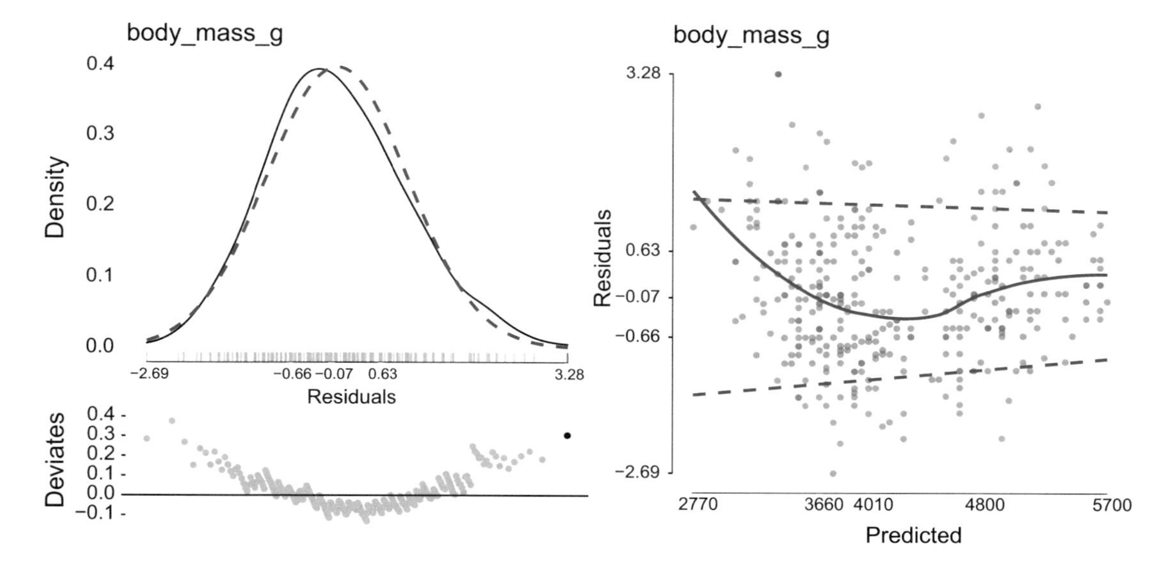

Figure 8-11. *Penguins data test for normality of residuals and homoscedasticity*

The right-side graph is new to us and helps us check the homoscedasticity assumption. The dotted lines on the right side are the 10th decile and the 90th decile of the predicted values. Remember homoscedasticity simply says that we expect the residuals to be fairly uniform about the regression line as in the B graph of Figure 8-9. Thus, we would expect the prediction vs. the residuals on the right side of Figure 8-11 to look more uniform. In other words, we expect those dashed lines to be roughly parallel to each other and trace out a rectangular shape. While they are not perfectly parallel, they are quite close. Thus, the model passes the homoscedasticity assumption on residuals.

The last assumption, independence, is a philosophical check. The assumption is that the penguins are independent of each other. Now, if they all came from the same family or the same species or the same island, we might worry. While the original research paper would need to be read to verify if a random sample or convenience sample was used for `palmerpenguins`[11], still, the fact that the penguins come from different locations is enough to suppose some level of independence:

```
penguinsData[order(island, species),
             .N,
             by = .(island, species)]
```

```
##       island   species   N
## 1:    Biscoe    Adelie  44
## 2:    Biscoe    Gentoo 124
## 3:     Dream    Adelie  56
## 4:     Dream Chinstrap  68
## 5: Torgersen    Adelie  52
```

With the four assumptions confirmed, we are now able to use this linear model. Using the familiar `summary()` function on our model variable `mPenguin`, quite a lot of data is printed to the screen. For now, our goal is to find the coefficients for our linear formula. In the section labeled `Coefficients`, the `Estimate` column provides the data needed. The `intercept` is $b_0 = -5780.83$, while our x variable, `flipper_length_mm`, is $b_1 = 49.69$:

```
summary(mPenguin)
```

```
##
## Call:
## lm(formula = body_mass_g ~ flipper_length_mm, data = penguinsData)
##
## Residuals:
##      Min       1Q   Median       3Q      Max
```

```
## -1058.8  -259.3   -26.9   247.3  1288.7
##
##  Coefficients:
##                   Estimate Std. Error t value Pr(>|t|)
## (Intercept)       -5780.83     305.81   -18.9   <2e-16 ***
## flipper_length_mm    49.69       1.52    32.7   <2e-16 ***
## ---
## Signif. codes:  0 '***' 0.001 '**' 0.01 '*' 0.05 '.' 0.1 ' ' 1
##
## Residual standard error: 394 on 340 degrees of freedom
##   (2 observations deleted due to missingness)
##  Multiple R-squared: 0.759, Adjusted R-squared: 0.758
##  F-statistic: 1.07e+03 on 1 and 340 DF, p-value: <2e-16
```

This yields the prediction formula

$$y_i = -5780.83 + 49.69 * x_i + \varepsilon_i$$

Of course, we do not know the error, ε_i. That said, for flipper lengths within the range of data the linear model was trained on, we now have a formula that could give us our prediction for any given flipper length. Suppose we measure a penguin with flipper length 201 mm. We might expect a weight of 4207 g based on our model:

```
range(penguinsData$flipper_length_mm,
      na.rm = TRUE)
```

```
## [1] 172 231
```

```
xiPenguin <- 201
yiPenguin <- -5780.83 + (49.69*xiPenguin) yiPenguin
```

```
## [1] 4207
```

Earlier, in Chapter 2, you installed the `visreg` package [7]. We now use a function from that package to *vis*ualize the penguin *reg*ression fitted by `lm()`. Because the regression line needs to be the line of best fit, one sometimes speaks of fitting a line to the data points (or fitting a model to the data). Thus, `visreg()` takes an argument named `fit =` that is given a model. In this case, we use the penguin model `mPenguin`. For now, do not worry about the second argument shown, band = FALSE. We explain that band in Chapter 9:

```
visreg(fit = mPenguin,
       band = FALSE)
```

Check where a flipper length of 201 would be on the graph Figure 8-12. Does the *y* value height appear to be approximately 4207? Does that seem fairly close to the scatter plot dots of actual penguins with that length of flipper?

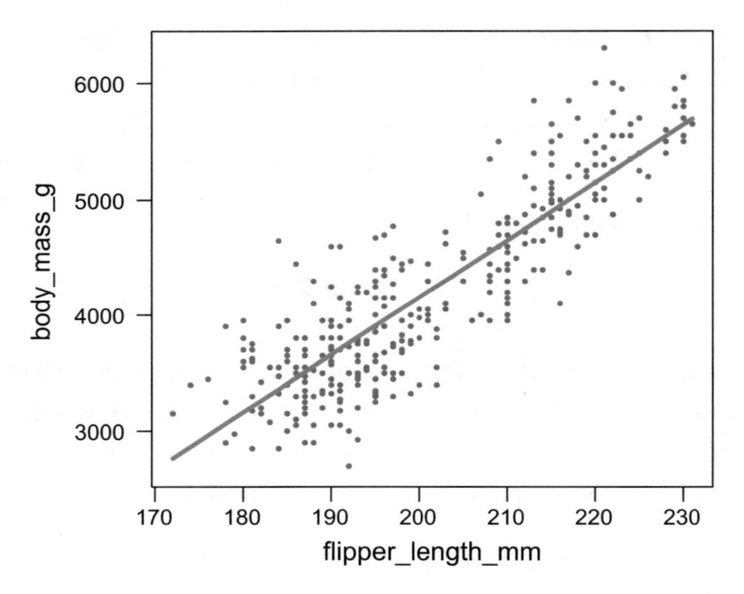

Figure 8-12. *Linear regression visual for penguinData*

The last thing we look at is the r^2. Notice that the summary() function gave us two values for the variance explained. One was 0.759 and was titled Multiple R-Squared. The other was an adjusted figure (although in both cases they round to 0.76). This figure – called the coefficient of determination – tells us 0.76 or 76% of the change/variation in outcome of body mass comes from the predictor of flipper length. You can see this in Figure 8-12. For any single flipper length, it does seem there are several different mass dots on the scatter plot. Using the probability idea of complement, 1 – 0.76 = 0.24 or 24% of body mass is not determined by our predictor of flipper length. Something else helps determine penguin mass.

In the end, we have a good model. It closely follows our four assumptions. Most of the change in the outcome variable of body mass seems to be predicted by our independent variable of flipper length. This tells us measuring flippers is a very good proxy for weighing penguins (and, if our original thoughts are correct, easier on researchers and penguins). While there may be improvements to be made to the model in future chapters, for now, we have a good start.

Example

For our last example, consider the `acesData` along with the `STRESS` vs. `NegAff`. Can we shorten our nightly survey for study participants and use `STRESS` as the independent variable or predictor and have `NegAff` as the dependent variable or outcome?

This would require we find the linear regression line and equation. In other words, we must calculate the intercept b_0 and the coefficient b_1 for STRESS and negative affect. To do this, we must first confirm the data fit the four assumptions of linear regression.

The linearity assumption is verified via a visual inspection of the scatter plot in Figure 8-13. These data are less clearly linear. We proceed with caution; a linear model may not be a great fit:

```
ggplot(data = acesData,
       mapping = aes(x = STRESS,
                     y = NegAff)) +
  geom_point()
```

```
## Warning: Removed 2 rows containing missing values (geom_point).
```

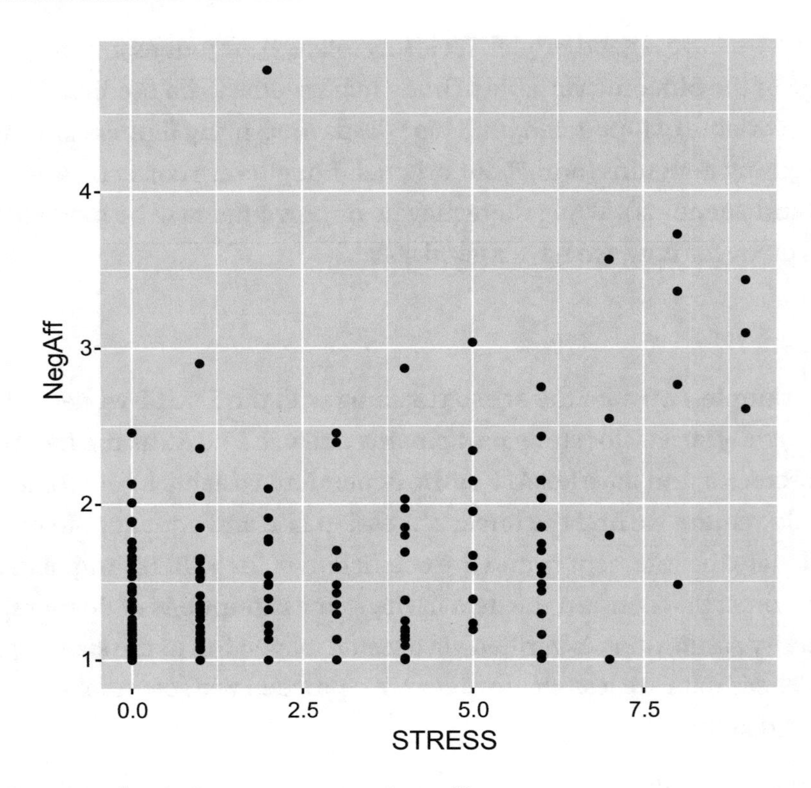

Figure 8-13. *Aces data. Stress vs. negative affect*

To verify the normality assumption on the residuals requires us to compute the linear regression line. Compare and contrast the similarities and changes to `lm()` between this example and the prior example:

```
mAces <- lm(formula = NegAff ~ STRESS,
            data = acesData)
```

To test both the normality of residuals and homoscedasticity assumptions, we use `modelDiagnostics()` again:

```
mAcesTests <- modelDiagnostics(mAces)

## Warning in .local(x, ...): singularity problem

## Warning in rq.fit.sfn(x, y, tau = tau, rhs = rhs, control = control,
...): tiny diagonals replaced with Inf when calling blkfct

## Warning in .local(x, ...): singularity problem
```

```
## Warning in rq.fit.sfn(x, y, tau = tau, rhs = rhs, control = control,
...): tiny diagonals replaced with Inf when calling blkfct

## Warning in rq.fit.br(x, y, tau = tau, ...): Solution may be nonunique

plot(mAcesTests,
     ncol = 2)

## 'geom_smooth()' using formula 'y ~ x'
```

The resulting Figure 8-14 shows that the distribution of residuals is somewhat normal, though not as clear-cut as our prior example. The deviates also are often close to the line, yet there is at least one point that is fairly far away. Real life is not perfect; the density plot is roughly normal shaped. We again proceed with caution. This does not mean we cannot use our linear fit. It does mean we must be more cautious in interpreting our results.

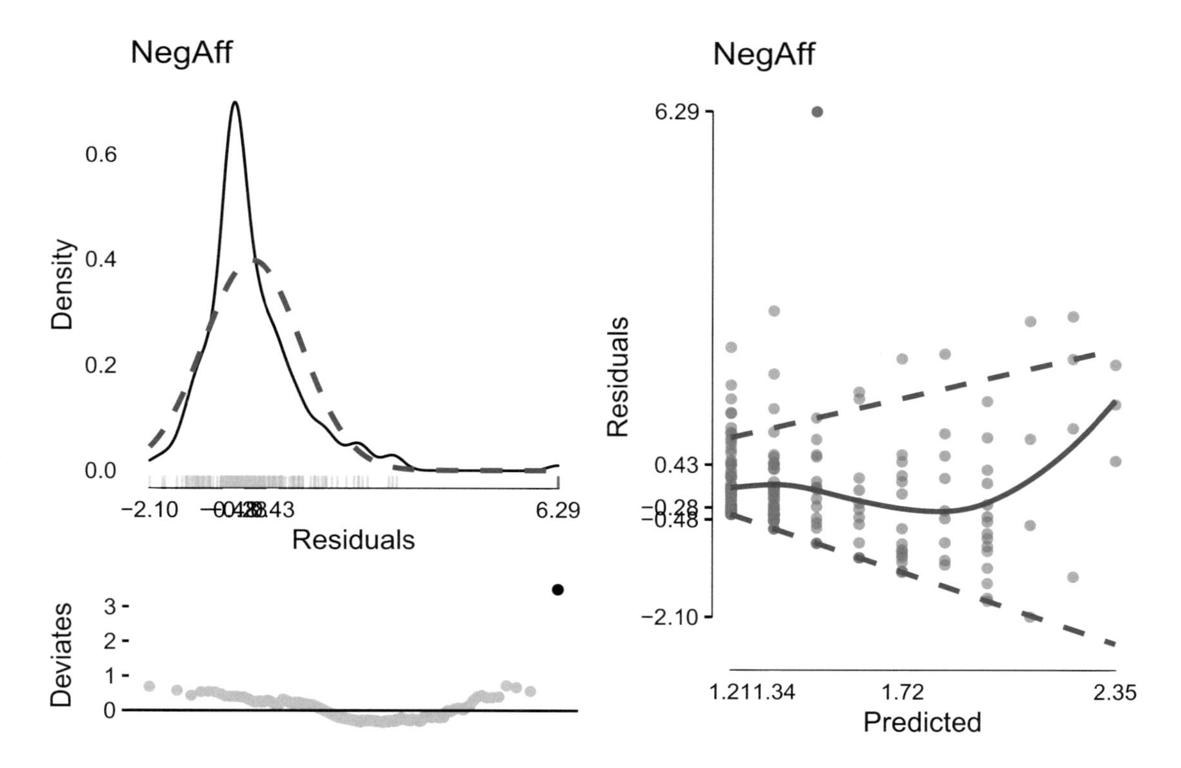

***Figure 8-14.** Aces one-day data test for normalcy of residuals and homoscedasticity*

The dashed lines on the right graph of Figure 8-14 are very clearly not parallel. There's a definite pinching in on the left side of the data that spreads wider by the right side. This is reminiscent of Figure 8-9 Panel A – heteroscedasticity. This assumption is either violated or close to it.

At this point we have a choice. There seems to be a fair bit of evidence that a linear fit in this case is not going to be as good as the linear fit on penguins was. In some cases, it may be acceptable to proceed. If we do move forward, we would want to clearly inform end users (even if that end user is ourselves) that this model is not ideal. If we choose to stop now and say that a linear fit is not indicated due to assumption violation including homoscedasticity, that could make sense.

In the interest of giving you some additional code experience, we continue forward.

The last assumption, independence, is a philosophical check. The assumption is that the participants are independent of each other. In the case of this data, we shall suppose the study was designed well enough to meet this assumption.

Moving onward despite one or more assumptions close to breaking, we inspect the linear model with the familiar summary() function:

```
summary(mAces)

##
## Call:
## lm(formula = NegAff ~ STRESS, data = acesData)
##
## Residuals:
##    Min     1Q Median     3Q    Max
## -1.095 -0.252 -0.146  0.228  3.318
##
## Coefficients:

##             Estimate Std. Error t value Pr(>|t|)
## (Intercept)   1.2097     0.0515   23.50  < 2e-16 ***
## STRESS        0.1265     0.0160    7.91  2.2e-13 ***
## ---
## Signif. codes: 0 '***' 0.001 '**' 0.01 '*' 0.05 '.' 0.1 ' ' 1
##
## Residual standard error: 0.529 on 185 degrees of freedom
##   (2 observations deleted due to missingness)
```

```
## Multiple R-squared:  0.253,  Adjusted R-squared:  0.249
## F-statistic: 62.6 on 1 and 185 DF, p-value: 2.23e-13
```

This yields the prediction formula

$$y_i = 1.2097 + 0.1265 * x_i + \varepsilon_i$$

Rather than computing an example, we inspect the graph in Figure 8-15. It is easy to see why this is either not linear or only barely linear. For a STRESS = 2, there is a fairly narrow range of NegAff ranging from 1 to a bit more than 2. On the other hand, by STRESS = 7, there is a wide range of NegAff ranging from 1 to about 3.5 it seems. At different points on our graph, the possible values for the outcome, NegAff, have different ranges:

```
visreg(fit = mAces,
       band = FALSE)
```

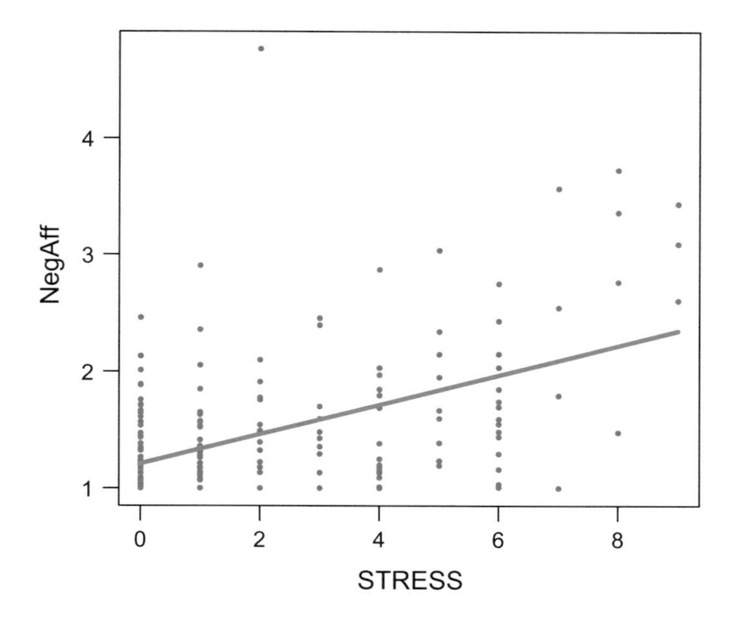

***Figure 8-15.** Linear regression visual for aces data*

Additionally, looking at $r^2 = 0.25$ gives another cause for concern. Even if this was a great linear model (it is not; the homoscedasticity is quite rough at best), only 25% of NegAff outcome is predicted by STRESS - in other words, the exact opposite of the penguin scenario. Here, 75% of the variance in outcome is *not* from our predictor.

This model is fairly weak at best. In any case, it seems wise to keep the both parts of the survey on the nightly questionnaire.

8.4 Summary

This chapter explored the relationship between two variables. In the case of linearity, a parametric correlation could be used. If the assumption of linearity does not hold up in the scatter plot, then non-parametric correlations can be used. From there, provided the four assumptions of linearity, normality of residuals, homoscedasticity, and independence hold, you learned how to fit a linear model. Table 8-2 should be a good reference of the key ideas you learned in this chapter.

Table 8-2. *Chapter Summary*

Idea	What It Means
`geom_point()`	`ggplot()` visual geometry function for scatter plots.
Parametric	Criteria or assumptions required for a technique to work.
Non-parametric	Statistical techniques that do not require assumptions.
Linear	Appearing to form a straight line.
Positive correlation	As x increases/decreases, y increases/decreases (same behavior).
Negative correlation	As x increases/decreases, y decreases/increases (opposite behavior).
`cor()`	Takes two variables and a `method` = " " argument (Pearson, Spearman, or Kendall).
Residuals	The difference between a linear regression and the actual data points.
Linearity	Visually tested via scatter plot.
Normality of residuals	Visually tested via `modelDiagnostics()`.
Homoscedasticity	Visually tested via `modelDiagnostics()`.
Independence	Of each observation. May be philosophically tested via unbiased study design.
R^2	Coefficient of determination; how much is variance of y predicted by x?
`lm()`	Linear model function. Takes formula and data arguments.
`modelDiagnostics()`	Tests normality and homoscedasticity.
`visreg()`	Visualize the regression.

8.5 Practice for Mastery

Check your progress and grow through practice by working through some exercises. Comprehension checks ask critical thinking questions that may be best answered with a written or verbal response. Part of the art of statistics is successfully communicating results to your stakeholders or audience. Sometimes that audience is highly technical and other times very much not technical. Exercises are more direct applications of the concepts explored in the chapter.

Comprehension Checks

1. Rephrase, in writing, in your own words, the four assumptions of a linear model. If you are taking this as part of a class, check your words with a classmate. If you are learning on your own, use a search engine on some of your words and see if you can crowd-source a check. If in doubt, tweet one of us; we will do our best to reply.
2. Thinking about the two linear regression model examples and homoscedasticity graphs, do you have a clear visual in your head of what the "cut-off" is for assumption met or assumption failed? If not, copy and paste that code against as many variables in the aces or penguins data as you can tolerate. Compare and contrast your results, and see if you can get examples of mostly parallel lines vs. not really at all parallel. Remember most real-world data do not perfectly fit the assumptions. Any collection of data can have a linear model fit; the difference is not all linear regressions are worth fitting. Assumptions help us sort out worthwhile models from badly damaged to broken to useless ones.

Exercises

1. Using the correlation examples, consider the aces data restricted to a single day and a single, evening survey. Looking at the scatter plot of `PosAff` and `NegAff`, is there a linear pattern? If there is, use Pearson's correlation. If there is not, use both Spearman's and Kendall's correlations. Is your correlation positive or negative? Strong or weak?

2. Using the correlation examples, consider the aces data restricted to a single day and a single, evening survey. Looking at the scatter plot of `Age` and `STRESS`, is there a linear pattern? If there is, use Pearson's correlation. If there is not, use both Spearman's and Kendall's correlations. Is your correlation positive or negative? Strong or weak?

3. Using the linear regression examples as a pattern, test the four assumptions for `penguinsData` using a predictor of `bill_length_mm` for an outcome of `bill_depth_mm`. If the assumptions hold, create the linear model and use `visreg()` to graph. If the assumptions hold, use the model to predict the bill depth given a bill length of 41.2 mm.

4. Using the linear regression examples as a pattern, test the four assumptions for `penguinsData` using a predictor of `bill_length_mm` for an outcome of `flipper_length_mm`. If the assumptions hold, create the linear model and use `visreg()` to graph. If the assumptions hold, use the model to predict the flipper length given a bill length of 41.2 mm.

CHAPTER 9

Confidence Intervals

How sure are we in our regression models from Chapter 8? For that matter, how confident are you a *sample* mean statistic is a good estimate of the population mean parameter? How might we build in some "wiggle room" into such estimates to consistently convey the level of accuracy that might be expected?

Most likely, you have seen an estimate given with a single number, followed by a plus or minus quantity. The single-number estimate is called a **point estimate**, while the range of the plus or minus is called a **confidence interval** (CI). For example, we can confidently tell you that this book is recommended by 50% of statisticians worldwide, ± 50%. While the point estimate of a global popularity rating of 50% may be impressive, you quickly realize the confidence interval ranges from 0% to 100%. Not so impressive.

NOTE A confidence interval is commonly abbreviated as **CI**. The confidence interval, or CI, defines an interval and is made up of two parts: a lower confidence limit and an upper confidence limit. For example, in the earlier example, the confidence interval was 0%–100%. Then 0% is the lower confidence limit, and 100% is the upper confidence limit. These often get abbreviated as **LL** for lower limit and **UL** for upper limit. Sometimes you also will see them abbreviated as **2.5** and **97.5** which are the percentiles of the most common type of confidence interval, a CI with a 95% confidence level (**CL**). We will learn more about CIs in this chapter, but hopefully this helps orient you to some common terminology.

The goal then is to develop a statistical framework that allows you to create a workable confidence interval. A good confidence interval should communicate the level of uncertainty. A good data sample will be robust enough to shrink the range of the confidence interval down to a workable level.

M. Wiley and J. F. Wiley, *Beginning R 4*, https://doi.org/10.1007/978-1-4842-6053-1_9

By the end of this chapter, you should be able to

- Understand the theory of confidence intervals (CIs) via the z-score.
- Understand the application of t-score confidence intervals.
- Create confidence intervals using the t-score method.
- Evaluate two samples for population similarity/dissimilarity.

9.1 R Setup

As usual, to continue practicing creating and using projects, we start a new project for this chapter.

If necessary, review the steps in Chapter 1 to create a new project. After starting RStudio, on the upper-left menu ribbon, select *File* and click *New Project*. Choose *New Directory* ➤ *New Project*, with *Directory name* `ThisChapterTitle`, and select *Create Project*. To create an R script file, on the top ribbon, right under the word *File*, click the small icon with a plus sign on top of a blank bit of paper, and select the *R Script* menu option. Click the floppy disk–shaped *save* icon, name this file `PracticingToLearn_XX.R` (where XX is the number of this chapter), and click *Save*.

In the lower-right pane, you should see your project's two files, and right under the *Files* tab, click the button titled *New Folder*. In the *New Folder* pop-up, type `data` and select *OK*. In the lower-right pane, click your new *data* folder. Repeat the folder creation process, making a new folder titled `ch09`.

Remember all packages used in this chapter were already installed on your local computing machine in Chapter 2. There is no need to re-install. However, this is a new project, and we are running this set of code for the first time. Therefore, you need to run the following `library()` calls:

```
library(data.table)
```

```
## data.table 1.13.0 using 6 threads (see ?getDTthreads). Latest news:
r-datatable.com
```

```
library(ggplot2)
library(visreg)
library(palmerpenguins)
library(JWileymisc)
```

For this chapter, we continue our work with just the evening survey data from one day of the ACES daily dataset, so each person contributes just one observation. For no reason in particular, we selected March 03, 2017:

```
acesData <- as.data.table(aces_daily)[SurveyDay == "2017-03-03" &
SurveyInteger == 3]
str(acesData)
```

```
## Classes 'data.table' and 'data.frame':      189 obs. Of 19 variables:
##  $ UserID            : int  1 2 3 4 5 6 7 8 9 10 ...
##  $ SurveyDay         : Date, format: "2017-03-03" ...
##  $ SurveyInteger     : int  3 3 3 3 3 3 3 3 3 3 ...
##  $ SurveyStartTimec11: num  0.518 0.428 0.529 0.423 0.391 ...
##  $ Female            : int  0 1 1 1 0 1 0 0 0 1 ...
##  $ Age               : num  21 23 21 24 25 22 21 21 22 25 ...
##  $ BornAUS           : int  0 0 1 0 0 1 0 1 0 1 ...
##  $ SES_1             : num  5 7 8 8 5 5 6 7 4 6 ...
##  $ EDU               : int  0 0 0 0 1 0 0 0 0 1 ...
##  $ SOLs              : num  NA 33.1 29.63 3.41 100.47 ...
##  $ WASONs            : num  NA 0 0 2 2 1 NA 1 1 1 ...
##  $ STRESS            : num  2 1 4 4 0 0 0 4 0 1 ...
##  $ SUPPORT           : num  5.93 3.94 8.84 4.45 2.1 ...
##  $ PosAff            : num  1.55 3.17 3.51 1.7 2.22 ...
##  $ NegAff            : num  1.39 1 1.69 1.38 1.08 ...
##  $ COPEPrb           : num  2.05 1.93 3.21 2.12 1 ...
##  $ COPEPrc           : num  2.31 1.69 3 1.7 1.1 ...
##  $ COPEExp           : num  2.24 1.8 2.05 2.93 1 ...
##  $ COPEDis           : num  1.94 2.13 2.53 1.21 1.86 ...
##  - attr(*, ".internal.selfref")=<externalptr>
```

Of course, we also use our familiar penguin and car data sets:

```
penguinsData <- as.data.table(penguins)
mtcarsData <- as.data.table(mtcars, keep.rownames = TRUE)
```

You are now free to learn more about statistics!

9.2 Visualizing Confidence Intervals

Why might it be foolish to trust one single sample?

To answer this question, let us use R to create 10,000 data points to represent a familiar, normal population. We called it the *perfect* data set. Using `set.seed(1234)` to make sure your "random" data matches ours, create a new `data.table()` that has two columns. One is named `Data` and has 10,000 random, normal numbers via `rnorm()` with a mean of 2.5 and a standard deviation of 0.75. In other words, `Data` is *N*(2.5, 0.75). Because these data are already pulled in a random fashion, we can use Kth sampling to put the data points into 100 groups as shown in the `Group` column. With $\frac{10,000}{100} = 100$, we have 100 points in 100 groups. So, while we used Kth sampling for speed, in this case, it is technically a random sample. Notice we called our population parameters `muP` and `sigmaP` for μ and σ *P*erfect:

```
set.seed(1234)
muP <- 2.5
sigmaP <- 0.75
perfectData <- data.table(Data = rnorm(n = 10000,
                                        mean = muP,
                                        sd = sigmaP),
                          Group = 1:100)
```

A sample size of 100 may not seem like much. However, imagine surveying 100 people; that could take some time!

Go ahead and compute the arithmetic `mean()` of the data for each group. Since we have 100 groups, there should be 100 rows or observations in the resulting data set. Calculating a new value is a column operation; thus, in a data table, it belongs in the jth column operation position. We are finding the arithmetic mean of `Data` elements. Because we want `mean(Data)` for each group, that is a by operation. We need a `by = Group`.

Notice, while the following line of code achieves these goals, it leaves something to be desired. First of all, notice in the output that the column holding the arithmetic mean has a not so great name, `V1`. That is not descriptive at all; future researchers (and even our own future selves) will have no idea what that column holds. Second, the following code does *not* get saved anywhere. The means are printed to the console; the middle 90 groups are truncated and hidden in our output. Also, the results are not stored anywhere in R, so we cannot use them for further analysis. While this code might be good enough for a quick check, it is not enough for our current needs:

```
perfectData[, mean(Data),
            by = Group]

##        Group    V1
##    1:      1 2.518
##    2:      2 2.449
##    3:      3 2.460
##    4:      4 2.442
##    5:      5 2.527
## ---
##   96:     96 2.437
##   97:     97 2.626
##   98:     98 2.396
##   99:     99 2.519
##  100:    100 2.554
```

It is clear we must brand the column with a better name and save our results for future use. There are two ways to name a column. You could create a new column in `perfectData` using the assignment operator `groupMean := mean(Data)`. If you do this, the `perfectData` will gain a third, new column named `groupMean`. However, remember that `perfectData` has 10,000 observations. *Every* observation gets that new column; we now have 10,000 items added to our table which is now 50% bigger in memory. To make this even worse, every group has only one mean; so there are 100 duplicates of each mean. That is a lot of extra information copies.

Another way to brand a column is to give it a name directly; this does not save any values to memory. Rather than use an assignment operator, we use the column selection operator `.()`. The following code is actually the same as the preceding output. Once again, nothing is saved to memory in your global environment; the only difference is the ambiguous `V1` is replaced for an easier-to-understand `groupMean` column name:

```
perfectData[,
            .(groupMean = mean(Data)),
            by = Group]
```

```
##       Group groupMean
##    1:     1     2.518
##    2:     2     2.449
##    3:     3     2.460
##    4:     4     2.442
##    5:     5     2.527
## ---
##   96:    96     2.437
##   97:    97     2.626
##   98:    98     2.396
##   99:    99     2.519
##  100:   100     2.554
```

Of course, we do actually want to save and use our values, so we use the external assignment operator: `<-`. This creates a new `data.table`, in this case named `perfectMeans`, that only has 100 rows/observations. Because the assignment captures the output, the results are not printed to the console again. However, you can check that the data you want is saved by using `head(perfectMeans)` if you wish:

```
perfectMeans <- perfectData[,
                            .(groupMean = mean(Data)),
                            by = Group]
```

Both methods of creating and saving column information have their uses. If your new column is unique to each row of the raw data, then an *internal* assignment such as `groupMean := mean(Data)` is a good way to go. On the other hand, when, like in this case, the new column is collapsing many rows of raw data into a single point estimate, `.(groupMean = mean(Data))` and an external assignment `<-` to a new data table may be preferred.

In any case, now that we have 100 sample means, let us take a look at those means. To keep the points distinct, we set the y-axis to be a discrete variable of the sample group identifier. Additionally, we use `geom_vline()` to set a vertical line with at `xintercept = muP` which is the population mean:

```
ggplot(data = perfectMeans,
       mapping = aes(x = groupMean, y = Group))+
  geom_point()+
  geom_vline(xintercept = muP)
```

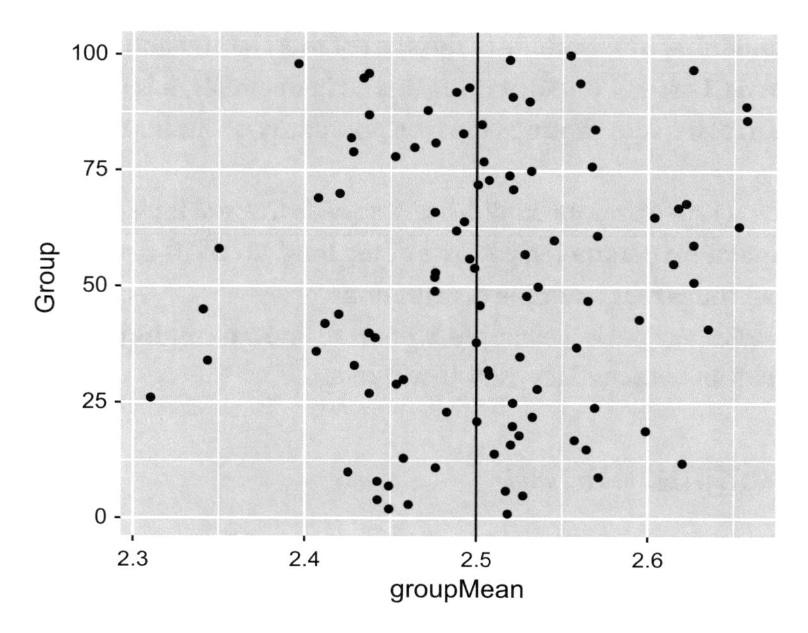

Figure 9-1. *Sample means with a vertical line for the population mean, mu = 2.5*

Now, keep in mind our population has a distribution of $N(2.5, 0.75)$. Each dot in Figure 9-1 is a mean from a sample of size $n = 100$. Now, you know the central limit theorem - and it is clear these means are looking pretty normal around 2.5 (like they should). When a sample is collapsed to a single data point (such as a mean), this is called a **point estimate**. While most of these point estimates are close to the parameter of 2.5, only a handful touch that 2.5 line.

Rather than a single point estimate, it might be nice to get some interval around any one sample statistic that probably captures the population parameter. In other words, we need to create a **confidence interval**. Confidence intervals take the form

$$(Point\ Estimate - Error\ Bound, Point\ Estimate + Error\ Bound)$$

How do we compute an error bound? Well, that is going to come down to some ideas about probability. Referring back to Figure 9-1, remember you are looking at 100 point estimates. In other words, we took 100 random samples, and some were very good at estimating close to $\mu = 2.5$ and others were not as good. Imagine if, instead of a dot, each one of those was a line stretching the length of some confidence interval. If we pick a large error bound, then maybe even those rare samples on the far left and far right of our graph will have their confidence interval line cross the 2.5 mark. However, that is of limited use.

After all, remember our example at the start of this chapter about how popular this book is that has a CI of (0%, 100%); writing this CI in decimals, it is (0.00, 1.00). Our confidence interval for sure captures the true popularity of our book. It is not a useful CI, however.

What we need is to compute confidence intervals that are long enough to probably capture the population parameter, yet are not too long. To do that, we need to think about the probabilities of the sample distributions.

Let us start with our perfect example, where we know everything. From there, we can explore a way to think about likely, real-world cases.

Example: Sigma Known

If we know the population parameters, in particular, sigma, we can create a probability framework for error bound. Keep in mind this almost never is the case in real life. After all, if we knew the population data, we would not be in the business of sampling! That said, to understand what is happening (and to explore the probabilities), it helps to start from known quantities.

So, for our perfect data set, we created the population ourselves from known parameters *N*(2.5, 0.75). Thus, we can calculate standard error:

```
standardErrorP <- sigmaP / sqrt(100)
```

Again referring to Chapter 7, the sampling distribution is generally

$$N\left(\mu, \frac{\sigma}{\sqrt{n}}\right)$$

In our particular case, *N*(2.5, 0.075) is our sampling distribution using the `standardErrorP` we just computed. Suppose we want a confidence *interval* (CI) that has a **confidence *level*** (CL) of 95%. In other words, can we make our error bound big enough that there is a 95% chance a random sample's confidence interval captures the parameter?

The answer is yes, and the formula for the **error bound of the mean** (EBM) follows:

$$EBM = (Z - Score) * (StandardError)$$

In more formal notation, the z-score quantile is written as

$$EBM = \left(Z_{\frac{\alpha}{2}}\right) * \left(\frac{\sigma}{\sqrt{n}}\right)$$

To understand the z-score quantile, we have to think about the normal curve and probability some more. As is often the case with probability, the complement is an interesting idea to consider. If we have a 95% confidence level, then we are said to have a 5% *alpha* or $\alpha = 0.05$. **Alpha** is the chance we pull a random sample from our population and it is so far from the middle that our confidence interval does not capture the population parameter.

Remember the normal curve has an area under the curve of 1 (this is not magic; mathematicians built it that way so it was easy to compute probabilities). Also recall the empirical rule from Chapter 6 (also called the 68-95-99.7 rule). That rule stated that the middle 68% was contained in a range of plus or minus one standard deviation. The middle 95% is captured (roughly) in plus or minus two standard deviations. The middle 99.7% is in plus or minus three standard deviations. These are very close approximations relating the area under the curve (which equals the probability) to the number of standard deviations (which is also called the z-score).

A way to visualize the empirical rule is in Figure 9-2 [6].

Notice in Figure 9-2 that 95% of values in the middle is the uppermost horizontal line. There are two dashed lines dropping down vertically that take you to -2σ on the left and $+2\sigma$ on the right. Follow that down a bit further, and you see on the z-score row those are just called -2 and +2. However, if you looked very closely, you would see that those vertical, dashed lines do not quite perfectly intersect with ± 2. Notice on either side of the horizontal 95% of values line in the middle of the normal curve, it shows -1.96σ and $+1.96\sigma$. These are the actual values of the sigma (and thus ± 1.96 is the z-score).

Remember, if 95% of the values are in the middle, then there is 5% of the values contained under the far-left and the far-right tails of the normal curve. It is that 5% that is called `alpha`. However, if we want to figure out the leftmost z-score (which is the number of standard deviations we are to the left of the absolute middle), we need to split that 5% in half (it gets split between both tails). This is the $\frac{\alpha}{2}$ mentioned in the preceding formula.

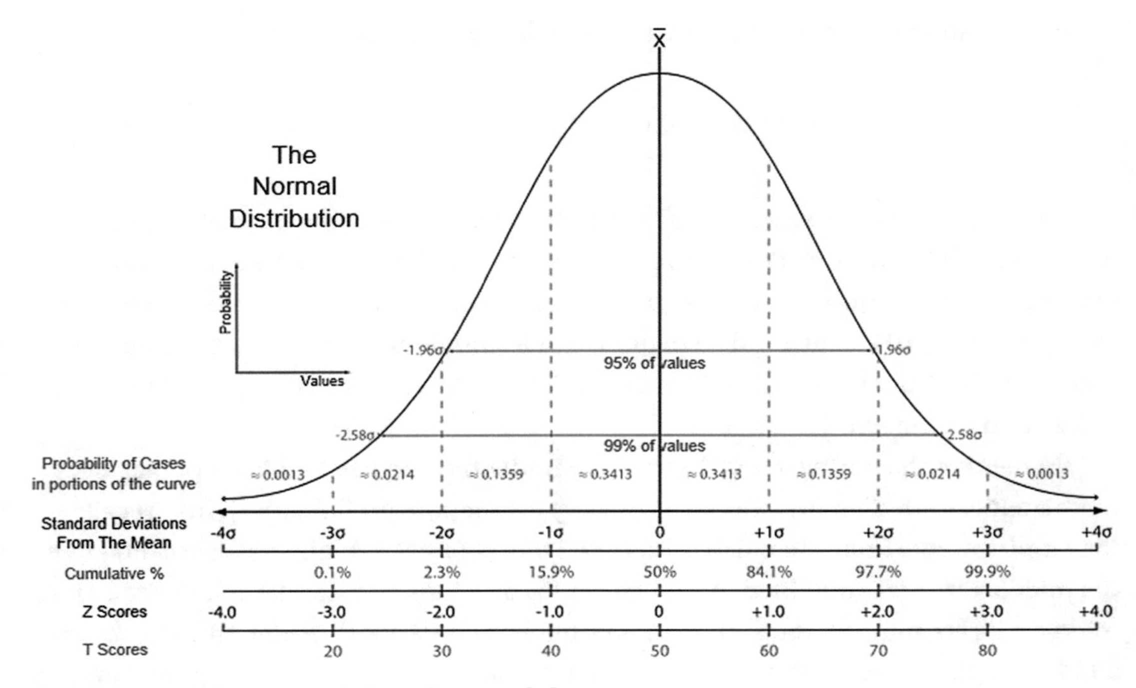

Figure 9-2. *The normal distribution [6]*

The qnorm() function takes an area (which is the same as a probability remember) and gives an output of the correct z-score for that area. In other words, going back to Figure 9-2, qnorm() takes the cumulative area % line as an input and outputs the z-score:

```
ConfidenceLevel <- 0.95
alpha <- 1 - ConfidenceLevel

qnorm(alpha/2)

## [1] -1.96
```

Notice that if we round this answer to two decimal places, one gets the -1.96σ Figure 9-2 suggested:

```
round(qnorm(alpha/2), 2)

## [1] -1.96
```

For the formula, $EBM = \left(Z_{\frac{\alpha}{2}}\right) * \left(\frac{\sigma}{\sqrt{n}}\right)$, we want a positive z-score. To do that, we want to start our cumulative percent not from the left side of the normal curve (also called the lower tail). Instead, we want to start from the right side (also called the upper tail).

By setting `lower.tail = FALSE` in the `qnorm()` function, one gets the positive side:

```
round(qnorm(alpha/2,
            lower.tail = FALSE),
      2)
```

```
## [1] 1.96
```

If one wishes a 99% confidence interval instead, only a few changes are required. The empirical rule approximated that three standard deviations would be about 99.7%. Thus, we would expect a 99% confidence interval to be more than 1.96 standard deviations and less than 3. Figure 9-2 suggests it is 2.58. Changing our confidence level to 0.99 in the following code shows this is correct:

```
ConfidenceLevel <- 0.99
alpha <- 1 - ConfidenceLevel

qnorm(alpha/2,
      lower.tail = FALSE)
```

```
## [1] 2.576
```

As you can see, this method allows you to compute the exact z-score required for any confidence level you need. In practice, 95% is the most frequent choice, and we carry on our example using that number.

Filling into our formula for the error bound for the mean for an arbitrary value of `perfectMeans`, Figure 9-1 yields

$$\begin{aligned} EBM &= \left(Z_{\frac{\alpha}{2}}\right) * \left(\frac{\sigma}{\sqrt{n}}\right) \\ &= (1.96) * \left(\frac{0.75}{\sqrt{100}}\right) \\ &= (1.96) * (0.075) \\ &= 0.147 \end{aligned}$$

The preceding final calculation is shown in R:

```
ebmP <- 1.96*standardErrorP
ebmP
```

```
## [1] 0.147
```

We now add the lower and upper error bounds to the perfectMeans data. This only requires creating a new column, and we want every row to have that column (since every row is one of our samples). Thus, we will use the internal assignment operator := in the jth column operation position:

```
perfectMeans[,
            lowerEBM := groupMean - ebmP]

perfectMeans[,
            upperEBM := groupMean + ebmP]
```

Recycling our ggplot() code from before, we add one more phrase of code. Line segments are created by geom_segment(). There is a lot going on in that segment; while it is not vital you understand it, we talk you through it because graphs are useful. Each segment needs to start and end at the lower and upper EBMs that were just added to perfectMeans. These need to be horizontal lines to stay in the sample group ID, so the y values are kept fixed. Lastly, we use the alpha = 0.4 to control the *transparency* of those lines to make them less bold. Notice that this has nothing to do, other than shared name, with the probability complement of the confidence level (CL; which in this case is 95%) which is also named alpha (i.e., $\alpha = 0.05$ that we often set)!

```
ggplot(data = perfectMeans,
       mapping = aes(x = groupMean, y = Group)) +
  geom_point() +
  geom_vline(xintercept = muP) +
  geom_segment( mapping = aes( x = lowerEBM, xend = upperEBM,
                               y = Group, yend = Group),
               alpha = 0.4)
```

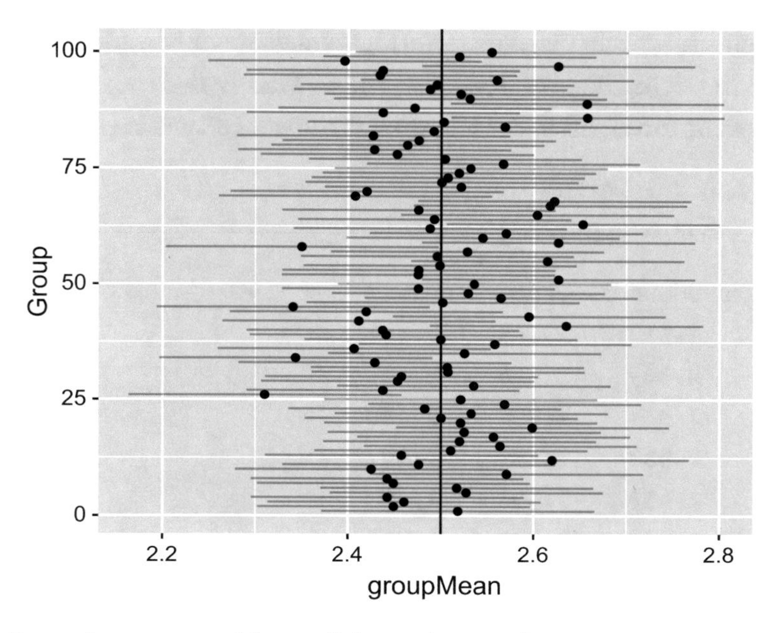

Figure 9-3. *Sample means with confidence intervals*

Can you spot any confidence interval lines that do not intersect our vertical line in Figure 9-3? In other words, can you spot some samples that were so extreme their confidence intervals miss the mark?

Remember we picked a confidence level (CL) of 95%. This means, 95% of the time, we expect our confidence interval (CI) to capture our population mean of 2.5. The formal way to say this is "95% of the confidence intervals ought to capture the population parameter $\mu = 2.5$."

Visually, it seems the theory and the practice were a perfect match. However, just because we picked 100 random samples does not mean that we will only have five of them fail. There is always noise in random, and we may have either gotten lucky or unlucky this round.

To quickly check, we can use data table logic rules. If our confidence interval (CI) starts too big to the right, then the `lowerEBM` will be *greater than* the mean. On the other hand, if our CI starts too small to the left, then the `upperEBM` will be *less than* the mean:

```
perfectMeans[muP < lowerEBM|
               muP > upperEBM]

##    Group groupMean lowerEBM upperEBM
## 1:    26     2.310    2.163    2.457
## 2:    34     2.343    2.196    2.490
## 3:    45     2.341    2.194    2.488
## 4:    58     2.350    2.203    2.497
## 5:    63     2.653    2.506    2.800
## 6:    86     2.657    2.510    2.804
## 7:    89     2.657    2.510    2.804
```

In this case, seven of our samples were off. This is quite close to the predicted quantity of five. Again, when working with probabilities, one expects them to hold true *in general*. Any specific case, such as this one, will usually be close, yet is not expected to match exactly. One way to check if the probability is followed over time is to change the `set.seed(1234)` to a different number. This example is a great showcase for how even good models can occasionally get the real-world outcome "wrong."

Where do we go from here? The next step is to avoid using σ. Most real-world use cases are calculated from sample statistics, not population parameters. Therefore, we need to find some ways to understand data when sigma is not known. What might be a consequence of not knowing sigma? Well, logically, when information is reduced, uncertainty increases. Thus, we might expect the confidence intervals to get a bit longer if we must use the sample's standard deviation to approximate the distribution.

Admittedly, a wider confidence interval is a step backward. The narrower that range, the more actionable a decision can be achieved. Again, remember our silly case of claiming this book was endorsed by half of all statisticians, plus or minus half. That is far too wide to be of any use. There is, however, a way to make the confidence interval narrower. In fact, there are two ways.

The easiest way is to reduce the confidence level required. In other words, in this example, we chose a 95% confidence level (CL). This is a common CL to choose; it is not, however, magic. You can choose any confidence level you want. In social and behavioral science work, 95% CL is a fairly common standard. However, other fields may prefer higher confidence (in which case the confidence interval gets wider) or may not need such a high confidence (in which case the CI gets narrower). In our silly example, that was a 100% CL. The risk of higher confidence *level* is that your confidence *interval* gets so large it is no longer useful.

The better way to make a confidence interval (CI) narrower is to increase the number of elements in your sample. Remember the z-score formula for sampling distribution of the mean includes the standard error (review Chapter 7 if needed). Notice the square root of n, where n is the sample size, is in the denominator of the fraction. If n were to *increase*, then the standard error would *decrease*:

$$z = \frac{\overline{x} - \mu}{\frac{\sigma}{\sqrt{n}}}$$

If the standard error decreases, that decreases the error bound! Notice the error bound of the mean is two numbers multiplied by each other. If we increase our confidence level, that makes $Z_{\frac{\alpha}{2}}$ bigger. To cut that back down to size, we must make the remaining term of our formula smaller. In other words, we have to increase *n*:

$$EBM = \left(Z_{\frac{\alpha}{2}}\right) * \left(\frac{\sigma}{\sqrt{n}}\right)$$

This is the perpetual challenge of statistics. How do we find ways to make n as large as possible? It can be expensive to survey people; it can be expensive to conduct tests. It can even be ethically risky to conduct tests (it might not be ethical to give everyone a new pharmaceutical just to increase sample size).

Thus, even as we explore how to adapt these formulas to a world without σ, keep in mind this lack of information comes at a cost of wider confidence intervals and that we may wish to boost (when possible) our sample size.

Example: Sigma Unknown

To recap, while z-scores work when sigma is known, most often researchers work with an unknown sigma. To help compare and contrast the two methods, we use the exact same data set as in the prior example. However, we shall pretend we do not know sigma and instead simply have 100 samples.

When sigma is unknown, instead of using the z-score and the normal distribution, we use the **Student's t-distribution**. From a distance, this distribution looks highly similar to the normal distribution – it is just a bit flattened. This has the net result of putting more area in the tails and less area in the middle (in other words, shorter and longer, the area is still 1). Additionally, the t-distribution depends on the number of elements in the sample. As n gets larger, the t-distribution gets taller and narrower. As n gets very large, the t-distribution eventually gets quite close to the normal distribution. Remember, as n gets larger, we would expect our standard error to get narrower, so this fits well with that idea. The formulae follow:

$$t = \frac{\overline{x} - \mu}{\frac{s}{\sqrt{n}}}$$

$$EBM = \left(t_{\frac{\alpha}{2}} \right) * \left(\frac{s}{\sqrt{n}} \right)$$

Because the t-score and t-distribution approach or converge to the z-score as n gets large, most modern software simply uses the t-distribution on sample data. The functions in R are no exception. Behind the scenes, the software adjusts for the size of *n*.

The function to compute the t-score and confidence interval is called `t.test()`. It takes several arguments, and for now, two are needed. The first argument is the sample data, and the second is the confidence level desired. For now, we keep the `ConfidenceLevel` we used in the prior section. To see how the function works, we take only the first sample from our `perfectData`. That first sample is in `Group == 1`, which is a row operation. We access the `Data` column information using the `$` operator. As mentioned, we set the second formal argument, `conf.level`, to our variable `ConfidenceLevel` which is currently set to 0.95:

```
tTestResults <- t.test(perfectData[Group == 1]$Data,
                       conf.level = ConfidenceLevel)
tTestResults

##
##       One Sample t-test
##
## data: perfectData[Group == 1]$Data
## t = 33, df = 99, p-value <2e-16
```

```
## alternative hypothesis: true mean is not equal to 0
## 99 percent confidence interval:
##  2.317 2.720
## sample estimates:
## mean of x
##     2.518
```

The result of this function gives us a lot of information. The t-score is 33; that could be computed manually via the preceding formula for t-score. The next is `df = 99`.

It is worth discussing **degrees of freedom** which is often abbreviated **df**. Imagine you knew you had two people in a sample and that their average salary was $100,000. Suppose you also know that one person has a salary of $50,000. The only way to take $50,000 and get an average of $100,000 is if the other person makes $150,000. So, while the sample size was $n = 2$, we say there is $n - 1 = 1$ degree of freedom because once you know what the first person makes and you know the average, the second person is not free to vary.

Traditionally, the t-distribution was built on those degrees of freedom (remember, as n increases, the t-distribution changes to look ever closer to the normal distribution). For a single sample like we have here, the formula is $df = n - 1$.

For now, ignore the p-value; we shall discuss that in the next chapter. The next thing of interest is that there is a 95% confidence interval. Rather than compute EBM by hand, `R` gives us the range. Lastly, the sample mean of 2.518 is given.

Without showing the code (because it gets needlessly messy), we use that same `t.test()` on all 100 samples. Letting the z-score confidence intervals stay as solid black lines, we superimpose the t-score confidence intervals as blue, dashed lines in Figure 9-4. As you can see, the sample means are the same; that is at it should be. The only part that changes from a z-score to a t-score is the switch from σ and normal distribution to s and t-distribution. As promised, the blue, dashed lines from the t-distribution are all just a bit longer than the z-score solid lines.

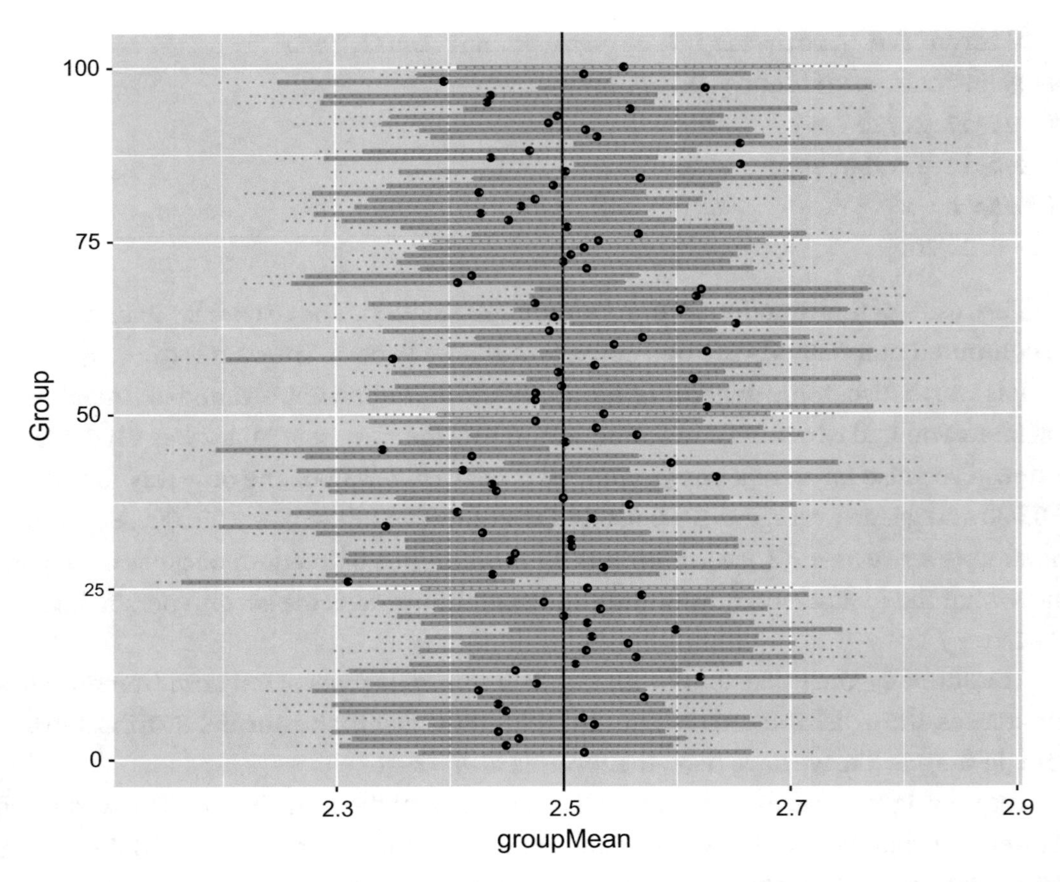

Figure 9-4. *Sample means with confidence intervals. Solid lines are z-values and blue, dashed lines are t-values*

Now that you have a good understanding and a good reference image for what confidence intervals are, you are ready to use them on real-life data samples where we do not yet know the population deviation. Additionally, you have seen how simple it is for software to compute the confidence interval. Using some examples from prior chapters, let us return to those and expand those analyses.

Example

Consider the penguin data set. That data set is a sample, and no one has (yet) measured every penguin on earth. Thus, the population standard deviation is not known. Using `t.test()`, one can compute a confidence interval that is likely to capture μ.

Our research question might be: Based on the sample of penguins collected, calculate at the 98% confidence level a confidence interval for μ, the population mean of flipper lengths. The data of flipper lengths goes as the first function argument in `t.test()`. This time, we set the `conf.level = 0.98`. Now is a great time to mention that the default CL is 95% in most software packages. This is so often the case that one can often think there is something sacred about 95%. In part, this was historically true as a way to simplify some of the techniques taught (by having a single default setting, there was less variability). Coding makes it easy enough to change confidence levels, and it is usually as simple as in the following code:

```
t.test(penguinsData$flipper_length_mm,
       conf.level = 0.98)

##
##      One Sample t-test
##
## data:  penguinsData$flipper_length_mm
## t = 264, df = 341, p-value <2e-16
## alternative hypothesis: true mean is not equal to 0
## 98 percent confidence interval:
##  199.1 202.7
## sample estimates:
## mean of x
##     200.9
```

You may recall our initial sample mean for flipper lengths has been 201; in all prior chapters, we had rounded that value to a whole number. Thus, the 200.9 is indeed the same as it has always been. The degrees of freedom are the 342 penguins that have flipper lengths recorded less one. Remember, even though `penguinsData` has 344 observation rows, two of those penguins are missing flipper length data. The CI is (199.1, 202.7).

In English, you would say you are 98% confident that the true population mean of penguin flippers is between 199.1 and 202.7. Due to the large sample size, the CI is fairly narrow.

Now, as you saw before in Figure 9-3, it is possible that not every sample's confidence interval successfully captures the population parameter. This can happen even with a perfectly random data pull. And, as discussed back in Chapter 5, sample bias can happen. Perhaps it is easier to catch penguins with long flippers because they are easier to see. Or maybe it is tougher because their longer flippers power them away more quickly. Nevertheless, as long as the data collection was done well, we now can be rather confident in a range for the population of penguins' flipper length "average."

Example

Confidence intervals do not only apply to single point estimates. They also apply to the regression lines calculated in Chapter 8. A confidence interval on a line looks like a band on either side of the line. Thus, it can also be called a confidence *band*. You may recall back in Chapter 8 we added an argument to `visreg()` which was `band = FALSE`. We repeat the analysis from before, this time looking closely at confidence levels and intervals.

First, we copy and run without any changes to the code to fit the linear model:

```
mPenguin <- lm(formula = body_mass_g ~ flipper_length_mm,
               data = penguinsData)

summary(mPenguin)
```

```
##
## Call:
## lm(formula = body_mass_g ~ flipper_length_mm, data = penguinsData)
##
## Residuals:
##     Min      1Q  Median      3Q     Max
## -1058.8  -259.3   -26.9   247.3  1288.7
##
## Coefficients:
##                   Estimate Std. Error t value Pr(>|t|)
## (Intercept)       -5780.83     305.81   -18.9   <2e-16 ***
## flipper_length_mm    49.69       1.52    32.7   <2e-16 ***
## ---
## Signif. codes:  0 '***' 0.001 '**' 0.01 '*' 0.05 '.' 0.1 ' ' 1
##
## Residual standard error: 394 on 340 degrees of freedom
##   (2 observations deleted due to missingness)
```

```
## Multiple R-squared:  0.759,  Adjusted R-squared:   0.758
## F-statistic: 1.07e+03 on 1 and 340 DF,  p-value: <2e-16
```

In the summary() output, notice that in the linear model coefficients area, there is a column titled Std. Error. That is, of course, standard error. Keeping in mind the error bound of the mean formula for t-scores, $EBM = \left(t_{\frac{\alpha}{2}}\right) * \left(\frac{s}{\sqrt{n}}\right)$, think back to our discussion of the z-score process and the code. The standard error is independent of your desired confidence level. Instead, the confidence level is used to calculate the t value or z value in that first term. Thus, by giving only the standard error in the Std. Error column, the linear model gives us everything we would need to calculate an error bound for any confidence level we later choose.

Additionally, the model tells us that the degrees of freedom are 340. Again, for the penguin data, there are 342 observations that do not have missing values. Unlike a one sample t-test where $df = n - 1$, in this linear model we have both flipper lengths and body mass. Thus, the formula is $df = n - 2 = 342 - 2 = 340$. That is also given in the summary output.

All this is to say that visreg() has access, from the linear model saved to the mPenguin variable, to a fair bit of information. Using this information, the function has a "confidence band" shaded on either side of the linear model as seen in Figure 9-5. That is, for every predicted point on the line, the confidence interval is shown through shading. Because the predictions are a line, the confidence intervals end up forming a "band" around the line:

```
visreg(fit = mPenguin)
```

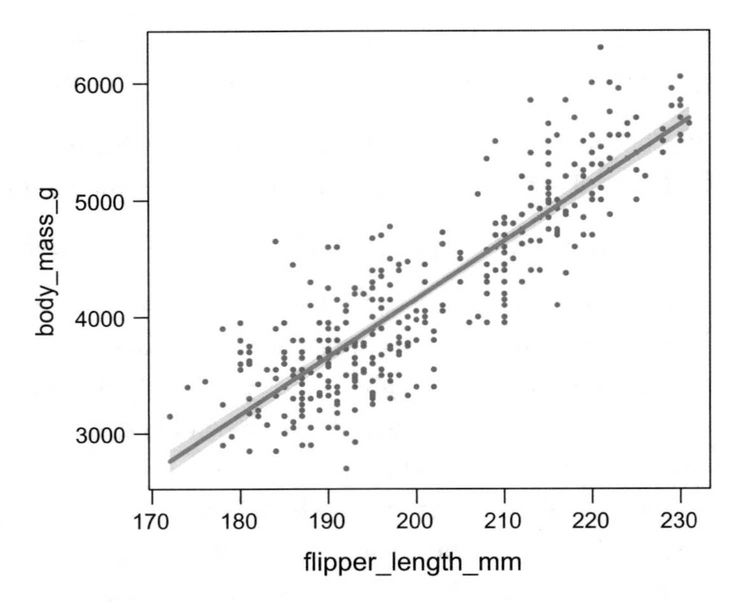

***Figure 9-5.** Linear regression visual for penguinData with confidence interval*

One thing to notice about that confidence band is it gets wider near the outer edges of the model. Recall the x-axis data, `flipper_length_mm`, ranges from 172 to 231:

```
range(penguinsData$flipper_length_mm,
      na.rm = TRUE)
```

```
## [1] 172 231
```

Using flipper length to predict body mass can likely be fairly accurate inside that range. This was a decent model fit. Using a model to predict the body mass of a new penguin who has a flipper length inside the range is called **interpolation**. The prefix *inter-* means "inside." Generally, interpolation, for a good model fit, can be fairly accurate. On the other hand, using the model to predict beyond the range of the independent variable is called **extrapolation**. Extrapolation is always less certain; it goes beyond the model. Many applications of data science call for extrapolation. Indeed, it is not unreasonable to suppose that a flipper length of 300 mm might yield a fairly weighty penguin. All the same, the error band getting wider as we approach the border between interpolation and extrapolation serves as a warning that extrapolating too far past our model is inherently less certain.

9.3 Understanding Similar vs. Dissimilar Data

Another way the confidence interval can be understood is to see if the confidence intervals of some different groups overlap. After all, if they do overlap, there is a chance the two groups share the same population parameter. And if they share a population parameter, they may well be from the same population. On the other hand, if the confidence intervals do not overlap, they may well be different.

Example

Suppose we did not know each penguin species group was in fact a different species. One way to realize there is something different about the species is to see that not only is the group average different but even the confidence intervals do not overlap! Thus, we are fairly sure the population mean for each species is significantly different from other species.

In this case, we choose a 95% confidence level (simply because it is the most typical). While Adelie penguins have a confidence interval (CI) of 188.9–191.0, the larger Chinstrap penguins have a CI of 194.1–197.5:

```
t.test(penguinsData[species == "Adelie"]$flipper_length_mm,
       conf.level = 0.95)
```

```
##
##      One Sample t-test
##
## data:  penguinsData[species == "Adelie"]$flipper_length_mm
## t = 357, df = 150, p-value <2e-16
## alternative hypothesis: true mean is not equal to 0
## 95 percent confidence interval:
##  188.9 191.0
## sample estimates:
## mean of x
##       190
```

```
t.test(penguinsData[species == "Chinstrap"]$flipper_length_mm,
       conf.level = 0.95)
```

```
##
##      One Sample t-test
##
## data:  penguinsData[species == "Chinstrap"]$flipper_length_mm
## t = 226, df = 67, p-value <2e-16
## alternative hypothesis: true mean is not equal to 0
## 95 percent confidence interval:
##  194.1 197.5
## sample estimates:
## mean of x
##     195.8
```

Notice this confidence interval for the population mean has little to do with the size of any single Adelie or Chinstrap. The longest Adelie flipper length is 210 mm. And the shortest Chinstrap flipper length is 178 mm.

```
range(penguinsData[species == "Adelie"]$flipper_length_mm,
      na.rm = TRUE)
```

```
## [1] 172 210
```

```
range(penguinsData[species == "Chinstrap"]$flipper_length_mm,
      na.rm = TRUE)
```

```
## [1] 178 212
```

All the same, using `t.test()` which considers *all* elements of each sample, we estimate with 95% confidence level that the population arithmetic mean for each species is different from the other species. We know it is *different* because the two confidence intervals do not overlap.

Example

As a contrasting example, consider the ages of participants in our ACES daily study. In particular, consider the ages of those not born in Australia vs. those who were born in Australia:

```
t.test(acesData[BornAUS == 0]$Age)

##
##      One Sample t-test
##
## data:  acesData[BornAUS == 0]$Age
## t = 112, df = 121, p-value <2e-16
## alternative hypothesis: true mean is not equal to 0
## 95 percent confidence interval:
##  21.22 21.98
## sample estimates:
## mean of x
##      21.6

t.test(acesData[BornAUS == 1]$Age)

##
##      One Sample t-test
##
## data:  acesData[BornAUS == 1]$Age
## t = 72, df = 64, p-value <2e-16
## alternative hypothesis: true mean is not equal to 0
## 95 percent confidence interval:
##  21.11 22.31
## sample estimates:
## mean of x
##     21.71
```

In this case, notice the ranges of the confidence intervals overlap. So even though the sample means are slightly different, there is a chance the population parameter of mean age is the same for both groups.

We will build a better vocabulary for this in Chapter 10. For now, we leave you with observation that if the confidence intervals of two samples overlap, there is a chance the population measure being estimated could be the same for both samples. And if both samples share a population parameter, then it is possible both samples came from essentially the same population.

9.4 Summary

This chapter explored nuances between point estimates and confidence intervals. In particular, by using the t-distribution, one can calculate a confidence interval range around sample statistics for any desired confidence level about the population parameters. An important idea from this chapter is that even random samples will, on occasion, "miss the mark." Large sample sizes can help reduce this risk. A consequence of this idea is that it is not reasonable to expect statistics to always be accurate for any *single* case. Indeed, individual data points can easily have a high level of variance. And, despite the central limit theorem, even reasonable sample sizes vary in their ability to predict the true parameters. This is not to say that statistics are bad. Rather, it is a caution that one must be reserved when interpreting statistics as estimates of parameters.

Another feature of this chapter is that two separate samples can be compared to each other by observing their confidence intervals at some confidence level. If the confidence intervals do not overlap, then you suspect there is no evidence they come from the same population. In other words, they come from different populations. On the other hand, if they do overlap, the two samples may well have been pulled from the same population (at least in the terms of that measure).

Table 9-1 should be a good reference of the key ideas you learned in this chapter.

Table 9-1. *Chapter Summary*

Idea	What It Means
Point estimate	A single sample statistic used to estimate population parameter.
Confidence interval (CI)	A range around a sample statistic that likely captures the population parameter.
`.(colName = Stuff)`	In `data.table` names and calculates a column without assignment.
`geom_vline()`	Vertical line on a `ggplot()`; only sets `xintercept` =.
Confidence level (CL)	CL of the CIs computed ought to contain the population parameter.
Error bound of the mean	Based on the CL times standard error, builds the confidence interval.
Alpha	The complement of CL; the chance our CI does not capture the parameter.

(continued)

Table 9-1. *(continued)*

Idea	What It Means
`qnorm()`	Quantile of the normal distribution; computes the z-score based on the CL.
`geom_segment()`	Creates a line segment on a `ggplot()`.
Student's t distribution	Used when σ is unknown. For large n, gets ever closer to normal distribution.
`t.test()`	Computes CI and mean from a sample using t-distribution.
Degrees of Freedom (df/DF)	n-1 for one-variable, n-2 for two-variable scatter plot–style data.

9.5 Practice for Mastery

Check your progress and grow through practice by working through some exercises. Comprehension checks ask critical thinking questions that may be best answered with a written or verbal response. Part of the art of statistics is successfully communicating results to your stakeholders or audience. Sometimes that audience is highly technical and other times very much not technical. Exercises are more direct applications of the concepts explored in the chapter.

Comprehension Checks

1. Thinking about the error bound, suppose all else is equal (e.g., everything else is the same) and you doubled the sample size. Would the error bound get wider or narrower?

2. Thinking about the error bound, suppose all else is equal (e.g., everything else is the same). Which confidence level (CL) would give a wider confidence interval: 90% CL or 99% CL?

Exercises

1. Using the `penguinsData` and a confidence level of 90%, compute the confidence interval for the mean for `bill_length_mm` using the methods described in this chapter.

2. Using the `penguinsData` and a confidence level of 95%, compute the confidence interval for the mean for `bill_length_mm` using the methods described in this chapter. Compare this result to your prior result.

3. Recall your code from Chapter 8 exercises that used a predictor of `bill_length_mm` for an outcome of `bill_depth_mm`. This time, graph using `visreg()` and allow the band of the CI to show in the graph. Based on that band, does your prediction of bill depth given a bill length of 41.2 mm seem likely to be very accurate or to have a lot of room for being only an estimate for likely real-life bill depth?

4. Using the `penguinsData` and a CL of 90%, compare body mass for male and female penguins using two separate `t.test()` function calls. Do the two confidence intervals overlap? Keeping in mind this chapter's discussion of similar vs. dissimilar data, does there seem to be a difference in mass for female vs. male penguins?

CHAPTER 10

Hypothesis Testing

You have reached a milestone. You are able to set up and use one of the most popular statistical programming languages, R (Chapters 1 and 2). In particular, you can import data to R from multiple sources (Chapter 3) and manipulate foreign data into analysis-ready formats (Chapter 4). Seeking to discover population parameters, you understand the collection of unbiased samples (Chapter 5). You have a collection of exploratory data analysis tools to deploy to understand your sample data (Chapter 6). More recently, you learned about the theories of population distributions (Chapter 7), analyzed the relationships between variables (Chapter 8), and evaluated confidence intervals – projecting sample statistics back onto population parameters (Chapter 9). You even visually evaluated two samples for whether or not they came from the same population. Admittedly that evaluation was visual; it is time to refine that idea and make it rigorous.

While there will always be more techniques and skills to learn for statistical research/ data science, being able to decide if a sample likely comes from a specific population is the last major philosophical thought for this book. Being able to determine if a sample likely comes from a specific population allows you to answer a variety of questions. Does the new website drive a higher level of client engagement? Does "single-click" purchasing increase the dollar value of each purchase? Do penguins on Dream Island have the longest flippers? Did the new syllabus format improve student performance?

Answering these inferential statistics questions requires hypothesis testing, and by the end of this chapter, you should be able to

- Create a null hypothesis and determine the alternative hypothesis.
- Understand Type I and Type II errors.
- Apply the steps of null hypothesis significance testing (NHST).
- Evaluate the results of NHST.
- Create simple, English statements that explain the results of NHST to a relevant audience.

M. Wiley and J. F. Wiley, *Beginning R 4*, https://doi.org/10.1007/978-1-4842-6053-1_10

10.1 R Setup

As usual, to continue practicing creating and using projects, we start a new project for this chapter.

If necessary, review the steps in Chapter 1 to create a new project. After starting RStudio, on the upper-left menu ribbon, select *File* and click *New Project*. Choose *New Directory* ➤ *New Project*, with *Directory name* `ThisChapterTitle`, and select *Create Project*. To create an R script file, on the top ribbon, right under the word *File*, click the small icon with a plus sign on top of a blank bit of paper, and select the *R Script* menu option. Click the floppy disk–shaped *save* icon, name this file `PracticingToLearn_XX.R` (where XX is the number of this chapter), and click *Save*.

In the lower-right pane, you should see your project's two files, and right under the *Files* tab, click the button titled *New Folder*. In the *New Folder* pop-up, type `data` and select *OK*. In the lower-right pane, click your new *data* folder. Repeat the folder creation process, making a new folder titled `ch10`.

Remember all packages used in this chapter were already installed on your local computing machine in Chapter 2. There is no need to re-install. However, this is a new project, and we are running this set of code for the first time. Therefore, you need to run the following `library()` calls:

```
library(data.table)
```

```
## data.table 1.13.0 using 6 threads (see ?getDTthreads). Latest news:
r-datatable.com
```

```
library(ggplot2)
library(visreg)
library(palmerpenguins)
```

```
library(JWileymisc)
```

For this chapter, we continue our work with just the evening survey data from one day of the ACES daily dataset, so each person contributes just one observation. For no reason in particular, we selected March 03, 2017:

```
acesData <- as.data.table(aces_daily)[SurveyDay == "2017-03-03" &
SurveyInteger == 3]
```

Of course, we also use our familiar penguin and car data sets as well.

```
penguinsData <- as.data.table(penguins)
mtcarsData <- as.data.table(mtcars, keep.rownames = TRUE)
```

You are now free to learn more about statistics!

10.2 H0 vs. H1

A **hypothesis** is a claimed reason or fact that can be checked for truth (or at least for sensibility). In formal statistical testing, we traditionally start with the **null** hypothesis. The **null hypothesis** is often abbreviated H0 or H_0 which is sometimes pronounced "H Zero" or "H Nought." H_0 is the claim of no difference or no change. Consider the penguins data you have been exploring. An example of H_0 might be "The Adelie penguins on Biscoe Island have the same flipper length to Adelie penguins in general."

It is important that this "no difference" feature of H_0 be a *testable* claim. In the Adelie example, the claim is indeed measurable. Another feature of this H_0 claim is that it includes the sense of equality. In the Adelie example, this is the idea that the mean Adelie penguin flipper length on Biscoe Island is equal to the mean Adelie penguin flipper length anywhere.

If H_0 is the claim of no difference or equality, then there is only one other possibility. That other possibility is the outcome of some sort of change or difference or inequality and is called the **alternative hypothesis**. The alternative hypothesis is often abbreviated H1, H_1, or H_a. There are a few possibilities for H_1 in the Adelie example. If we know nothing about these penguin flipper lengths on various islands, we may simply go with the broadest H_1: "Adelie penguins on Biscoe Island have *different* flipper lengths than Adelie penguins in general." This would be an example of what is called a two-tailed test – the non-equality might be a left-tailed *less than* or a right-tailed *more than*.

On the other hand, maybe we happen to know Biscoe Island has a lot of food; thus, we may suspect those penguins ought to have longer flippers. In this case, we may want to make H_1 be "Adelie penguins on Biscoe Island have *longer* flipper lengths than Adelie penguins in general." This would be a scenario where H_1 was single tailed, right tailed in fact, and greater than.

Of course, if we instead knew that Biscoe had suffered a famine recently, we might suspect H_1 to be "Adelie penguins on Biscoe Island have *shorter* flipper lengths than Adelie penguins in general." This of course would lead H_1 to be single, left-tailed, and a less than scenario.

The key feature of H_0 vs. H_1 is they are complements of each other; deciding H_0 determines what H_1 will be.

Example

Recall in Chapter 8 we did a correlation between the miles per gallon (mpg) and the horsepower (hp) from the `mtcarsData`. At the time, we were simply interested in the negative correlation of that data. A more rigorous question might have included phrasing that exploration in terms of null and alternative hypotheses. As an example of a research question using those data:

H_0: The correlation between mpg and hp is equal to 0. In other words, there is neither positive nor negative correlation between mpg and hp.

From there, based on our limited knowledge of physics, we might wonder if more horsepower leads to less efficient engines. Thus, we might develop an alternative hypothesis that does not include equality – and in particular expect that the correlation ought to be negative. This would be an example of a left-tailed or "less than" H_1.

H_1: The correlation between mpg and hp is less than 0. In other words, there is evidence of a negative correlation between mpg and hp.

Example

Recall in Chapter 8 we did a correlation between `STRESS` and `NegAff` from the `acesData`. At the time, we were simply interested in the positive correlation of that data. A more rigorous question might have included phrasing that exploration in terms of null and alternative hypotheses. As an example of a research question using those data:

H_0: The correlation between `STRESS` and `NegAff` is equal to 0. In other words, there is neither positive nor negative correlation between `STRESS` and `NegAff`

From there, based on our knowledge of stress, we might wonder if more stress leads to more negative emotions (e.g., negative affect). Thus, we might develop an alternative hypothesis that does not include equality – and in particular expect that the correlation ought to be positive. This would be an example of a right-tailed or "greater than" H_1.

H_1: The correlation between STRESS and NegAff is greater than 0. In other words, there is evidence of a positive correlation between STRESS and NegAff.

10.3 Type I/II Errors

We must now discuss some more probability. The null and alternative hypotheses are – to use the probability phrase you learned in Chapter 7 – mutually exclusive. As you learned in the preceding section of this chapter, they are built that way on purpose.

Presumably, we are getting ready to share how to use statistics to decide whether H_0 or H_1 is more likely. However, recall from Chapter 9 the example of the 95% confidence level with 100 samples from the "perfect" distribution that ended up having seven of the confidence intervals miss the mark instead of the predicted five. Any finite set of real-world examples are never guaranteed to match the probability. A single flip of a coin is not half and half (outside of a truly extraordinary landing-on-edge rare event). Instead, one flip of a coin will give 100% heads or tails. Thus, when we have two possible scenarios, H_0 or H_1, we must be prepared for the best decision based on probability to sometimes be wrong.

It helps then to think through how things can go wrong. That way, we can understand how likely we are to be wrong and at least make sure that, if we are wrong, we are wrong in the "best" way.

Now, the English will seem a trifle clunky. In the statistical method you are about to learn, everything is framed on H_0. That is, the two possible options are you will (based on the computed probability) either reject H_0 or not reject H_0. Regardless of what the probability says is *likely*, in real, actual fact, either H_0 is true or H_1 is true.

Ideally, when you reject H_0, then H_1 will turn out to have actually been true. Similarly, you hope that when you do not reject H_0, H_0 turns out to be true.

However, sometimes, you will, based on the statistical probability, reject H_0, and in real life, it will turn out that H_0 was, in fact, the true choice. This is called a **Type I error** which occurs when, based on the data, you reject H_0 and yet H_0 was the actual truth.

Other times, you will, based on the statistical probability, not reject H_0, and in real life, it will turn out that H_1 was in fact the true choice. This is called a **Type II error** which occurs when, based on the data, you do not reject H_0 and yet H_1 was the actual truth.

Table 10-1 shows the relationship between rejecting/not rejecting H_0 and whether H_0 is true. It may be worth noting that I and II are Roman numerals for 1 and 2. It is also worth noting that the decision we are preparing to make (belief in either H_0 or H_1) is based on a "playing" the odds. Thus, in formal statistical language, the phrasing is always cautious. One never says "We are going with H_1" or "We choose H_0." Rather, one only ever rejects H_0 or does not reject H_0 – whichever is more probable.

***Table 10-1.** Type I and II Errors*

	H_0 Is Actually True	H_0 Is Actually False
Reject H_0	Type I error	Statistical probability Matched real-life outcome
Not reject H_0	Statistical probability Matched real-life outcome	Type II error

Example

Going back to the Adelie penguins on Biscoe Island, let us lock down our null and alternative hypotheses and consider what the *consequences* of a Type I or Type II error might be.

While this is not quite right, suppose for the sake of this example that the *population* mean of Adelie penguin flipper lengths is 190 mm:

```
summary(penguinsData[species == "Adelie"]$flipper_length_mm)
```

```
##    Min. 1st Qu. Median    Mean 3rd Qu.    Max.    NA's
##     172     186    190     190     195     210       1
```

Then, our formal null and alternative hypotheses for a two-tailed test about Biscoe Island might be as follows:

H_0: Adelie penguins on Biscoe Island have a mean flipper length statistically **equal** to Adelie penguins in general, 190 mm.

H_1: Adelie penguins on Biscoe Island have a mean flipper length statistically **not equal** to Adelie penguins in general, 190 mm.

The statistical test you will soon see will either tell you to reject H_0 or not reject H_0.

Suppose NHST tells you to reject H_0. Thus, you believe Adelies on Biscoe Island are *different* from other Adelies. You write a grant, get funding, and spend a year on Biscoe Island in the (presumable) cold. If the test told the truth, you will discover the difference (and hopefully what causes it). Your name goes down forever in the halls of penguinology. On the other hand, what if the test led you wrong? This is a Type I error; you rejected H_0, you believed H_1, and yet actually those Biscoe Adelies are the same as every other Adelie. You suffered a year of cold for nothing.

Suppose NHST tells you to not reject *H*0. Thus, you believe Adelies on Biscoe Island are *not* different from other Adelies. There is no point in researching this, and you never write that grant. Perhaps you decide all possible penguin research has been completed. You quit marine biology, go back to school to become a chef, and open a successful sandwich shop. On the other hand, what if the test led you wrong? This is a Type II error; you did not reject H_0, you believed H_0, and yet actually those Biscoe Adelies were different. You could have climbed to the height of penguinology and opened new avenues of research, and, while successful, you are still making sandwiches for the people who still study penguins.

Which error is worse? As you can see from our fanciful story, the answer is "it depends." Indeed, each combination of H_0 and H_1 can yield different consequences for Type I or Type II errors. In the next section, we will discuss some ways to control to think about the probability of those errors. For now, we continue with our other two examples.

Example

In the mtcarsData example, we settled on the following, left-tailed or "less than" research questions.

H_0: The correlation between mpg and hp is equal to 0 (namely, no correlation).

H_1: The correlation between mpg and hp is less than 0 (namely, negative correlation).

Suppose NHST tells you to reject H_0. Thus, you believe H_1 is correct - that as mpg increases, horsepower decreases. If H_1 is not correct, you are making a Type I error, which may lead you to buy a low-mpg car in the hopes of getting more horsepower for a faster ride.

Suppose NHST tells you to not reject H_0. Thus you believe H_0 is correct - that as mpg increases, there is no change in horsepower. If H_0 is not correct, you made a Type II error. You may find yourself trying to race a hybrid car at your local race track.

Example

In the `acesData` example, we settled on the following, right-tailed or "greater than" research questions.

H_0: The correlation between `STRESS` and `NegAff` is equal to 0.

*H*1: The correlation between `STRESS` and `NegAff` is greater than 0; an increase in stress results in an increase in negative mood.

Suppose NHST tells you to reject H_0. Thus, you believe H_1 is correct – that as stress increases, negative mood increases. If H_1 is not correct, you made a Type I error. Your new stress relief ball will not help your bad mood after your boss yells at you.

Suppose NHST tells you to not reject H_0. Thus, you believe H_0 is correct – that as stress increases, there is no impact on negative mood. If H_0 is not correct, you made a Type II error. One day your boss yells at you, and you think there is nothing to be done. Instead, you might have bought a stress relief ball.

10.4 Alpha and Beta

In the preceding examples, you may have found either Type I or Type II errors more concerning depending on the scenario. Of course, this is not to say that errors are ever what we want. Instead, you may find that one type of error might be worse than the other. An example of this might be in deciding whether to deploy a medical treatment. If H_0 is "The patient is not sick" and H_1 is "The patient is sick; we must do surgery!" there is a cost to both Type I errors (surgery on a healthy patient) and Type II errors (letting a sick patient go without surgery).

How do we understand and set the risk level for each type of error?

Recall Chapter 9 stated $\alpha = 1 - CL$; this same α is the probability of making a Type I error. In the "perfect" example from Chapter 9, suppose we had a random sample like one of those seven confidence *intervals* which did not intersect the known population mean parameter. If we did the testing we are about to discuss using one of those seven samples (as might happen in real life), it would lead us into a Type I error. Increasing the confidence level, which also increases the confidence interval, *decreases* α.

So why do we ever use anything other than a 99% confidence level? After all, that would limit us to $\alpha = 0.01$ which is only 1%. Well, recall a Type I error occurs after rejecting H_0. If we go with a superwide confidence interval, we will almost never reject H_0. Because you must choose, in the end, either H_0 or H_1, by default, if you never reject

H_0, then you live perpetually in a world of possible Type II error. Thus, it is not practical to simply avoid Type I errors altogether. As H_0 always includes equality, it can often be translated into English as "there is no difference between our current scenario and business as usual." Thus, a high confidence level can easily turn into inaction or paralyzed decision making.

This is the artistic side of statistics, thinking through the possible error risk, and the consequences of making such errors, and deciding acceptable risk levels for each. This turns our attention to Type II errors and those probabilities.

The chance of a Type II error is called β (Greek letter pronounced "beta"). The complement of β is called the "power of the test." Without knowing the true population standard deviation, it is impossible to compute β exactly. Instead, we often use estimates. Just like a 95% confidence interval is considered standard across many disciplines (mostly because of tradition - after all, 96% would not be all that different really), so too a power of the test set to 80% is typical. Thus, as $\beta = 1 - PoweroftheTest$, 20% β is traditional.

The power of the test, and thus β, is also in part determined by how intense the effect is. Ergo, if the Biscoe Adelies have double the flipper length of other Adelies, that ought to be easy to spot. However, if the difference is only 1 mm or so, that will require more penguins in the sample to confirm. For estimate purposes, one traditionally uses **Cohen's d** [8] which gives a value for what we might think effect level would be. Table 10-2 contains the estimated effect level and the number one uses for each.

Table 10-2. *Cohen's d*

Effect Size	*d*
Small	0.2
Medium	0.5
Large	0.8

To use this power of the test, R has a function named `power.t.test()`. It takes seven arguments. If you enter data for any of the first four of those arguments, it will estimate the remaining fifth. Most often, we use this idea of power to estimate the sample size n required to get the power of the test we want. In other words, for a given level of β we are willing to accept, we use this function to calculate the required sample size to keep our Type II error restricted to some level.

The argument n is for sample size; this is most often set to NULL to allow the function to give us the estimated sample size required.

Set the delta argument to the correct value for the effect level you believe you will find based on Table 10-2. Remember, when choosing the alternative hypothesis, H_1, you could choose two tailed/not equal, left tailed/less than, or right tailed/greater than. While the default when you do not know the correct choice is two tailed, in some cases (e.g., mpg or stress), you might already reasonably suspect less than or greater than. So too, with Cohen's d, you will suspect that the effect you are observing or the treatment you are deploying will either have a small, medium, or large effect.

When using this function in this fashion, the standard deviation is set to the standard normal curve of sd = 1. The mathematics behind changing this are more suited to a statistical theory course than a statistical methods book, so we leave this set to 1 for now.

The sig.level is the α that is the complement of our confidence level. You can of course choose a CL other than 95% if you desire. Similarly, by choosing β, that determines the power of the test.

The last two arguments let the function know what type of test we are doing. The type = argument takes either one.sample for the type of data we work on in this chapter or two.sample if you are comparing two samples to each other. The alternative = argument cares if you are using a two-tailed test two.sided or one of the left-/right-tailed tests one.sided.

Most commonly, this type of power test is used to estimate sample size rather than estimate the power of a test. For example, in the case of the penguin, mtcars, or aces data sets, there is little point, now, in estimating the power of those tests. The error will either occur or not, and we must use the data already collected. Additionally, there is minimal value in estimating sample size for those – that data collection is over.

However, suppose you were getting ready to start a new research project. You decide to go with the usual CL of 95%. Additionally, you choose the usual beta of 20%. Lastly, because you are making a decision about a moderately expensive project that you intend to have a medium effect level, you choose Cohen's d to be 0.5. You know you only have time to collect one sample, and your intervention is untested, so you go with a two-sided H_1 to be safe:

```
confidenceLevel <- 0.95
alpha <- 1 - confidenceLevel

beta <- 0.20
powerOfTest <- 1 - beta
```

```
#here we choose medium effect level.
cohensD <- 0.5

power.t.test(n = NULL ,
             delta = cohensD,
             sd = 1,
             sig.level = alpha,
             power = powerOfTest,
             type = "one.sample",
             alternative = "two.sided")

##
##      One-sample t test power calculation
##
##               n = 33.37
##           delta = 0.5
##              sd = 1
##       sig.level = 0.05
##           power = 0.8
##     alternative = two.sided
```

As you see, the result is that you would need to have at least 34 people in your sample to be confident. That is not a lot of people; a medium effect size is expected to have such clear results we do not need to look too far to notice what is happening.

A small effect size, with all else equal, would require almost 200 people in the study:

```
#here we choose small effect level.
cohensD <- 0.2

power.t.test(n = NULL ,
             delta = cohensD,
             sd = 1,
             sig.level = alpha,
             power = powerOfTest,
             type = "one.sample",
             alternative = "two.sided")
```

```
##
##      One-sample t test power calculation
##
##               n = 198.2
##           delta = 0.2
##              sd = 1
##       sig.level = 0.05
##           power = 0.8
##     alternative = two.sided
```

You have your null and alternative hypotheses, you know the types of errors you might make, and you have even estimated the sample size you must collect to keep those errors to a minimal level. What assumptions or biases might we be facing before we actually complete a null hypothesis significance test?

10.5 Assumptions

You have your null and alternative hypotheses. You selected a confidence level, which determined α – your risk of a Type I error. You even estimated the sample size you must collect to keep your Type II error to a comfortable level.

What else is required for a **null hypothesis significance test** (NHST)? As has been the case for many statistical tests, there are some assumptions about the data set.

The sample ought to be from a normally distributed population; this assumption is entirely due to this beginning text using normal distributions. There are highly similar methods that can be used for other distributions. Remember normality can be estimated using some of the techniques from Chapter 7.

The elements of the sample ought to be independent of each other. This is often achieved via a random sample.

10.6 Null Hypothesis Significance Testing (NHST)

You are now ready to conduct a **null hypothesis significance test**, often abbreviated NHST. Philosophically, when we talk about the null hypothesis H_0 being "no different," what we are really asking is "How likely is it that our sample came from the expected population?" In other words, we are asking if our sample statistic's confidence interval contains the population parameter.

Thus, when you choose a confidence level, which determines α, you are setting the benchmark by which you will measure the probability of a given sample coming from the extreme tails of the population's normal distribution.

Remember, for any given population, if you take a sample, the central limit theorem (CLT – mentioned in Chapter 7) tells us it is "unlikely" the sample will be "too extreme." However, random does mean random. There exists a random sample of 100 people in the world where the sample is all billionaires. Is that likely? Of course not! Randomly calling 100 people and getting all billionaires would be an extremely rare event – even more rare than you and your two friends flipping a fair coin three times in a row and getting TTT each game (review the faire coin example in Chapter 7).

The **p-value** is the sample-estimated probability of getting your sample from your H_0 population. If that p-value is small, then it is not likely that your sample was pulled from the H_0 population. Thus, you reject H_0. On the other hand, if the p-value is large, then it is likely your sample came from the H_0 population. Thus, you do not reject H_0.

The p-value is not the chance that H_0 or H_1 is correct! The p-value simply tells us how likely it would be to randomly capture a sample from the H_0 population that has our sample's statistic (or a statistic further from the parameter than our statistic). Remember the statistic is the sample measure (e.g., mean or correlation) and the parameter is the H_0 population's measure.

How small is small enough to reject H_0? Smaller than α. That was the purpose of choosing a confidence level to begin with it turns out. If the p-value is smaller than alpha, your risk of a Type I error will be no larger than alpha.

The steps for NHST are listed for your reference:

1. Determine your research question. It should be clear and measurable.

2. From the research question, build a null hypothesis H_0 which includes equality in the measure. From H_0, build the alternative hypothesis H_1 which is either not equal, strictly less than, or strictly greater than your measure from H_0.

3. Select a confidence level (most often 95%) which determines α. This α is your Type I error. Thus, your choice of α is in part informed by your tolerance for committing a Type I error.

4. If your research is to use newly collected data, based on your selection of β (Type II error) and your estimate of effect level (Cohen's d), determine the sample size required using `power.t.test()`.

5. Confirm that your sample, whether newly collected or historical data, is random and independent and from a normal enough population.

6. Compare the calculated p-value to α. For $p < \alpha$, reject H_0. For $p >= \alpha$, do not reject H_0.

7. Phrase your choice to reject or not reject H_0 in simple English so a nonstatistical audience can understand. Explain which of Type I or Type II error is at risk, and (ideally) state some consequences of that risk.

You now have all the background needed to conduct NHST.

Example

1: Research Question

The Adelie penguins live on three islands and, overall, seem to have a mean flipper length of 190 mm. Perhaps due to prior research, perhaps a psychic guess, you wonder if the Biscoe Island Adelies are different. Before you write a grant and travel out to Biscoe Island yourself for a year as a penguinologist, you decide to do some background research:

```
penguinsData[species == "Adelie",
             .N,
             by = island]

##       island  N
## 1: Torgersen 52
## 2:    Biscoe 44
## 3:     Dream 56

mean(penguinsData[species == "Adelie"]$flipper_length_mm, na.rm = TRUE)

## [1] 190
```

2: H_0 and H_1

From this research question came your null hypothesis, H_0, and your alternative hypothesis, H_1.

H_0: Adelie penguins on Biscoe Island have a mean flipper length statistically **equal** to Adelie penguins in general, 190 mm.

H_1: Adelie penguins on Biscoe Island have a mean flipper length statistically **not equal** to Adelie penguins in general, 190 mm.

3: Select α

You decide to use a standard confidence level of 95%, which you know also determines $\alpha = 0.05$. You know that a Type I error may occur if you reject H_0. The consequences of a Type I error include spending a year on a (presumably) cold island only to find the Biscoe Adelies are like all the other Adelies.

4: Consider β

This research is depending on data already collected; there is no need to consider `power.t.test()`. The odds of a Type II error are outside your ability to control in this case. Of course, you might go ahead and calculate how many penguins you will sample once you are there to confirm your results. However, that is something to do at another time.

5: Assumption Checks

Based on the `palmerpenguins` [11] research, you believe the data was properly collected. There may be some limitations on the purely random nature of the penguin sampling (the article does mention weather conditions preventing access to all selected nests). All the same, the data ought to be mostly random and mostly independent. Prior analysis suggests flipper length is mostly normal.

6: Compute and Compare p to α

Using the `R` function `t.test()`, you conduct NHST. You have one data set, so you set the first data argument `x =` to the flipper lengths of the Adelie penguins at Biscoe Island. Your alternative hypothesis, H_1, is "not equal" which is `two.sided`. In this case, you suppose that the $\overline{x} = 190$ for the entire Adelie species in the sample might be close enough to the population parameter to risk as a stand-in for `mu =`. Lastly, you already chose 95% as the `conf.level =`:

```
t.test(x = penguinsData[species == "Adelie" & island == "Biscoe"]$flipper_
length_mm,
       alternative = "two.sided",
       mu = 190,
```

```
        conf.level = 0.95)

##
##      One Sample t-test
##
## data:  penguinsData[species == "Adelie" & island == "Biscoe"]$flipper_
   length_mm
## t = -1.2, df = 43, p-value = 0.2
## alternative hypothesis: true mean is not equal to 190
## 95 percent confidence interval:
##  186.7 190.8
## sample estimates:
## mean of x
##     188.8
```

Reading through the output of the t-test, you see several things that make sense. However, you focus on the end of the second line `p-value = 0.2`. The p-value is greater than alpha (which was 0.05 in this case). Thus, you do *not reject* H_0. Of course, you are savvy enough to know there is a chance you are making a Type II error. However, you are also experienced enough to know that chance does not change the results of the test. If the consequences of a Type II error are large, you will simply need to conduct your own research test.

7: Communicate Results

While you are experienced enough at statistics to know what this means, you must translate for your family and friends who are excited to see you travel to Biscoe Island. You might tell them

"To be worth spending a year in the field, we needed make sure our sample mean of 188.8 (1.2 mm shorter than the average Adelie) was statistically significant. Using a t-test analysis, we found there was a 20% chance a sample at least as extreme as our data set came from the known population of Adelie penguins. That is likely enough to happen that we have no reason to believe the Biscoe Island Adelies are different from regular Adelie penguins. This means we will not be going to Biscoe Island. There is a chance our analysis led us to make the wrong choice, in which case we are passing up on a fascinating trip."

It is a skill all of its own to successfully communicate the results of statistical analytics to stakeholders. In some cases, that stakeholder may be you. In others, it may be experienced research types, and you can mention specific tests by name. In many cases, however, your stakeholders are experienced in their fields of expertise (e.g., a CEO or a board member) yet need your area of statistical expertise explained.

Example

1: Research Question

Miles per gallon of petrol (gasoline) and horsepower are suspected to be correlated. In fact, because it makes physical sense that increases in horsepower might create a less efficient car, one suspects a negative correlation. In a classic example of correlation is not causation, Figure 10-1 is an example that just because mpg increases and hp decreases, it does not make sense that mpg increase all by itself must decrease hp. In fact, it is likely the other way around:

```
ggplot(data = mtcarsData,
       mapping = aes(x = mpg,
                     y = hp)) +
  geom_point()
```

In any case, while you have seen in Chapter 8 that these appear to be correlated, the question becomes, “Is the correlation statistically significant?” We can test this by creating null and resulting alternative hypotheses.

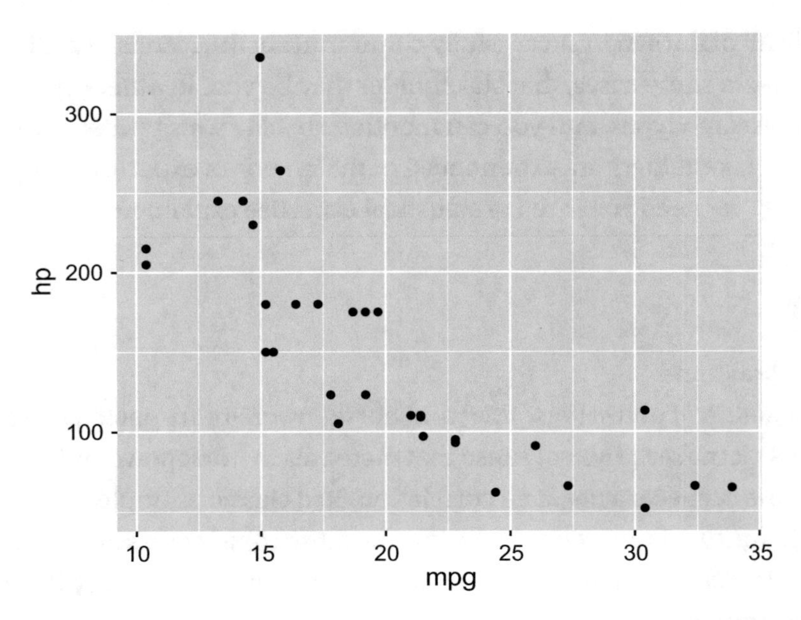

***Figure 10-1.** Scatter plot of miles per gallon (mpg) and horsepower (hp) in the mtcars dataset*

2: H_0 and H_1

H_0: The correlation between mpg and hp is equal to 0 (namely, no correlation).

H_1: The correlation between mpg and hp is less than 0 (namely, negative correlation).

3: Select α

While it may be sacrilegious, car mpg vs. hp is not high stakes. It may make sense to relax the confidence level to 90%. This naturally increases alpha to 0.10.

4: Consider β

You are using data already collected from `mtcars` [1] which is from 1974 (you know these data are dated). There is no need to perform a `power.t.test()`.

5: Assumption Checks

You do not have a copy of that magazine and will simply hope they conducted a random sample of cars. Most likely this is not true. While technically you know you should stop here, you wish to practice your newfound skills and continue.

6: Compute and Compare p to α

Unlike the prior example, this is a *correlation*. Therefore, you know to use the `cor.test()` function from `R`. You set the first data value, `x =`, to mpg and the second data value, `y =`, to hp. In this case, H_1, your alternative hypothesis, was *less than*. Thus, you set `alternative = "less"` rather than `"two.sided"` or `"greater"`. From Chapter 8, you know the connection was consistent and thus met the assumptions for Pearson's correlation. Thus, you set `method = "pearson"` rather than `"kendall"` or `"spearman"`. Lastly, you decided that cars from 1974 were for fun rather than strenuous; you set `conf.level = 0.90`:

```
cor.test(x = mtcarsData$mpg,
         y = mtcarsData$hp,
         alternative = "less",
         method = "pearson",
         conf.level = 0.90)

##
##      Pearson's product-moment correlation
##
## data:  mtcarsData$mpg and mtcarsData$hp
## t = -6.7, df = 30, p-value = 9e-08

## alternative hypothesis: true correlation is less than 0
## 90 percent confidence interval:
##  -1.0000 -0.6627
## sample estimates:
##     cor
## -0.7762
```

Reading through the output of the Student's t-distribution test, you see several things that make sense. However, you focus on the end of the second line `p-value = 9e-08`. You know this is scientific notation for eight decimal places to the left including the leading 9, which is `p-value = 0.00000009`. Thus, the p-value is definitely smaller than $\alpha = 0.10$. Therefore, you reject H_0. Of course, you are savvy enough to know there is a chance you are making a Type I error. However, you are also experienced enough to know that chance does not change the results of the test. If the consequences of a Type I error are bad in the real world (likely not for 1974 cars), you will simply need to conduct your own research with a larger sample size (and ensure a proper random sample).

7: Communicate Results

All the same, you are ready to opine to your friends about cars and are confident to tell them that after a statistical test, miles per gallon appears to be negatively correlated with horsepower. While there is a chance the correlation is nonexistent between mpg and hp, for a cordial chat among friends after a week's work, the consequences of being wrong seem slight.

Example

1: Research Question

Stress and negative affect are suspected to be correlated. In fact, because it makes psychological sense that increases in stress might cause negative emotions, one suspects a positive correlation.

2: H_0 and H_1

H_0: The correlation between `STRESS` and `NegAff` is equal to 0.

H_1: The correlation between `STRESS` and `NegAff` is greater than 0; an increase in stress results in an increase in negative mood as seen in Figure 10-2 and generated by the following code:

```
ggplot(data = acesData,
       aes(x = STRESS,
           y = NegAff)) +
  geom_point()

## Warning: Removed 2 rows containing missing values (geom_point).
```

3: Select α

Generally, for psychology research, the confidence level is 95%, and thus the alpha is 0.05.

4: Consider β

You are using data already collected from `acesData`; there is no need to perform a `power.t.test()`.

5: Assumption Checks

These data are simulated in `JWileymisc` [21] from a proper study conducted in Melbourne, Australia [24]. The individuals in the study are independent, and you can count them as such.

6: Compute and Compare p to α

As you are testing correlation, you know to use the `cor.test()` function from R. You set the first argument to `x = STRESS` and the second argument to `y = NegAff`. In this case, H_1, your alternative hypothesis, was *greater than*. Thus, you set `alternative = "greater"` rather than `"two.sided"` or `"less"`. From Chapter 8, you know the connection was somewhat iffy and thus a non-parametric option such as Kendall's was best for correlation. Thus, you set `method = "kendall"` rather than `"pearson"`. As mentioned, you set `conf.level = 0.95`:

```
acesData[, cor.test(x = STRESS,
                    y = NegAff,
                    alternative = "greater",
                    method = "kendall",
                    conf.level = 0.95
                    )]
```

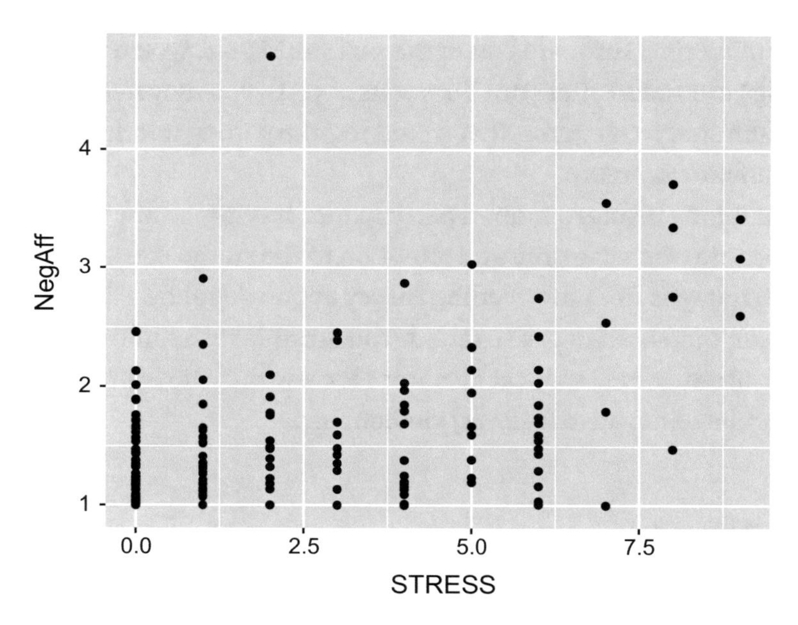

***Figure 10-2.** Scatter plot of stress and negative affect (mood) in the ACES daily dataset*

```
##
##      Kendall's rank correlation tau
##
## data:  STRESS and NegAff
## z = 6.1, p-value = 4e-10
## alternative hypothesis: true tau is greater than 0
## sample estimates:
##    tau
## 0.3342
```

Reading through the output of the Student's t-distribution test, you see several things that make sense. However, you focus on the end of the second line `p-value = 4e-10`. You know this is scientific notation for ten decimal places to the left, which is `p-value = 0.0000000004` (ten decimal places total including the leading 4). Thus, the p-value is definitely smaller than $\alpha = 0.05$. Therefore, you reject H_0. Of course, you are savvy enough to know there is a chance you are making a Type I error. However, you are also experienced enough to know that chance does not change the results of the test. If the consequences of a Type I error are bad in the real world (e.g., worse than a stress relief ball not helping your mood after your boss yells at you), you may need to increase the sample size of the study. Of course, that would require getting another grant.

7: Communicate Results

Because this study involved human participants, it is important to share out the results to the people who gave time and attention to the study. You are able to share that a statistical analysis from the evening survey of one day shows that increased stress correlates with an increased negative mood. You are able to include some information on coping with stress, as well as local resources for anyone who is finding coping with stress (and the related negative feelings) challenging.

10.7 Summary

This chapter explored the concepts and steps of null hypothesis significance testing (NHST). Table 10-3 should be a good reference of the key ideas you learned in this chapter.

Table 10-3. *Chapter Summary*

Idea	What It Means
Hypothesis	A measurable claim.
Null hypothesis (H_0)	Sample is not different from the population.
Alternative hypothesis (H_1)	Sample is different from the population.
Type I error	H_0 true, yet due to $p - value < \alpha$, rejected H_0.
Type II error	H_0 false, yet due to $p - value >= \alpha$, did not reject H_0.
Alpha	If H_0 is true, probability of making a Type I error.

(*continued*)

Table 10-3. *(continued)*

Idea	What It Means
Beta	If H_0 is false, probability of making a Type II error.
Power of the test Cohen's d	A set of traditional values for small, medium, and large effects .
`power.t.test()`	Using Cohen's d, can estimate required sample size.
NHST	Null hypothesis significance test
p-value	The probability of a sample like ours (or more extreme) from the H_0 population.
`t.test()`	Calculates p-value for a sample $\overline{x}$ vs. a population μ.
`cor.test()`	Calculates p-value for correlation of two samples.

10.8 Practice for Mastery

Check your progress and grow through practice by working through some exercises. Comprehension checks ask critical thinking questions that may be best answered with a written or verbal response. Part of the art of statistics is successfully communicating results to your stakeholders or audience. Sometimes that audience is highly technical and other times very much not technical. Exercises are more direct applications of the concepts explored in the chapter.

Comprehension Checks

1. Does a p-value less than α *prove* the alternative hypothesis is true?
2. In the example about the Adelie penguins of Biscoe Island, we estimated the population mean at 190 mm and the sample mean on Biscoe Island at 188.8 mm. Now, in this case, the p-value was 0.20. However, suppose the p-value was 0.02 instead. That would be enough to suppose there was a *statistically* significant difference in flipper lengths. Does statistical significance always translate into real-world significance? In other words, does 1.2 mm make an appreciable difference? Why or why not?

Exercises

1. Explore the Adelie penguins of Biscoe Island some more. This time, instead of flipper lengths, consider body mass. Perhaps there is more food or less food on Biscoe Island? Work your way through all the steps of NHST.
2. Explore the `mtcars` data set more. Swap out horsepower for weight (`wt`). What are your findings? Are you understanding the steps of NHST better?
3. You studied `STRESS` and `NegAff`. Rework the data set for `STRESS` and `PosAff`. Do you need to change H_0? What are the results of NHST?

CHAPTER 11

Multiple Regression

The last three chapters, this one included, are admittedly complicated. If this book is your first learning of both statistics and R, it may be wise to take a moment here to look back on what you have already learned. If there is a chapter that you know was challenging, this is an opportunity to reread that chapter. This all may sound a bit ominous; do not be overly worried! In every learner's life, there comes a time when you have learned so much that it is time to climb up to the next level. You have now arrived at your time to climb.

When you first met regression in Chapter 8, you learned it created a relationship between an x value input/predictor and a y value output/response. In that scenario, there were only two values in the relationship. Adding in more inputs/predictors can create a more complete output/response. However, more values in a relationship complicate things.

By the end of this chapter, you should be able to

- Understand linear regression better by more deeply exploring the theory.
- Apply the background/theory principles of linear regression to achieve multiple regression.
- Create and fit a multiple regression model on several inputs/predictors.
- Evaluate nonnumeric or categorical predictors.

M. Wiley and J. F. Wiley, *Beginning R 4*, https://doi.org/10.1007/978-1-4842-6053-1_11

11.1 R Setup

As usual, to continue practicing creating and using projects, we start a new project for this chapter.

If necessary, review the steps in Chapter 1 to create a new project. After starting RStudio, on the upper-left menu ribbon, select *File* and click *New Project*. Choose *New Directory* ➤ *New Project*, with *Directory name* `ThisChapterTitle`, and select *Create Project*. To create an `R` script file, on the top ribbon, right under the word *File*, click the small icon with a plus sign on top of a blank bit of paper, and select the *R Script* menu option. Click the floppy disk–shaped *save* icon, name this file `PracticingToLearn_XX.R` (where XX is the number of this chapter), and click *Save*.

In the lower-right pane, you should see your project's two files, and right under the *Files* tab, click the button titled *New Folder*. In the *New Folder* pop-up, type `data` and select *OK*. In the lower-right pane, click your new *data* folder. Repeat the folder creation process, making a new folder titled `ch11`.

Remember all packages used in this chapter were already installed on your local computing machine in Chapter 2. There is no need to re-install. However, this is a new project, and we are running this set of code for the first time. Therefore, you need to run the following `library()` calls:

```
library(data.table)
```

```
## data.table 1.13.0 using 6 threads (see ?getDTthreads). Latest news:
r-datatable.com
```

```
library(ggplot2)
library(palmerpenguins)
library(visreg)
library(emmeans)

library(JWileymisc)
```

For this chapter, we continue our work with just the evening survey data from one day of the ACES daily dataset, so each person contributes just one observation. For no reason in particular, we selected March 03, 2017:

```
acesData <- as.data.table(aces_daily)[SurveyDay == "2017-03-03" &
SurveyInteger == 3]
```

While this was not relevant earlier in the book, the STRESS variable in this data set ranges from 0 to 9.

```
range(acesData$STRESS, na.rm = TRUE)
```

```
## [1] 0 9
```

A stress level above a 2 is of interest in some research (and in some models we will do in this chapter). Thus, we **recode** the data to just three levels of 0, 1, and 2 or higher (coded as 2+). Often, real-world data requires this type of pre-analysis work or **recoding** to be "analysis-ready" data.

One way to do this is via `factor()`. You already met factors in this book; the penguin species are an example of factor data (see Chapter 5). A **factor** simply puts a numeric value behind the scenes in R on character or *ordinal* data. In this case, our goal is to collapse the STRESS values that are 2 or higher down to a single category named 2+. This is a column operation and belongs in the jth column operation position. If the STRESS values are greater than or equal to 2, we want that to be stored as a 2+ and given an ordinal factor value behind the scenes. Otherwise, if the values are less than 2 (e.g., 0, 1, or NA), we wish them to stay as they are. This is achieved via the fast if-else function, `fifelse()`. The results of the fast if-else are wrapped in the `factor()` function and assigned to a new column named StressCat in a single operation:

```
acesData[, StressCat := factor(fifelse(STRESS >= 2,
                                       "2+",
                                       as.character(STRESS)))]
```

By using the structure function `str()`, we show what these three columns look like. Notice the `Factor w/ 3 levels` on the new column, StressCat:

```
str(acesData[, .(UserID, STRESS, StressCat)])
```

```
## Classes 'data.table' and 'data.frame':     189 obs. Of 3 variables:
##  $ UserID   : int 1 2 3 4 5 6 7 8 9 10 ...
##  $ STRESS   : num 2 1 4 4 0 0 0 4 0 1 ...
##  $ StressCat: Factor w/ 3 levels "0","1","2+": 3 2 3 3 1 1 1 3 1 2 ...
##  - attr(*, ".internal.selfref")=<externalptr>
```

Using unique(), we dive deeper into the contents of this new column:

```
unique(acesData$StressCat)
```

```
## [1] 2+   1    0    <NA>
## Levels: 0 1 2+
```

We also use the familiar penguin data set; we include the structure again so you can observe the factors:

```
penguinsData <- as.data.table(penguins)
str(penguinsData)
```

```
## Classes 'data.table' and 'data.frame':      344 obs. of 8 variables:
##  $ species          : Factor w/ 3 levels "Adelie","Chinstrap",..: 1 1 1 1
1 1 1 1 1 1 ...
##  $ island           : Factor w/ 3 levels "Biscoe","Dream",..: 3 3 3 3 3
3 3 3 3 3 ...
##  $ bill_length_mm   : num 39.1 39.5 40.3 NA 36.7 39.3 38.9 39.2 34.1 42
...
##  $ bill_depth_mm    : num 18.7 17.4 18 NA 19.3 20.6 17.8 19.6 18.1 20.2
...
##  $ flipper_length_mm: int 181 186 195 NA 193 190 181 195 193 190 ...
##  $ body_mass_g      : int 3750 3800 3250 NA 3450 3650 3625 4675 3475
4250 ...
##  $ sex              : Factor w/ 2 levels "female","male": 2 1 1 NA 1 2 1
2 NA NA ...
##  $ year             : int 2007 2007 2007 2007 2007 2007 2007 2007 2007
2007 ...
##  - attr(*, ".internal.selfref")=<externalptr>
```

11.2 Linear Regression Redux

Previously we learned about linear regression with a single predictor variable, yet at the time had not yet learned about confidence intervals and hypothesis testing. Now that those topics are familiar, we can refresh and expand our coverage of linear regressions adding confidence intervals and hypothesis testing.

Linear regression in R is generally conducted using the `lm()` function, which fits a **l**inear **m**odel. It uses a formula interface to specify the desired model, with the format *outcome predictor*. We can store the results of the linear regression in an object, `m`, and then call various functions on this to get further information. For example, we can use the `summary()` to get a quick summary including hypothesis tests for the regression coefficients and the overall model. We can also use the `confint()` function to get confidence intervals for the regression coefficients.

To begin with, we revisit a simple linear regression and look at the results from the `summary()` function. When we use `summary()`, we get quite a bit of output. The headings are as follows:

- **Call**: This is the code we used to call R to produce the model; it is a handy reminder of the variables and outcome especially if you are running and saving many models.
- **Residuals**: This section provides some descriptive statistics on the residuals including the minimum (0th percentile), first quartile (25th percentile), median (second quartile, 50th percentile), third quartile (75th percentile), maximum (100th percentile).
- **Coefficients**: This is really the main table normally reported for regressions. The column "Estimate" gives the model parameter estimates or regression coefficients (the *bs*). The column "Std. Error" gives the standard error (SE) for each coefficient, a measure of uncertainty. The column "t value" gives the t-value for each parameter, used to calculate the p-values and defined as $\frac{b}{SE}$. Finally, the column "$Pr(> |t|)$" gives the probability value, the p-value for each coefficient. The significance codes are added at the bottom to indicate what the asterisks mean.
- Next is the residual standard error, σ_ε, and the residual degrees of freedom (used with the t-value to calculate p-values).
- Next are the overall model R^2, the proportion of variance in the outcome explained by the model in this specific sample data, and the adjusted R^2, a better estimate of the true population R^2.

- Finally the F-statistic and degrees of freedom and p-value are provided. The F-test is an overall test of whether the model is statistically significant overall or not. It will test all the predictors simultaneously. Because we only have one predictor right now, the p-value for the F-test is identical to the p-value for the t-test of the `stress` coefficient. In multiple regression models, they will not be identical.

Example

To see these in action, we look at positive affect as the outcome and stress as a predictor:

```
m <- lm(formula = PosAff ~ STRESS,
        data = acesData)

summary(m)

##
## Call:
## lm(formula = PosAff ~ STRESS, data = acesData)
##
## Residuals:
##     Min      1Q  Median      3Q     Max
## -2.0750 -0.7924  0.0934  0.7908  2.4828
##
## Coefficients:
##             Estimate Std. Error t value Pr(>|t|)
## (Intercept)    3.075      0.103    29.8  < 2e-16 ***
## STRESS        -0.141      0.032    -4.4  1.8e-05 ***
## ---
## Signif. codes: 0 '***' 0.001 '**' 0.01 '*' 0.05 '.' 0.1 ' ' 1
##
## Residual standard error: 1.06 on 185 degrees of freedom
##   (2 observations deleted due to missingness)
## Multiple R-squared:  0.0946, Adjusted R-squared:  0.0897
## F-statistic: 19.3 on 1 and 185 DF,  p-value: 1.85e-05
```

There are a lot of results. We go through each part, one by one. Before moving on, we highly recommend you make sure you understand each part and can map each piece of output back to what you understand about linear regression and a graph of linear regression.

On a revisited note, in R output, you will sometimes see scientific notation. Scientific notation is a compact way to write very large or very small numbers. For example 10,000 can be written "1e4"; the "e4" part means move the decimal point four places to the right. So instead of 1.0, you get 10000.0, which is 10,000. Very small numbers close to zero can be written too. For example, .0001 is 1e-4. The "e-4" part means move the decimal four places to the left, so that 1.0 becomes .0001. 2.3e-3 is equal to .0023 and so on. Finally, R relies on approximate representations of numbers and only can work with a certain number of decimal places effectively. Thus, sometimes it will give a boundary. For example, $< 2e - 16$ means less than .0000000000000002. R does not accurately represent numbers closer to zero than that, so it will just write out $< 2e - 16$ instead.

- **Call**: This is the code we used to call R to produce the model; it shows that we fit a linear model using `lm()` and that the formula defining the model had positive affect, `PosAff`, as the outcome and `STRESS` as a predictor. The dataset used to fit the data is called `acesData`.

- **Residuals**: This section provides some descriptive statistics on the residuals. We can see that minimum residual is -2.08, which is the most any observed positive affect score; the outcome (since this is negative) falls **below** the predicted positive affect score. Likewise, the maximum residual is 2.48, which is the most any observed positive affect score; the outcome (since this is positive) falls **above** the predicted positive affect score. Looking at the lowest and highest residuals can help give a quick sense of whether there are extreme values. In addition, the median should be close to the mean of zero, if the residuals follow a normal distribution. Thus, looking for whether the median residual is about zero is another quick helpful diagnostic.

- **Coefficients**: This is really the main table normally reported for regressions. The column "Estimate" gives the model parameter estimates or regression coefficients (the *b*s). In this model, the coefficients are 3.08 and -0.14. These give the intercept and slope of the regression line. The column "Std. Error" gives the standard error

(SE) for each coefficient. The standard error captures how much uncertainty there is in the estimate of each coefficient due to the fact that what we analyze is a sample taken from a larger population. That is, although we want to know the population coefficients, due to only having a (assumed random) sample from the population, there could be differences in the true population regression coefficients. The standard error quantifies how uncertain we are in using our sample estimates to infer what the population coefficients are. Here the standard errors are 0.10 and 0.03. The column "t value" gives the t-value for each parameter. Although this is a regression model, just as with t-tests, the regression model uses a t-value and looks it up against the t-distribution to calculate p-values. In fact, the t-value reported is $\frac{b}{SE}$, that is, the regression coefficients divided by their standard error. Here the t-values are 29.82 and -4.40. Finally, the column "*Pr*(> |*t*|)" gives the probability value, the p-value for each coefficient. Even if the predictors in the model are not statistically significant (i.e., are not less than the α (alpha) value specified), often the intercept (i.e., the expected value of the outcome when all the predictors are equal to zero) is very different from zero and so has a very small p-value close to zero. For this reason, p-values often are shown using scientific notation in R. In this case, the p-values are $p < .001$ and $p < .001$. Because p-values are not exactly zero, it is convention to write them as < .001 or < .0001 when they are very small, rather than rounding and writing, for example, "p = 0.000" which suggests that the p-value is exactly zero, which it likely is not. The significance codes at the bottom are used to indicate what the asterisks mean. *** indicates $p < .001$, ** indicates $p < .01$, * indicates $p < .05$, a full stop "." indicates $p < .10$, and a blank indicates $p > .10$.

- **Residual standard error**: σ_ε and the residual degrees of freedom (used with the t-value to calculate p-values). Here the residual standard error is 1.06. The residual degrees of freedom are calculated as $df = N - k$. Here, N is the number of rows of data or *observations,* while k is the number of coefficients. In this case, this is 187 – 2, 185. We also get a note in parentheses where applicable about any observations removed due to missing data. Here we get a note that two observations were removed due to missing data. Observations will be removed if they are missing data on any of the variables in the model, that is, on either the outcome or any of the predictor(s).

- **R-squared**: Next is the overall model R^2, the proportion of variance in the outcome explained by the model in this specific sample data. The R^2 also is the squared correlation between the outcome, here positive affect, `PosAff`, and the predicted positive affect, $b_0 + b_1 * STRESS$. How do we understand R^2?

  If the R^2 is 0, it indicates that none of the variability in the outcome can be explained by the predictor. For this model, it would mean that none of the variance in positive affect across people can be explained by their level of stress. Put differently, it would mean that the squared correlation between the model-predicted positive affect and observed positive affect was zero, a pretty poor model! If the R^2 was 1, it would indicate that 100% (all) of the variance in positive affect could be explained by stress or that the predicted and observed positive affect squared correlation was 1, which would mean a perfect correlation. That is, we can perfectly predict positive affect from stress. If the model is that good, the predictor is probably the same as the outcome. Normal models will have R^2 values somewhere between 0 and 1, often closer to 0.

  In this model, the R^2 is 0.0946. Additionally, the output includes the adjusted R^2, a better estimate of the true population R^2, which in this case is 0.0897. The idea behind the adjusted R^2 is that because we fit the model to our sample, it will always explain just a bit more variance in our sample data than it would in the true population and we can adjust for that. The differences between R^2 and adjusted R^2 will be larger in smaller sample sizes and tend to be very minor in very large samples.

- **F-statistic**: The overall model F-statistic and degrees of freedom and p-value are provided. The F-test is an overall test of whether the model is statistically significant overall or not. It will test all the predictors simultaneously. Because we only have one predictor right now, the p-value for the F-test is identical to the p-value for the t-test of the STRESS coefficient. In multiple regression models, they will not be identical because the F-test can be a multivariable test, whereas the t-test always tests one predictor at a time.

Using confint() gives us 95% confidence intervals for the parameter estimates, the regression coefficients. Recall that a 95% CI in the middle leaves 5% to be split into the two tails of the normal curve. Half of 5% is 2.5%, so the left side of the CI starts at 2.5%, and then, moving from left to right, 2.5% + 95% yields 97.5%. This is where the right tail of the normal curve starts:

```
confint(m)
```

```
##               2.5 %   97.5 %
## (Intercept)  2.872  3.27841
## STRESS      -0.204 -0.07762
```

The confidence interval results give us a small table with one row for each coefficient, including the intercept. The two columns give the 2.5th percentile and the 97.5th percentile which are the lower and upper limits of the 95% confidence interval. Although it is possible to get other kinds of confidence intervals (e.g., 99% intervals), 95% are almost universally the standard, so we focus on those.

The confidence interval is calculated as

$$b + SE * t_{crit}$$

where b is the regression coefficient, SE is the standard error for a regression coefficient, and t_{crit} is the critical t-value, based on the degrees of freedom and the desired confidence interval width. For example, a 95% confidence interval uses the bottom 2.5th and upper 97.5th percentiles of the t-distribution. In R we can calculate the critical t-value by using the qt() function which gives the t-value for a given probability, p, and degrees of freedom, df. Recall the qnorm() function from Chapter 9 (used to calculate the z-score). This is the same idea, only for t-scores. Here are the critical t-values for 185 degrees of freedom:

```
qt(p = .025, df = 185)
```

```
## [1] -1.973
```

```
qt(p = .975, df = 185)
```

```
## [1] 1.973
```

The lower confidence intervals for our coefficients have an intercept and slope 3.075 + 0.103 * -1.973 = 2.872 and -0.141 + 0.032 * -1.973 = -0.204.

The upper confidence intervals for our coefficients have an intercept and slope 3.075 + 0.103 * 1.973 = 3.278 and -0.141 + 0.032 * 1.973 = -0.078.

We could interpret the results of the regression as follows. There was a statistically significant association between stress and positive affect. A one-unit-higher stress score was associated with a -0.14 lower positive affect score [95% CI -0.20 to -0.08], $p < .001$. For someone with a stress score of zero, positive affect was expected to be 3.08 [2.87, 3.28], $p < .001$ (the intercept). Overall, stress explained 9.0% of the variance in positive affect.

We can also use the `visreg` package to help visualize the results from the regression to better understand what it means (see Figure 11-1). The defaults work fairly well in this case, giving us the regression line (in blue), and the shaded gray region shows the 95% confidence interval around the regression line. Partial residuals are plotted as well by default. We use the argument `gg = TRUE` to ask `visreg()` to make a `ggplot2` graph which lets us do things like customize the theme and work with the usual ggplot2 graphing framework we are accustomed to now.

```
visreg(m, xvar = "STRESS", gg = TRUE)
```

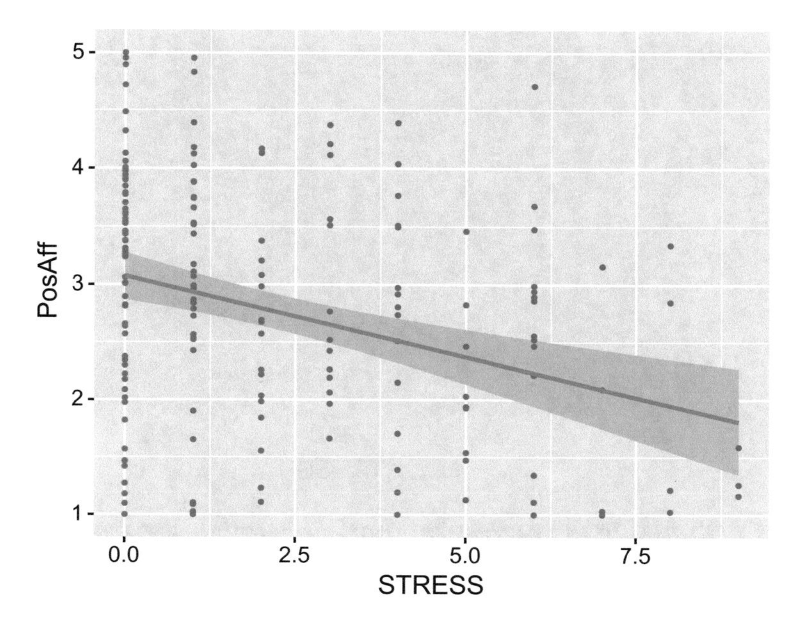

Figure 11-1. *Visualizing a regression model of positive affect predicted by stress using the visreg() function*

If you want just the regression line, you can set `parital = FALSE` and `rug = FALSE` to stop plotting partial residuals. As with other graphs in `ggplot2`, we can use + to add additional elements, such as controlling the title with the `ggtitle()` function. The result of this customization is in Figure 11-2:

```
visreg(m, xvar = "STRESS", partial = FALSE, rug = FALSE, gg = TRUE) +
  ggtitle("Linear regression of Positive Affect on Stress",
          subtitle = "Shaded region shows 95% confidence interval.")
```

You can use the `annotate()` function to add annotations to the graph, such as a label indicating the coefficient and p-value. Figuring out the best position and size to be attractive is typically a process of some trial and error in any graph. Play around with x and y coordinates and the size until you are happy with it. You can also use the `xlab()` function to add text to the x label and the `ylab()` function to add text to the y label.

The final graph is in Figure 11-3 and provides a good high-level summary of our regression model including key information.

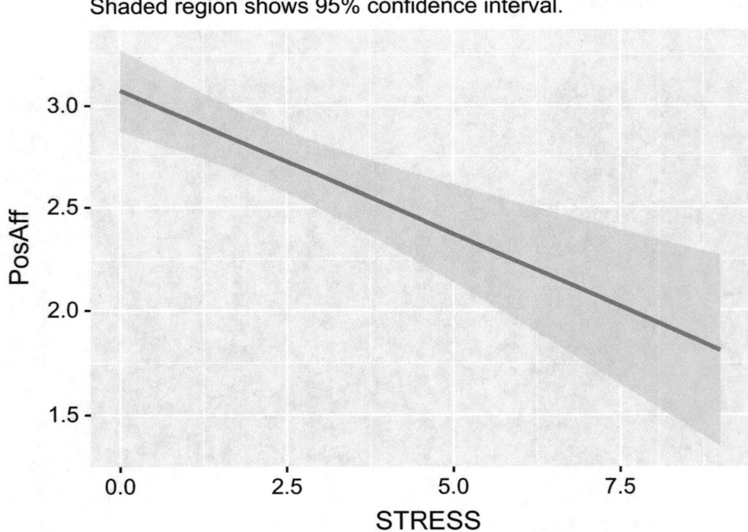

***Figure 11-2.** Customizing the regression model visualization from visreg()*

```
visreg(m, xvar = "STRESS", partial = FALSE, rug = FALSE, gg = TRUE) +
   ggtitle("Linear regression of Positive Affect on Stress",
```

```
        subtitle = "Shaded region shows 95% confidence interval.")+
  annotate("text", x = 6, y = 3, label = "b = -0.14, p < .001", size = 5) +
  xlab("Perceived Stress") + ylab("Positive Affect (Mood)")
```

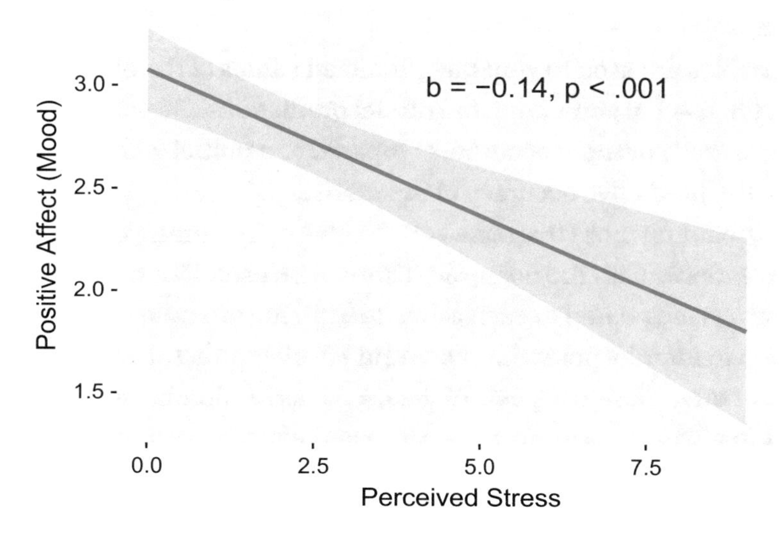

***Figure 11-3.** Adding annotations to show the slope in regression model visualization*

11.3 Multiple Regression

Now that you have delved into the mathematics behind Chapter 8 a bit more, you are ready to push forward into multiple regression. Remember that learning statistics is a process, not a single event. Your work in Chapter 8 helped you understand the preceding section. Each read through (and each round of experimentation with the code), you learn and understand more. There is no denying the following section is reasonably complex. After all, your goal is to leverage computers and mathematics to build a relationship between *multiple* variables (or dimensions) of data! That said, you do not need to understand everything right away. Run the code, compare and contrast between a single predictor and multiple predictors, and read through the text explanations a bit at a time. Every reread will help you understand better. No one understands multiple regression on their first try! At least, we certainly did not.

Implications of Multiple Predictors

Multiple linear regression works in principle basically the same way as does simple linear regression. The difference is more than one predictor (explanatory) variable is allowed in a single model. Compared to simple linear regression with a single predictor (explanatory) variable, including multiple predictors in the same regression has two main implications.

First, all variables are used to generate predicted values of the outcome. Because more variables are used in generating the model predictions, we would generally expect the model to be equally or more accurate compared to a model with only one predictor. This focuses on the predictive accuracy of the model.

Second, by including more than one variable, we can examine the *unique* association of a variable with the outcome. This can be useful for better understanding how and why things are related to each other. Identifying the unique association of a variable can help to identify potential cause and effect–type associations. Note that generally without other design aspects of data, such as randomization, it is very difficult to be confident that what is identified is a cause and effect association, yet multiple regression can help rule out likely alternatives.

We are going to focus on the second aspect of having multiple predictors: identifying the unique association of each variable. For example, suppose that you believe that high stress causes people to feel less happy. You might try to examine whether data support this belief by running a linear regression predicting positive affect from stress. You would expect a negative association between stress and positive affect. However, your friend has a different idea. Your friend thinks that when people are lonely, they feel more stressed and less happy. That is, your friend believes that loneliness or a lack of supportive friends causes both stress and unhappiness. Your friend reasons that it is natural that stress and happiness may appear to have a negative association, because both of them are caused by the same thing: lack of good friends.

How could you settle this question? One way would be to collect data on only people who had plenty of good friends or on people who all lacked good friends. Then you could see in only the people with good friends, if stress is associated with happiness or not. In practice, this may be difficult as it is not always easy to find and collect data on specific groups of people. Further, it can become impractical quickly. Suppose you have another friend who reasons it is neither stress nor friends; it's how far you live away from work. If you live too far away, you spend a long time commuting; so you do not have time for friends, feel more stressed, and are less happy. To collect data to address both these friends' arguments, you would need to get data on stress and happiness in people who all live the same distance from work and all have the same amount of good friends – a challenging task.

Multiple regression is a different approach to trying to tackle this problem. By including additional variables in the same model, you can model their associations and look at the **unique** association of each variable. That is, you can try to statistically control for the fact that different people may have a different length of commute to work or a different amount of support from friends. Then, statistically taking these other variables into account, you can see whether stress is associated with less happiness or not. This is really the "key" behind multiple regression compared to simple linear regression. In multiple regression, what each regression coefficient captures is not the total association between that variable and the outcome. Instead, what each coefficient captures is the **unique** association between a predictor and the outcome. In practice, you will see people talk and write about this same idea in many different ways. If you are new to regression, this can be confusing and may not be clear whether they all mean the same thing. Here are some common ways people describe results from multiple regression that are all trying to capture the same idea:

- The independent association of stress and happiness
- The unique association of stress and happiness
- The association of stress and happiness, controlling for support from friends
- The association of stress and happiness, covarying for support from friends
- The association of stress and happiness, adjusted for support from friends

These are not an exhaustive list of the ways people discuss results from multiple regression; hopefully they help give you some examples. In all cases, the key point is that the multiple regression *statistically* adjusts or takes into account the impact of other variables, so that an adjusted or independent association of two variables can be explored. If the model assumptions are met (e.g., there are linear associations), multiple regression can use statistics to control for the differences between people in other variables (like friend support and commute length to work) to create a sort of hypothetical world where differences on those variables have been removed to answer the question, "What is the unique or independent association of stress with happiness, if there was no variation in friend support or commute length?" Taking out the variation in those covariates, extraneous or alternative explanations for the association between stress and happiness can help clarify whether stress and happiness are truly related to each other or are only incidentally associated because of some third factor.

It is worth noting that even if we find that stress and happiness *are* independently associated with each other, it still does not prove that stress causes happiness. Without randomization, you could always have another friend who reasons some new variable may explain why stress and happiness are related, and you would endlessly need to measure variables and include them in the regression model. Nevertheless, multiple regression can allow us to adjust or control for other variables which can help rule out alternate explanations for the association we are interested in testing.

Multiple Regression in R

As before, we have an outcome, *y*, predicted by the combination of model parameters, the *b*s; yet instead of only one predictor, *x*, we can have an indeterminate number, *k*, of them. This *k* could be two predictors, three, twenty... The number does not change the underlying technique:

$$y = b_0 + b_1 * x_1 + \ldots + b_k * x_k + \varepsilon$$

The regression coefficients (sometimes they are more generally referred to as model parameters, where in this case the model is a regression) are interpreted fairly similarly to those in simple linear regression, yet with some extra requirements:

- b_0 is the intercept, the expected (model-predicted) value of *y* when all predictors are zero. Note you may see this sometimes written $E(y|x_1 = 0, \ldots, x_k = 0)$, which is just a way of saying the expected or predicted value of *y* when the predictors are zero. If people have many predictors or do not want to write them all out, sometimes they just write $E(y|X = 0)$ with the upper case "X" standing in for all the predictors.
- b_1 is the slope of the line and captures how much *y* is expected to change for a one-unit change in x_1, controlling for the other predictors.

- b_k is the slope of the line and captures how much y is expected to change for a one-unit change in x_k, controlling for all other predictors.
- ε is the residual/error term, the difference between the model-predicted value and the observed value.

Visually, it becomes more difficult to visualize, yet remains possible with a 3D graph, which we will explore a bit later.

Example

To practice with multiple regression, let us look at a real example. We will work with the hypothetical scenario discussed earlier. You believe stress makes people unhappy; your friend reasons lacking supportive friends results both in stress and unhappiness.

With the right data, we can use multiple regression to help determine which person is right. To do this, we add both stress and social support variables into a regression model predicting positive affect. This will let us separate the **unique** association of stress and positive affect as well as social support and positive affect. The model would look like this, written as an equation:

$$Positive\,Affect = b_0 + b_1 * STRESS + b_2 * SUPPORT + \varepsilon$$

You can conduct multiple linear regressions in R almost the same way as you would a simple linear regression. Additional predictors can be added by simply using + between them. The following code shows an example adding both STRESS and social SUPPORT as predictors simultaneously:

```
m2 <- lm(formula = PosAff ~ STRESS + SUPPORT,
         data = acesData)
```

Next, we run diagnostics and only after diagnostics take a look at the summary. We leave the `summary()` until after diagnostics on purpose as there is not much point interpreting the coefficients if the model assumptions are badly violated. We would want to first fix the model and then interpret. Conveniently, the model diagnostics for multiple regression are essentially the same as for simple linear regression. Specifically, we look at the distribution of the residuals to evaluate whether the normality assumption is met and identify potential extreme values. We also look at a scatter plot of the predicted positive affect vs. residual values to evaluate whether the amount of variability in the residuals is equal across the range of predicted values (homoscedasticity). We use the 0.5th and 99.5th percentiles of a normal distribution as the criteria for defining extreme values. The diagnostic graphs are shown in Figure 11-4:

```
m2d <- modelDiagnostics(m2, ev.perc = .005)
plot(m2d, ncol = 2, ask = FALSE)
```

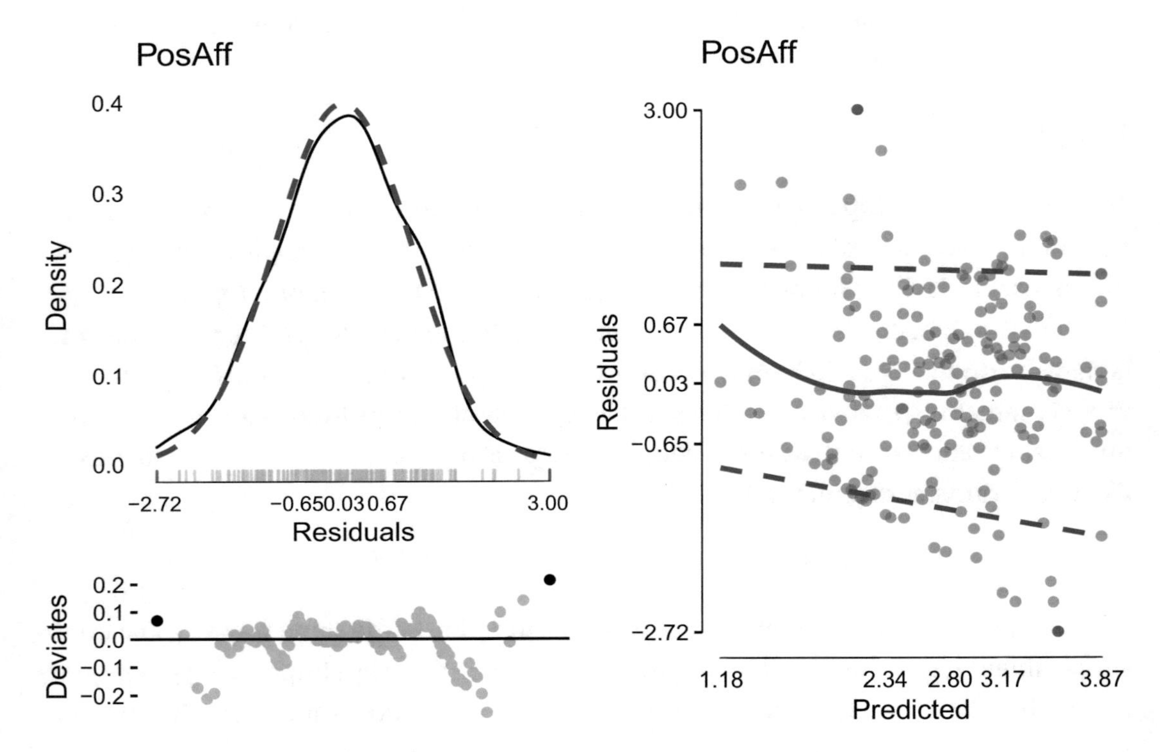

***Figure 11-4.** Multiple regression model diagnostics*

Note, if you get a warning about singularity problems or tiny diagonals, do not worry about them; they come from internal functions that estimate where to draw the blue lines on the residual vs. predicted value plot. Those blue lines are added just to help make it easier to see patterns in the data though are not needed to evaluate the diagnostics, and the warnings only pertain to those models, so you can safely ignore them. In the book, we do not print the warnings.

Looking at Figure 11-4, we can see a couple of extreme residual values, yet they are not too extreme. Additionally, there only seem to be two of them, so they are probably not going to make a large difference whether we leave or exclude them. The key behind evaluating extreme values is to try to ascertain whether the extreme values may indicate either problems/inaccurate data (e.g., if the values were impossible) or whether those few extreme values are extreme enough or there are enough of them to likely substantially influence the results of the model. The fewer extreme values there are relative to the rest of the sample size and the less extreme they are, the less likely they are to actually make much difference in the model results. It can be helpful to remember that being relatively extreme does not make those observations a "problem" or "wrong" or something to be "fixed." We want to examine them, but then can make a reasoned decision. There always has to be a lowest and highest value, but often taking a look at the extremes can help identify if there are problems.

Figure 11-4 looks pretty good, so we might decide to proceed without any changes. The next thing we could look at is a summary of the model results and confidence intervals for the results:

```
summary(m2)
```

```
##
## Call:
## lm(formula = PosAff ~ STRESS + SUPPORT, data = acesData)
##
## Residuals:
##     Min      1Q  Median      3Q     Max
## -2.5611 -0.6133  0.0285  0.6304  2.8065
##
## Coefficients:
##             Estimate Std. Error t value Pr(>|t|)
## (Intercept)   2.0881     0.1714   12.18  < 2e-16 ***
## STRESS       -0.1340     0.0287   -4.67  5.8e-06 ***
## SUPPORT       0.1780     0.0260    6.84  1.2e-10 ***
## ---
## Signif. codes:  0 '***' 0.001 '**' 0.01 '*' 0.05 '.' 0.1 ' ' 1
##
## Residual standard error: 0.949 on 184 degrees of freedom
##   (2 observations deleted due to missingness)
## Multiple R-squared:  0.278,  Adjusted R-squared:  0.27
## F-statistic: 35.4 on 2 and 184 DF,  p-value: 9.7e-14
```

```
confint(m2)

##                 2.5 %   97.5 %
## (Intercept)  1.7500  2.42619
## STRESS      -0.1906 -0.07737
## SUPPORT      0.1267  0.22943
```

We go through the output one part at a time to make sure it is very clear how to interpret each piece:

- **Call**: This is the code we used to call `R` to produce the model; it shows we fit a linear model using `lm()` and the formula defining the model had positive affect, `PosAff`, as the outcome and both `STRESS` and `SUPPORT` as predictors. The dataset used to fit the data is called `acesData`.
- **Residuals**: This section provides some descriptive statistics on the residuals. We can see that minimum residual is -2.56, which is the most any observed positive affect score; the outcome falls **below** the predicted positive affect score. Likewise, the maximum residual is 2.81, which is the most any observed positive affect score; the outcome falls **above** the predicted positive affect score. Because both stress and social support are predictors, the predicted positive affect score will be based off of both these predictors, yet the residuals are still just the difference between the observed positive affect and predicted positive affect scores. Looking at the lowest and highest residuals can help give a quick sense of whether there are extreme values. In addition, the median should be close to the mean of zero, if the residuals follow a normal distribution. Thus, looking for whether the median residual is about zero is another quick helpful diagnostic.
- **Coefficients**: This is really the main table normally reported for regressions. The column "Estimate" gives the model parameter estimates or regression coefficients (the *b*s). In this model, the coefficients are 2.09 and -0.13 and 0.18. These give the intercept and the expected value of positive affect for someone with zero stress and zero social support (i.e., zero on all predictors). The coefficients also contain the slope of the regression line for the unique association of stress and positive affect and of social support and positive affect.

The column "Std. Error" gives the standard error (SE) for each coefficient. The standard error captures how much uncertainty there is in the estimate of each coefficient due to the fact that what we analyze is a sample taken from a larger population. That is, although we want to know the population coefficients, due to only having a (assumed random) sample from the population, there could be differences in the true population regression coefficients. The standard error quantifies how uncertain we are in using our sample estimates to infer what the population coefficients are. Here the standard errors are 0.17 and 0.03 and 0.03.

The column "t value" gives the t-value for each parameter. Although this is a regression model, just as with t-tests, the regression model uses a t-value and looks it up against the t-distribution to calculate p-values. In fact, the t-value reported is $\frac{b}{SE}$, that is, the regression coefficients divided by their standard error. Here the t-values are 12.18 and -4.67 and 6.84.

Finally the column "$Pr(>|t|)$" gives the probability value, the p-value for each coefficient. Even if the predictors in the model are not statistically significant (i.e., are not less than the α (alpha) value specified), often the intercept (i.e., the expected value of the outcome when all the predictors are equal to zero) is very different from zero and so has a very small p-value close to zero. For this reason, p-values often are shown using scientific notation in R. In this case, the p-values are $p < .001$ and $p < .001$ and $p < .001$. Because p-values are not exactly zero, it is convention to write them as < .001 or < .0001 when they are very small, rather than rounding and writing, for example, "p = 0.000" which suggests that the p-value is exactly zero, which it likely is not. The significance codes at the bottom are used to indicate what the asterisks mean. *** indicates $p < .001$, ** indicates $p < .01$, * indicates $p < .05$, . indicates $p < .10$, and a blank indicates $p > .10$.

- **Residual standard error**: σ_ε and the residual degrees of freedom (used with the t-value to calculate p-values). Here the residual standard error is 0.95. The residual degrees of freedom are calculated as $df = N - k$. Here, N is the number of rows of data or *observations*, while k is the number of coefficients. In this case, this is 187 – 3, 184. We also get a note in parentheses where applicable about any observations removed due to missing data. Here we get a note that two observations were removed due to missing data. Observations will be removed if they are missing data on any of the variables in the model, that is, on either the outcome or any of the predictor(s).

- **R-squared**: Next is the overall model R^2, the proportion of variance in the outcome, positive affect, explained by the model (here our model has two predictors: stress and social support) in this specific sample data. The R^2 also is the squared correlation between the outcome, here positive affect, `PosAff`, and the predicted positive affect, $b_0 + b_1 * STRESS + b_2 * SUPPORT$.

  If the R^2 is 0, it indicates that none of the variability in the outcome can be explained by the predictor. For this model, it would mean that none of the variance in positive affect across people can be explained by their level of stress or social support. Put differently, it would mean that the squared correlation between the model-predicted positive affect and observed positive affect was zero. If the R^2 was 1, it would indicate that 100% (all) of the variance in positive affect could be explained by stress and social support. Again put differently, the predicted and observed positive affect squared correlation was 1, which would mean a perfect correlation. That is, we can perfectly predict positive affect from stress and social support.

  In this model, the R^2 is 0.278, which is the proportion of variance in positive affect in this sample explained by stress and social support. Additionally, the output includes the adjusted R^2, a better estimate of the true population R^2, which in this case is 0.2701. The idea behind the adjusted R^2 is that because we fit the model to our sample, it will always explain just a bit more variance in our sample data than it would in the true population and we can adjust for that.

- **F-statistic**: The overall model F-statistic and degrees of freedom and p-value are provided. The F-test is an overall test of whether the model is statistically significant overall or not. It will test all the predictors simultaneously; that is, it tests whether the coefficients for stress and social support are both 0 or at least one is not zero. Because we have multiple predictors, the p-value for the F-test will not match the p-value for the t-test of a specific coefficient, as it did for the simple linear regression.

We have seen a lot of the output from a multiple regression, but what does it look like? Figure 11-5 shows the regression surface predicted by a multiple regression of positive affect on both stress and social support. With two predictors, a multiple regression will make a three-dimensional (3D) plane or surface. The surface will be flat, like a piece of paper, because we are currently only modeling a linear association for each predictor.

A graph of the multiple regression is shown in Figure 11-5. The two different figures are just viewing the same surface from a different orientation to help see what the plane looks like. There are three axes: one for stress, one for social support, and one for predicted positive affect. It is important to note that what is being graphed is the *predicted* positive affect from the multiple regression model. Actual positive affect scores are not shown. Looking at the figure, you can see the slope of a line for any one variable is the same regardless of the other variable. That is, the slope of the association between stress and positive affect is the same regardless of the level of social support. Likewise, the slope of the association between social support and positive affect is the same regardless of the level of stress. The idea behind this is that multiple regression allows people to differ on a variable, say social support, but still assumes that the association of stress and positive affect is the same for everyone. People may start at different levels due to other variables, like social support, but the underlying assumption is the association is the same for everyone. By allowing the lines, the surface, to have different heights depending on other variables, like social support, it is possible to isolate the unique effect of stress, independent of social support. If we look at the graph from another angle, we could interpret it similarly, but focused on the unique effect of social support, controlling for stress or holding stress constant. It is one model, but we can talk about stress or social support.

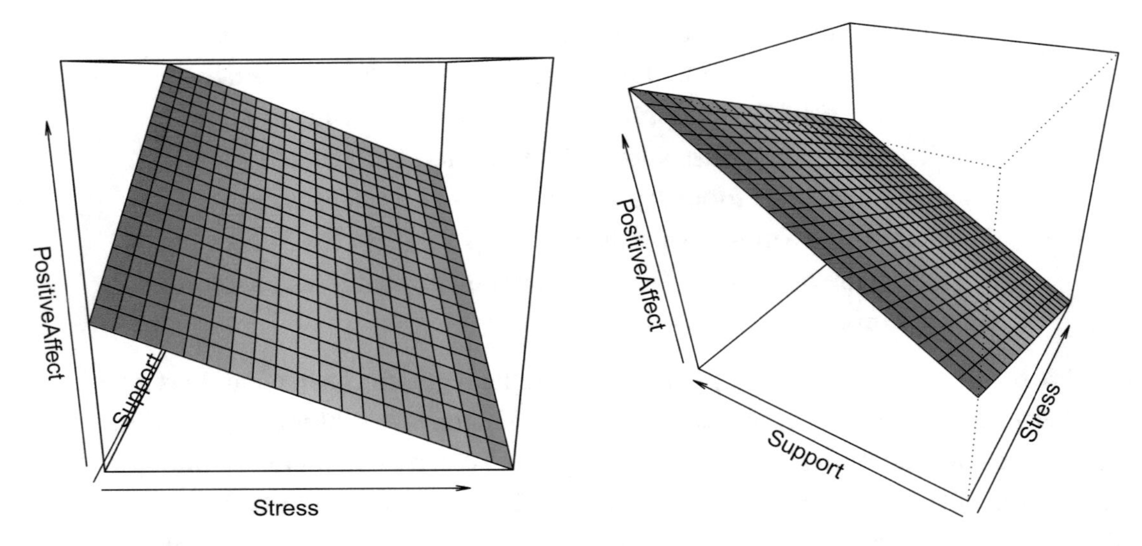

***Figure 11-5.** 3D graph of a multiple regression surface*

Multiple regression can include many predictors, but it is not very feasible to show a figure with more than two predictors. The ideas remain the same, however. However many predictors are added, what the multiple regression allows us to examine is the **unique** association of each predictor with the outcome, controlling for (or independent of) all the other predictors added into the model.

In practice, even for multiple regression with just two variables, we rarely create a 3D graph because it is not too easy to interpret or show a 3D graph in 2D like paper or screens. Rather, we often just plot the association of one variable at a time. This is okay because as you'll recall from the 3D graph, the slope capturing the association of stress and positive affect is the same regardless of level of social support. The height or absolute level of positive affect changes, but the slope of the line does not change. Thus, we can really just pick any level of social support and then plot the association between stress and positive affect. We can do this using the `visreg()` function, which is probably the easiest way to get a quick graph to help interpret and understand a regression model in R.

The following code creates a graph showing the association between stress and positive affect, holding social support constant. `visreg()` generally holds other predictors not graphed at their means as a default, although you can change this if desired. Note this code is actually the same as we used before when we had only a simple linear regression. What changed is our model now statistically controls for or adjusts for social support so what we are now graphing is the unique association of stress and positive affect. The plot is shown in Figure 11-6. We can see higher stress scores are associated with lower positive affect scores, independent of social support. The shaded gray region captures the 95% confidence intervals around the regression line:

```
visreg(m2, xvar = "STRESS", partial = FALSE, rug = FALSE, gg = TRUE) +
  ggtitle("Association of stress and positive affect, adjusted for social
  support")
```

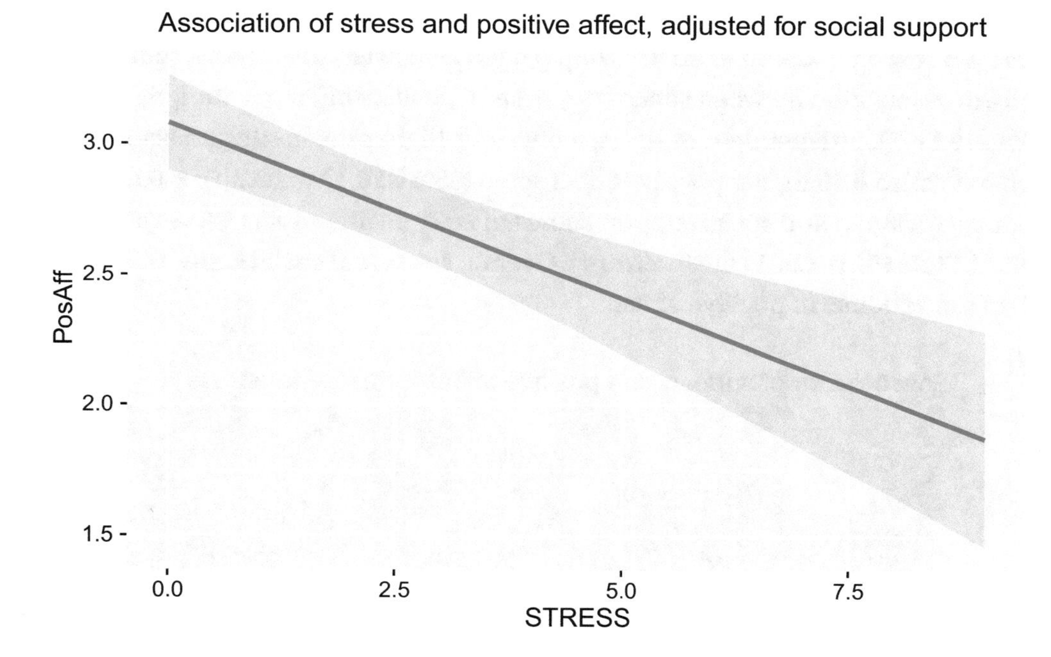

Figure 11-6. *Multiple regression of stress and social support predicting positive affect. This figure shows only the unique stress and positive affect association*

We can create the same sort of graph for social support instead of stress, in which case we see the unique association of social support and positive affect, controlling for stress. The plot is shown in Figure 11-7.

```
visreg(m2, xvar = "SUPPORT", partial = FALSE, rug = FALSE, gg = TRUE) +
  ggtitle("Association of support and positive affect, adjusted for stress")
```

From the figure, we can see social support has a positive association with positive affect: higher social support scores are associated with higher positive affect scores, independent of stress.

If we put everything we have learned together, we could interpret and write up the results of this multiple regression model as follows. Note this is quite a detailed write-up. In practice, you may only include parts of it rather than reporting everything.

A multiple linear regression model with positive affect as the outcome and stress and social support as predictors: There was a statistically significant association between stress and positive affect, controlling for social support (see Figure 11-6). Independent of social support, a one-unit-higher stress score was associated with a -0.13 lower positive affect score [95% CI -0.19 to -0.08], p <.001. Likewise, there was a statistically significant association between social support and positive affect, controlling for stress (see Figure 11-7). Independent of stress, a one-unit-higher social support score was associated with a 0.18 higher positive affect score [95% CI 0.13 to 0.23], p <.001. For someone with stress and social support scores of zero, positive affect was expected to be 2.09 [1.75, 2.43], p <.001 (the intercept). Overall, stress and social support explained 27.0% of the variance in positive affect.

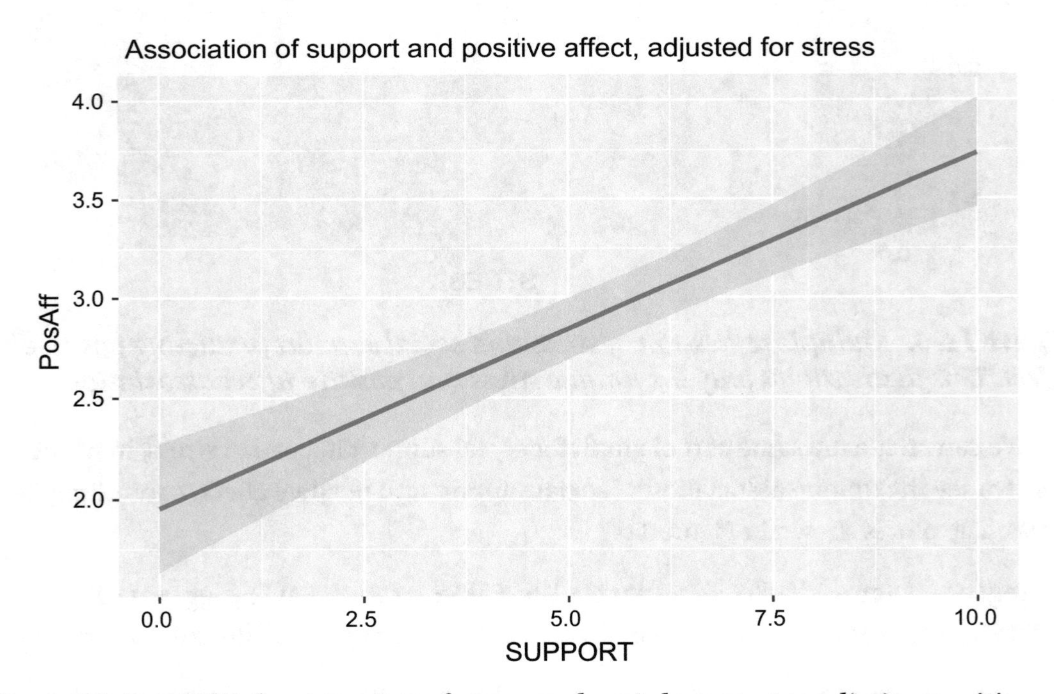

Figure 11-7. *Multiple regression of stress and social support predicting positive affect. This figure shows only the unique social support and positive affect association*

Effect Sizes and Formatting

We have already seen R^2 which captures the proportion of variance in the outcome accounted for by the model. Often, R^2 or another effect size is desirable. Effect sizes are standardized and allow comparing the **magnitude** of an effect across variables or studies. For example, because regression coefficients are in the units of the predictor and outcome, they are not comparable.

As a simple example, suppose one study measured time in seconds and another study measured time in minutes. Even if the magnitude of effects was identical in the two studies, the regression coefficients, the *b*s, would not be the same because time does not have the same scale (seconds vs. minutes). The goal of effect sizes is to somehow standardize so the magnitude of effect can be compared. R^2 is a standardized measure. No matter the outcome or predictors, it always ranges from 0 (or 0%) which means none of the variance is explained to 1 (or 100%) which means all of the variance is explained.

We can use R^2 to get an effect size estimate for the unique contribution of a predictor as well. The mathematics behind this is fairly straightforward. Suppose we have our multiple regression model, which we will call $Model_{AB}$:

$$Model_{AB} : Positive\ Affect = b_0 + b_1 * STRESS + b_2 * SUPPORT + \varepsilon$$

Suppose what we want to know is an effect size for the unique association of stress and positive affect. The R^2 for our regression model $Model_{AB}$ tells us the total variance in positive affect that can be explained by both stress and social support; this we can call R^2_{AB}. We could run *another* regression model with just social support as a predictor, which we call $Model_A$:

$$Model_A : Positive\ Affect = b_0 + b_1 * SUPPORT + \varepsilon$$

The R^2 for this model tells us the total variance in positive affect explained by social support, which we can call R^2_A. Now we have two R^2 values: R^2_{AB} is the total variance in positive affect explained by both stress and social support. R^2_A is the variance in positive affect explained by social support. The only difference between the two models, and thus the two R^2 values, is the addition of stress. Thus, the difference in the two R^2 values is the **unique** additional variance in positive affect that is explained uniquely by stress over and above social support:

$$R^2_{stress} = R^2_{AB} - R^2_A$$

We can apply this same process generally to any variable or set of variables to identify how much variance in an outcome they uniquely explain, as an effect size measure. The general process is as follows:

1. Fit a multiple regression model with all predictors included and note down the R^2 value; this is the full R^2_{AB}.
2. Fit a (multiple) regression model with all predictors **except** the predictor(s) you want to calculate an effect size for, and store the R^2 value; this is the reduced R^2_A.
3. Calculate the difference in the two R^2 values to calculate the unique variance explained by your predictor(s) of interest, and this is an effect size measure for them.

Example

Earlier we fit the full multiple regression model with both stress and social support as predictors of positive affect and stored the results in an object, `m2`. If we fit a regression model predicting positive affect from social support only, we will have all the pieces needed to identify the **unique** variance in positive affect explained by stress. The `R2()` function can be used to pull out just the R^2 values from a model without having to get the whole `summary()`. We use `R2()` on the full model to get the total variance explained by both stress and social support and then subtract the variance that can be explained by social support so that what is left is variance uniquely explained by stress:

```
model_a <- lm(PosAff ~ SUPPORT, data = acesData)

R2(m2) - R2(model_a)

##       R2   AdjR2
## 0.08557 0.08209
```

Often people stop at calculating the unique variance explained by a predictor. However, Cohen [8] defined another effect size measure, called f^2. The idea behind f^2 is that if you think about the unique or additional variance in an outcome explained by a predictor, it gets harder to explain variance the more has already been explained. For example, if a model $Model_A$ already explains 90% of the variance in an outcome, it is very difficult to explain an additional 9% (that would bring the model to nearly 100% variance explained). In contrast, if a model explains 0% of the variance, adding one more predictor may have an easier time explaining 9% of the variance. Thus Cohen's f^2 effect size accounts for this. Cohen's f^2 effect size for an overall model is defined as

$$f^2 = \frac{R^2}{1-R^2}$$

It is based on the model R^2 but divided by the variance not explained, $1 - R^2$.

In addition, Cohen's f^2 effect size for an individual predictor is defined as

$$f^2 = \frac{R_{AB}^2 - R_A^2}{1-R_{AB}^2}$$

We see the numerator is the unique variance explained by the predictor. The only addition for Cohen's f^2 is that the unique variance explained is divided by $1-R_{AB}^2$, the unexplained variance. Because of this, Cohen's f^2 can be any number greater than or equal to zero.

According to Cohen's [8] guidelines, there are some commonly used conventions for what is a small, medium, or large effect size:

- **Small** is $f^2 \geq 0.02$.
- **Medium** is $f^2 \geq 0.15$.
- **Large** is $f^2 \geq 0.35$.

Data scientists often use these guidelines or cut-offs to help interpret a f^2 effect size for a predictor to decide whether it has a small, medium, or large effect, based on the amount of unique variance it explains. We have seen how to calculate the difference in R^2 values using the `R2()` function. It is easy enough to divide by $1-R_{AB}^2$ to get Cohen's f^2:

```
(R2(m2) - R2(model_a)) / (1 - R2(m2))
```

```
##      R2  AdjR2
## 0.1185 0.1125
```

In this case, we see the f^2 values are in the "small" to "medium" range, as they are getting close to but do not quite reach the threshold for a "medium" effect size.

Although it is possible to do these calculations by hand, it becomes tedious as regression models have more predictors included. Each predictor requires two models, the full model and a model without that predictor, in order to calculate the effect sizes. In addition, although the `summary()` function gives us most the information we would want about a regression model, it leaves some out and the formatting often is not in the format that would make it easy to include in a report.

Although not strictly required for using and reporting multiple regression models, these next two functions can help get all the information you may want nicely formatted and in one place. The `modelTest()` function takes a model and automatically works out additional calculations for it, such as confidence intervals and effect sizes for each predictor. It does the same work we did to identify the unique effect size for stress, but `modelTest()` does that all automatically and it does it for every predictor in the regression model. In addition, the `APAStyler()` function can be used on the results from `modelTest()` to get a nicely formatted set of output with many things like rounding taken care of.

Note, we do recognize that "nicely formatted" can be a very personal choice, which is why we teach the general building blocks for fitting a regression and getting output, but in many applications, the format from APAStyler() is close to how results are commonly presented these days.

```
m2test <- modelTest(m2)
APAStyler(m2test)

##                   Term                          Est           Type
##  1:        (Intercept)  2.09*** [ 1.75, 2.43] Fixed Effects
##  2:             STRESS -0.13*** [-0.19,-0.08] Fixed Effects
##  3:            SUPPORT  0.18*** [ 0.13, 0.23] Fixed Effects
##  4: N (Observations)                        187 Overall Model
##  5:          logLik DF                        4 Overall Model
## ---
##  9:                 F2                     0.38 Overall Model
## 10:                 R2                     0.28 Overall Model
## 11:             Adj R2                     0.27 Overall Model
## 12:             STRESS    f2 = 0.12, p < .001 Effect Sizes
## 13:            SUPPORT    f2 = 0.25, p < .001 Effect Sizes
```

The table has three columns and three subsections:

- **Term**: The "Term" column tells us what we are seeing. For example, it indicates which regression coefficient or what effect size.
- **Est**: The "Est" column gives us the values, the actual number results from our model.
- **Type**: The "Type" column indicates the broad section or category of the table, covered next.

The three broad sections of the table occur on different rows. You can tell which section a row belongs to by looking at the column "Type." The sections are

- **Fixed Effects**: The Fixed Effects section has the regression coefficients, the *b*s, we learned about. In linear regression and multiple regression, these regression coefficients will always be fixed effects. `modelTest()` uses this name (fixed effects) because the same function can be used for other models where there may be random effects, effects that are allowed to vary randomly across different people. That is not important for now, but may help explain why things are labeled as they are.
- **Overall Model**: The Overall Model section has information or details that pertain to the overall model, such as the number of observations included, the overall model R^2, and so on.
- **Effect Sizes**: The Effect Sizes section has effect sizes for each predictor in our regression model. These are labeled basically the same way as the regression coefficients, but the effect sizes reported are f^2 values.

If you have sharp eyes, you may have noticed that the table printed earlier skipped a few rows jumping from row 5 to row 9. This happens because it is a `data.table` and we set up to only view the first and last five rows, which is helpful when showing a dataset so that hundreds of rows are not printed, but not always helpful for something like a table of model output where you may want to see it all. Different environment settings on different computers can yield different results. You can ensure every row is printed by using the `print()` function and specifying the maximum number of rows, shown in the following code:

```
print(APAStyler(m2test), nrow = 100)

##                   Term                           Est           Type
##  1:        (Intercept)  2.09*** [ 1.75, 2.43] Fixed Effects
##  2:             STRESS -0.13*** [-0.19,-0.08] Fixed Effects
##  3:            SUPPORT  0.18*** [ 0.13, 0.23] Fixed Effects
##  4: N (Observations)                      187 Overall Model
##  5:          logLik DF                      4 Overall Model
##  6:             logLik                -254.09 Overall Model
##  7:                AIC                 516.18 Overall Model
##  8:                BIC                 529.10 Overall Model
##  9:                 F2                   0.38 Overall Model
## 10:                 R2                   0.28 Overall Model
## 11:             Adj R2                   0.27 Overall Model
## 12:             STRESS   f2 = 0.12, p < .001  Effect Sizes
## 13:            SUPPORT   f2 = 0.25, p < .001  Effect Sizes
```

Now working with the full output, we go through each row to make sure it is clear exactly what the output means and how to practically interpret and use it:

1. **Term – (Intercept)**: This row has the regression coefficient, *b*, for the intercept; the asterisks are used to indicate the significance (p-value) just as in `summary()`. *** indicates $p < .001$, ** indicates $p < .01$, and * indicates $p < .05$. In this case, the intercept is significantly different from zero, $p < .001$. Next in square brackets is the 95% confidence interval for the intercept. All the output is rounded to two decimal places, as is commonly used in written reports. Here we can see that when stress and social support are zero, the model predicts a positive affect score of 2.09, and this is significantly different from zero.

2. **Term – STRESS**: This row has the regression coefficient, *b*, the slope for the association between stress and positive affect. As for the intercept, the asterisks indicate the significance level, and

the 95% confidence intervals are in square brackets. Here we can see that independent of social support, stress has a negative association with positive affect, $p < .001$.

3. **Term – SUPPORT**: This row has the regression coefficient, *b*, the slope for the association between social support and positive affect. As for the intercept, the asterisks indicate the significance level, and the 95% confidence intervals are in square brackets. Here we can see that independent of stress, social support has a positive association with positive affect, $p < .001$.

4. **Term – N (Observations)**: This row shows the total number of observations that were included in the regression model. If there are some observations excluded due to missing data, these are not counted. What is counted here is the number of observations on which data were available that were actually included in the analysis.

5. **Term – logLik DF**: This row and the next three rows can be used to test or compare different models. They are outside the scope of this beginning introduction to statistics in R, and you are free to ignore them.

6. **Term – logLik**: See the preceding text.

7. **Term – AIC**: See the preceding text.

8. **Term – BIC**: See the preceding text.

9. **Term – F2**: This row gives Cohen's f^2 for the overall model, that is, $\frac{R^2}{1-R^2}$. Here we can see that overall, stress and social support (i.e., this regression model, since those are the two predictors in the model) have a large effect size, indicating that stress and social support together explain what is considered a large amount of the variance in positive affect.

10. **Term – R2**: This row gives the proportion of variance in positive affect in this sample of data explained by the model overall (i.e., by stress and social support together).

11. **Term – Adj R2**: This row gives the estimated proportion of variance in positive affect in the population explained by the model overall (i.e., by stress and social support together). It is the adjusted R^2, where the adjustment is to try to give a more accurate population estimate rather than just an estimate in this sample of data.

12. **Term – STRESS**: This row gives the unique effect size for stress. It shows Cohen's f^2 for stress, independent of social support. It also presents the exact p-value, rather than "just" asterisks as we saw earlier for the Fixed Effects section. In this case, we see the same f^2 we calculated manually earlier. Independent of social support, stress has between a small and medium effect size on positive affect.

13. **Term – SUPPORT**: This row gives the unique effect size for social support. It shows Cohen's f^2 for social support, independent of stress. It also presents the exact p-value, rather than "just" asterisks as we saw earlier for the Fixed Effects section. In this case, we see that independent of stress, social support has a medium effect size on positive affect.

Beyond the individual pieces of information we can learn, one final bit of information can be cleaned. Unlike regression coefficients which are not directly comparable because "one unit" on stress may be very different from "one unit" on social support, effect sizes *are* comparable. Thus, although we do not have a significance test for this, we can say that social support has a larger unique effect size than does stress on positive affect in this model.

Example

It may seem that this is all a very long process, which, in one way, it is. On the other hand, once you get familiar with the code and the process, it can happen fairly quickly. As an example of this, let us consider if flipper length and bill depth are predictors for body mass.

Having worked through the prior section, you know everything starts with a linear model. In this case, `body_mass_g` is your outcome, and both `flipper_length_mm` and `bill_depth_mm` are your predictors. You store the model in `m2Penguin`:

```
m2Penguin <- lm(formula = body_mass_g ~ flipper_length_mm + bill_depth_mm,
                data = penguinsData)
```

You again run the modelDiagnostics() using the 0.5th and 99.5th percentiles of the normal distribution to shade in extreme values:

```
m2Penguind <- modelDiagnostics(m2Penguin, ev.perc = 0.005)
plot(m2Penguind, ncol = 2, ask = FALSE)
```

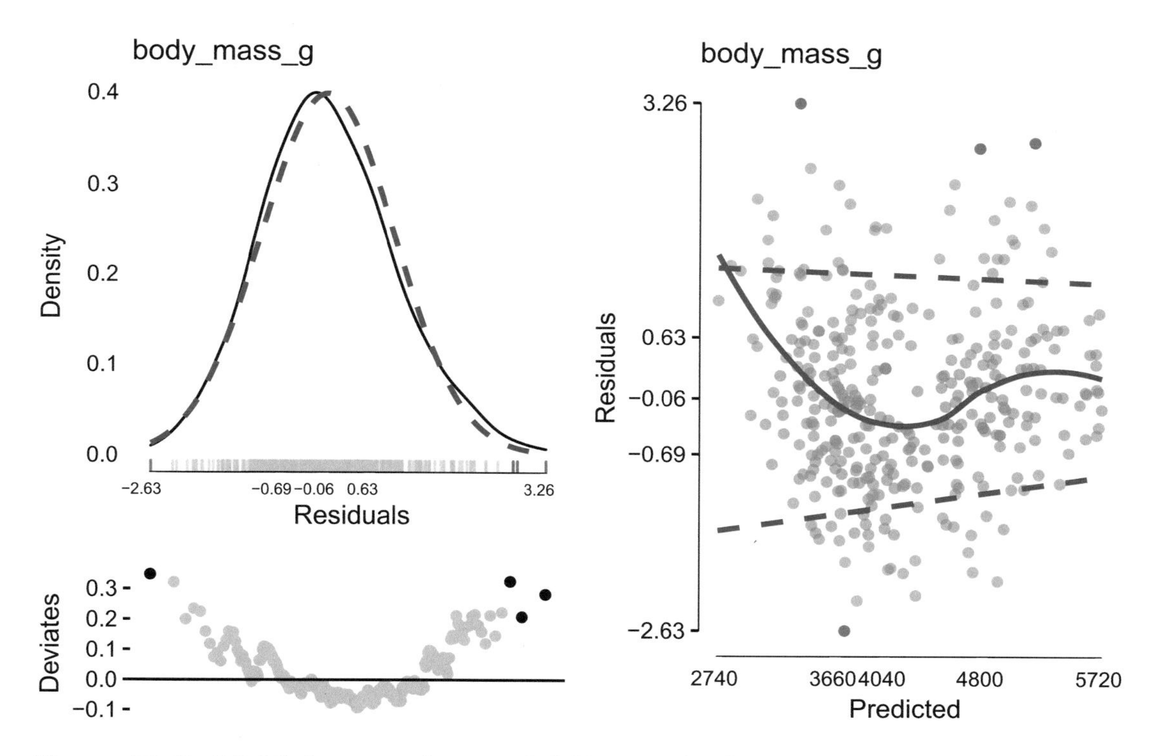

Figure 11-8. *Multiple regression model diagnostics for penguins*

In Figure 11-8, the distribution on the left looks mostly normal. The Deviates line only has a few extreme values, and you can see that the y-axis for the deviates on the lower left does not have a very wide range. On the right side, those dashed lines look fairly horizontal. Overall, this looks good enough to move forward.

Running the summary() function next, you step through the six areas we mentioned in the prior example:

```
summary(m2Penguin)

##
## Call:
## lm(formula = body_mass_g ~ flipper_length_mm + bill_depth_mm,
##     data = penguinsData)

##
## Residuals:
##      Min      1Q  Median      3Q     Max
## -1029.8 -271.5   -23.6   245.2  1276.0
##
## Coefficients:
##                   Estimate Std. Error t value Pr(>|t|)
## (Intercept)       -6541.91     540.75   -12.1   <2e-16 ***
## flipper_length_mm    51.54       1.87    27.6   <2e-16 ***
## bill_depth_mm        22.63      13.28     1.7    0.089 .
## ---
## Signif. codes: 0 '***' 0.001 '**' 0.01 '*' 0.05 '.' 0.1 ' ' 1
##
## Residual standard error: 393 on 339 degrees of freedom
##   (2 observations deleted due to missingness)
## Multiple R-squared:  0.761,  Adjusted R-squared:  0.76
## F-statistic:  540 on 2 and 339 DF,  p-value: <2e-16
```

- **Call**: As expected, this repeats the model we just built.
- **Residuals**: Looking at the lowest and highest residuals can help give a quick sense of whether there are extreme values. In addition, the median should be close to the mean of zero, if the residuals follow a normal distribution. Thus, looking for whether the median residual is about zero is another quick helpful diagnostic. In this case, -23.6 is not exactly 0. It is closer to 0 than it was back in Chapter 8 when we ran a similar model without the bill depth. You carry onward.

- **Coefficients**: In this model, the coefficients are -6541.91 and 51.54 and 22.63. These give the intercept, the expected value of penguin body mass in grams for a penguin with zero flipper length and zero bill depth (i.e., zero on all predictors). The coefficients also contain the slope of the regression line for the unique association of flipper length and body mass and of bill depth and body mass.

  The column "Std. Error" gives the standard error (SE) for each coefficient. The standard error captures how much uncertainty there is in the estimate of each coefficient due to the fact that what you analyze is a sample taken from a larger population. That is, although you want to know the population coefficients, due to only having a (assumed random) sample from the population, there could be differences in the true population regression coefficients. The standard error quantifies how uncertain you are in using your sample estimates to infer what the population coefficients are. Here the standard errors are 540.75 and 1.87 and 13.28.

  The column "t value" gives the t-value for each parameter. Although this is a regression model, just as with t-tests, the regression model uses a t-value and looks it up against the t-distribution to calculate p-values. In fact, the t-value reported is $\frac{b}{SE}$, that is, the regression coefficients divided by their standard error. Here the t-values are -12.10 and 27.64 and 1.70.

  Finally the column "*Pr*(> |*t*|)" gives the probability value, the p-value for each coefficient. Even if the predictors in the model are not statistically significant (i.e., are not less than the α (alpha) value specified), often the intercept (i.e., the expected value of the outcome when all the predictors are equal to zero) is very different from zero and so has a very small p-value close to zero. For this reason, p-values often are shown using scientific notation in R. In this case, the p-values are $p < .001$ and $p < .001$ and $p < .089$.

  The significance codes at the bottom are used to indicate what the asterisks mean. *** indicates $p < .001$, ** indicates $p < .01$, * indicates $p < .05$, . indicates $p < .10$, and a blank indicates $p > .10$.

- **Residual standard error**: σ_ε and the residual degrees of freedom (used with the t-value to calculate p-values). Here the residual standard error is 393.18. The residual degrees of freedom are calculated as $df = N - k$, which in this case is 342 – 3, 339. We also get a note in parentheses where applicable about any observations removed due to missing data. Here we get a note that two observations were removed due to missing data. Observations will be removed if they are missing data on any of the variables in the model, that is, on either the outcome or any of the predictor(s).

- **R-squared**: Next is the overall model R^2, the proportion of variance in the outcome, body mass in grams, explained by the model (here our model has two predictors: flipper length in millimeters and bill depth in millimeters) in this specific sample data. The R^2 also is the squared correlation between the outcome, here body mass, body_mass_g, and the predicted body mass, $b_0 + b_1 * flipper\_length\_mm + b_2 * bill\_depth\_mm$.

  If the R^2 is 0, it indicates that none of the variability in the outcome can be explained by the predictor. For this model, it would mean that none of the variance in body mass across penguins can be explained by their flipper length or bill depth. Put differently, it would mean that the squared correlation between the model predicted body mass and observed body mass was zero. If the R^2 was 1, it would indicate that 100% (all) of the variance in body mass could be explained by flipper length or bill depth. Again put differently, the predicted and observed body mass squared correlation was 1, which would mean a perfect correlation. That is we can perfectly predict body mass from flipper length or bill depth.

  In this model, the R^2 is 0.761, which is the proportion of variance in body mass in this sample explained by flipper length or bill depth. Additionally, the output includes the adjusted R^2, a better estimate of the true population R^2, which in this case is 0.7596. The idea behind the adjusted R^2 is that because we fit the model to our sample, it will always explain just a bit more variance in our sample data than it would in the true population and we can adjust for that. In this case, there is not much difference.

- **F-statistic**: The overall model F-statistic and degrees of freedom and p-value are provided. The F-test is an overall test of whether the model is statistically significant overall or not. It will test all the predictors simultaneously, that is, it tests whether the coefficients for flipper length and bill depth are both 0 or at least one is not zero. Because we have multiple predictors, the p-value for the F-test will not match the p-value for the t-test of a specific coefficient, as it did for the simple linear regression.

One thing to note is that it does not seem that adding in bill depth gained us all that much R^2 compared with the single predictor of flipper length back in Chapter 8. Another thing to notice is that bill depth was not as exciting in terms of p-value.

Looking at the 3D graphs in Figure 11-9 starts to show you why both those facts are true. You can see that while a greater bill depth does tilt that sheet up a little, this is not a major part of penguin body mass. Remember p-value measures how significant an effect is. If the total change is huge (e.g., I flip 20 tails in a row on a coin), a small sample size is enough for a significant p-value. On the other hand, even a smaller total change (e.g., just three tails in a row) is perhaps significant if I and two of my friends get that. In this case, the p-value is 0.089 which shows that there is a weaker total change and the sample size is not large enough to ensure that change is as significant as it might be (e.g., vs. flipper length).

Next, you investigate more closely, inspecting the two dimension graphs using `visreg()`. In Figure 11-10, you see comparatively wide 95% confidence interval bands. The range of the bill depth effects only a smallish change in body mass:

```
visreg(m2Penguin, xvar = "bill_depth_mm", partial = FALSE, rug = FALSE,
gg = TRUE) +
  ggtitle("Association of bill depth and body mass, adjusted for flipper
  length")
```

For comparison, you consider the two-dimensional graph for flipper length in Figure 11-11, and notice tighter confidence interval bands and more total movement on the y-axis:

```
visreg(m2Penguin, xvar = "flipper_length_mm", partial = FALSE, rug = FALSE,
gg = TRUE) +
  ggtitle("Association of flipper length and body mass, adjusted for bill depth")
```

The last bit of evidence that bill depth is probably not something we want to use in a model is shown when we look at the output from the `modelTest()`. The unique effect size for `bill_depth_mm` using Cohen's f^2 is 0.01. Recall that a small effect size starts at *greater* than 0.02. Bill depth just does not have much effect on penguin body mass:

```
m2PenguinTest <- modelTest(m2Penguin)
print(APAStyler(m2PenguinTest), nrow = 100)
```

```
##                   Term                                   Est          Type
## 1:         (Intercept) -6541.91*** [-7605.56, -5478.26] Fixed Effects
## 2: flipper_length_mm     51.54*** [   47.87,    55.21] Fixed Effects
## 3:       bill_depth_mm       22.63 [   -3.49,    48.76] Fixed Effects
## 4:  N (Observations)                                 342 Overall Model
## 5:          logLik DF                                  4 Overall Model
```

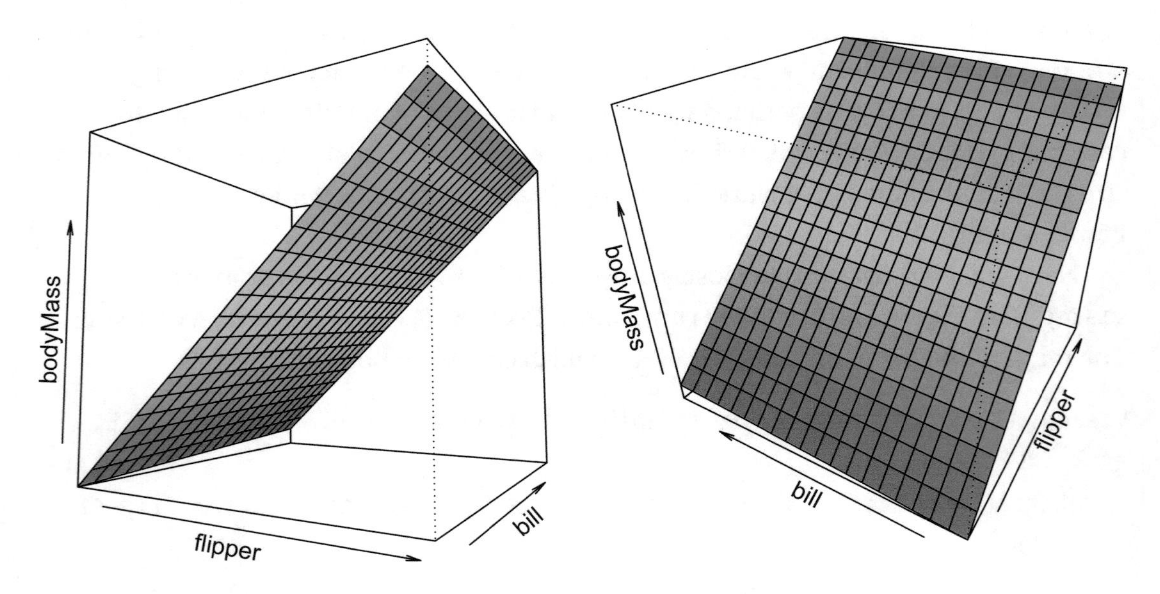

Figure 11-9. *3D graph of a multiple regression surface on penguins*

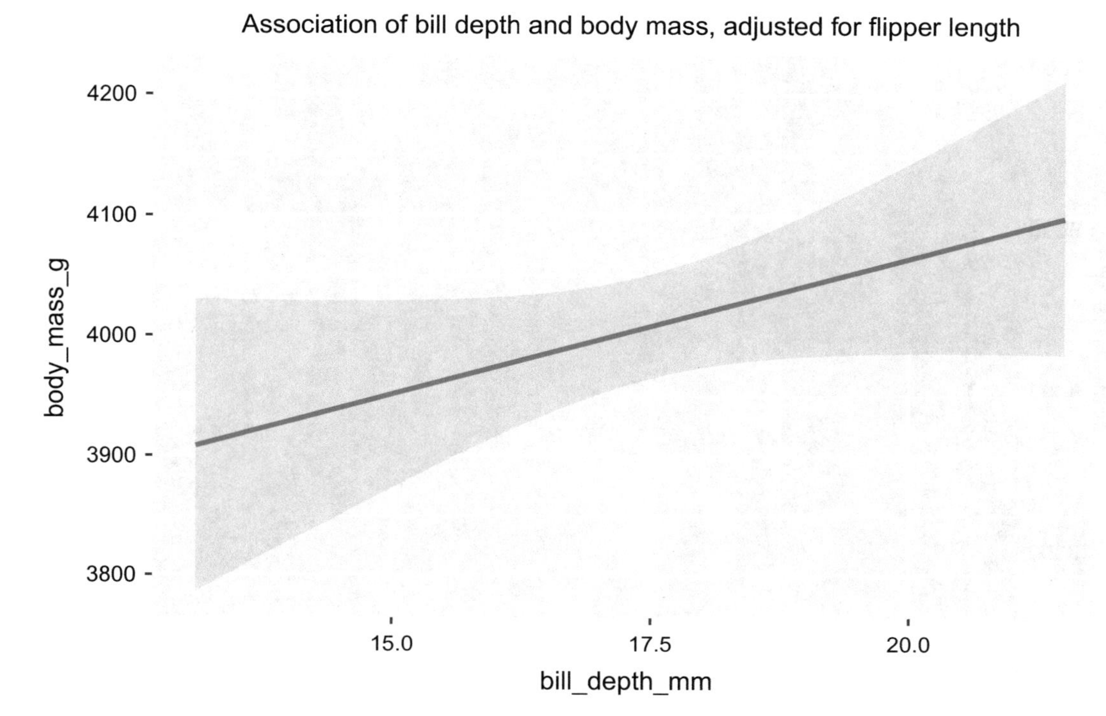

Figure 11-10. *Multiple regression of flipper length and bill depth predicting penguin body mass. This figure shows only the unique bill depth and body mass association*

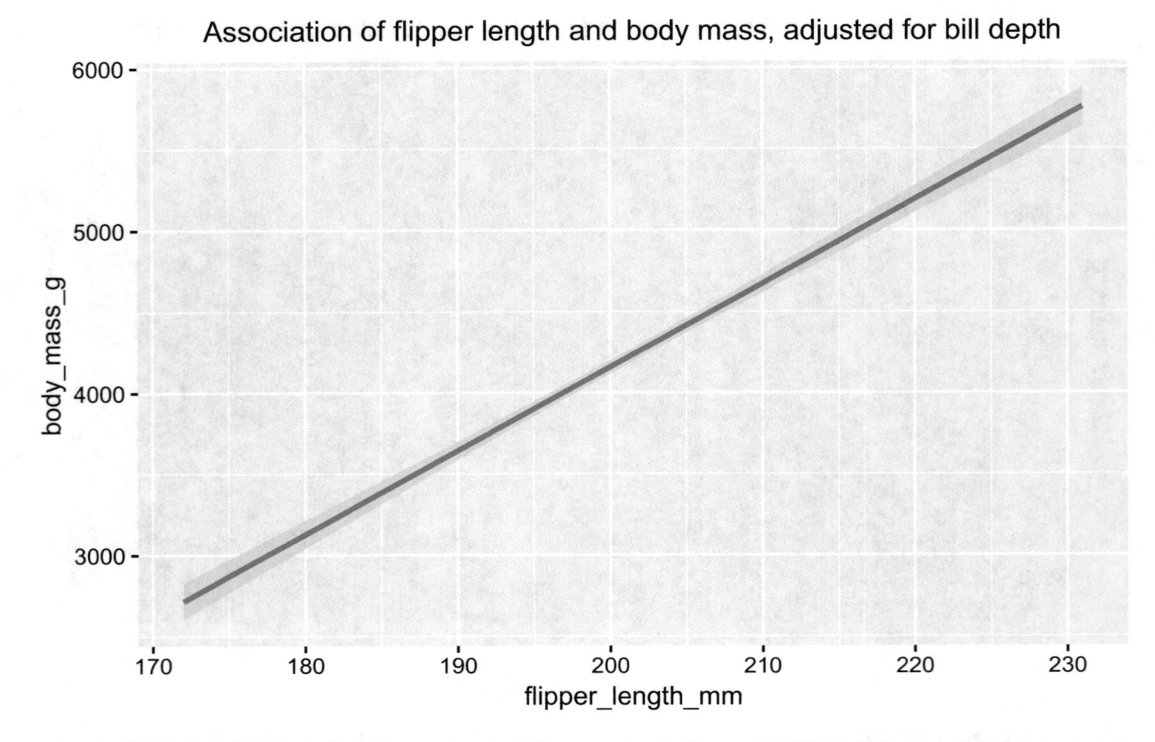

Figure 11-11. *Multiple regression of flipper length and bill depth predicting penguin body mass. This figure shows only the unique flipper length and body mass association*

```
##  6:            logLik                  -2526.97 Overall Model
##  7:               AIC                   5061.94 Overall Model
##  8:               BIC                   5077.28 Overall Model
##  9:                F2                      3.18 Overall Model
## 10:                R2                      0.76 Overall Model
## 11:            Adj R2                      0.76 Overall Model
## 12: flipper_length_mm  f2 = 2.25, p < .001  Effect Sizes
## 13:     bill_depth_mm  f2 = 0.01, p = .089  Effect Sizes
```

Now that you have seen this second example, notice how quickly you might experiment with different variables in a data set. Had you imagined that bill depth allows for more fish to be caught, you might suppose greater bill depth has some sort of outsize effect on penguin mass. It seems the effect is so slight; perhaps the only effect at all is the actual extra mass of the bill.

Assumption and Cleaning

Multiple regression has the same assumptions as linear regressions, so you can refer to Chapter 8 for more details.

The only diagnostics that typically vary between simple linear regression and multiple regression are (1) a linear association is assumed for all predictors, not just one, and (2) sometimes it can be helpful to check whether the predictors are highly correlated. The second issue, highly correlated predictors, is often called multicollinearity. That is a big word, although the idea is not hard.

Suppose you wanted to run a regression predicting the job salary of alumni from your school using grades from their classes. You have data on one class, which had a set of quizzes and a final exam. You have the total class score, the total quiz score, and the final exam score. You could include quiz and final exam scores or quiz and total scores or exam and total scores, but not all three. You could not include all three because the total class score is just the sum of the quiz and the exam scores. Even though you have three variables and no two variables are perfectly the same, two of the variables perfectly predict the third. This is what is meant by multicollinearity. There may appear to be a certain number of variables, but one or more of those variables do not really have unique information.

In practice, even if one of your predictors is not perfectly predicted by the rest, it can still be problematic. For example, you may have variables that are highly correlated, such as a correlation $r > .90$ or $r < -.90$ between two variables. The reason having very highly correlated variables in the same regression can be problematic is that if you recall, one use of multiple regression is to identify the **unique** contribution of a variable. However, if you have two predictors that are correlated $r = .95$, there is very little unique information in either one. Almost all their predictive performance will be overlapping, not unique. In the most extreme cases, like the total class score being exactly the sum of the final exam and quiz scores, there is no unique solution to the regression, so what R normally does is drop one predictor for you as the regression cannot be estimated otherwise.

Generally speaking, multicollinearity is not a major problem. Also, it often is possible to know whether it may be an issue or not simply by knowing what your predictors are. For example, if you had the classroom data, even without calculating any statistics, you could know that the quiz and final exam scores would be collinear with the total score. We find that in practice, beyond common sense about whether two predictors capture basically the same thing, multicollinearity is not a common problem and rarely needs to be addressed. If you do have it or are concerned about it, the most common approach to resolving it is simply to drop one or more predictors. There are some other options, but they are beyond the scope of a beginning book.

Example

To give an example of assumption checking, we revisit the multiple regression example we have been using in this chapter. We briefly looked at the assumptions, and they were not too bad. Now we go through them more carefully. To start off, we can quickly check whether stress and social support are very highly correlated. Multicollinearity is not likely a problem as stress and social support are different measures; however, if you were not certain, you could always look at the correlations:

```
acesData[, cor(STRESS, SUPPORT, use = "pairwise.complete.obs")]
```

```
## [1] -0.03476
```

The correlation between stress and social support is actually very close to zero, so multicollinearity will not be any issue. Next, we can use the `modelDiagnostics()` function to calculate diagnostics and the `plot()` function to create a visual display:

```
## assess model diagnostics
m2d <- modelDiagnostics(m2, ev.perc = .005)
plot(m2d, ncol = 2, ask = FALSE)
```

As noted before, the normality assumption is typically tested by comparing the residuals from the model against a normal distribution. The independence assumption cannot readily be tested, but we normally assume it is true if participants are independent of each other (e.g., they are not siblings, we do not have repeated measures on the same person, etc.).

The homogeneity of variance assumption is typically assessed by plotting the residuals against predicted values. The expectation is that across the range of predicted values, the spread (variance) of the residuals is about equal and that there should not be any systematic trends in the residuals.

Although not an assumption, we often want to identify outliers or extreme values that may unduly influence our findings. This is often done by examining the residuals from the model.

The **left panel** of Figure 11-12 shows a density plot of the residuals (solid black line) and for reference the density of what a perfectly normal distribution would look like with the same mean and residual variance (dashed blue line). The rug plot under the density plot (small, vertical lines) shows where raw data points fall. The x-axis labels (for the residuals) show the minimum, 25th percentile; median, 75th percentile; and maximum, providing a quick quantitative summary.

The lower side of the left panel shows a QQ deviates plot; this is related to what is often called a QQ plot. It plots the observed quantiles on the x-axis against the deviations between the observed and theoretical normal quantiles on the y-axis. If the points fall exactly on the line, that indicates they fall exactly where they would be expected under a normal distribution. Both the rug plot and the dots plotted in the QQ deviates plot are light gray if they are not extreme and are solid black if they meet the criteria for an extreme value, as specified in the call to `modelDiagnostics()`. In the case of Figure 11-12, that lower-left panel has two such extreme values.

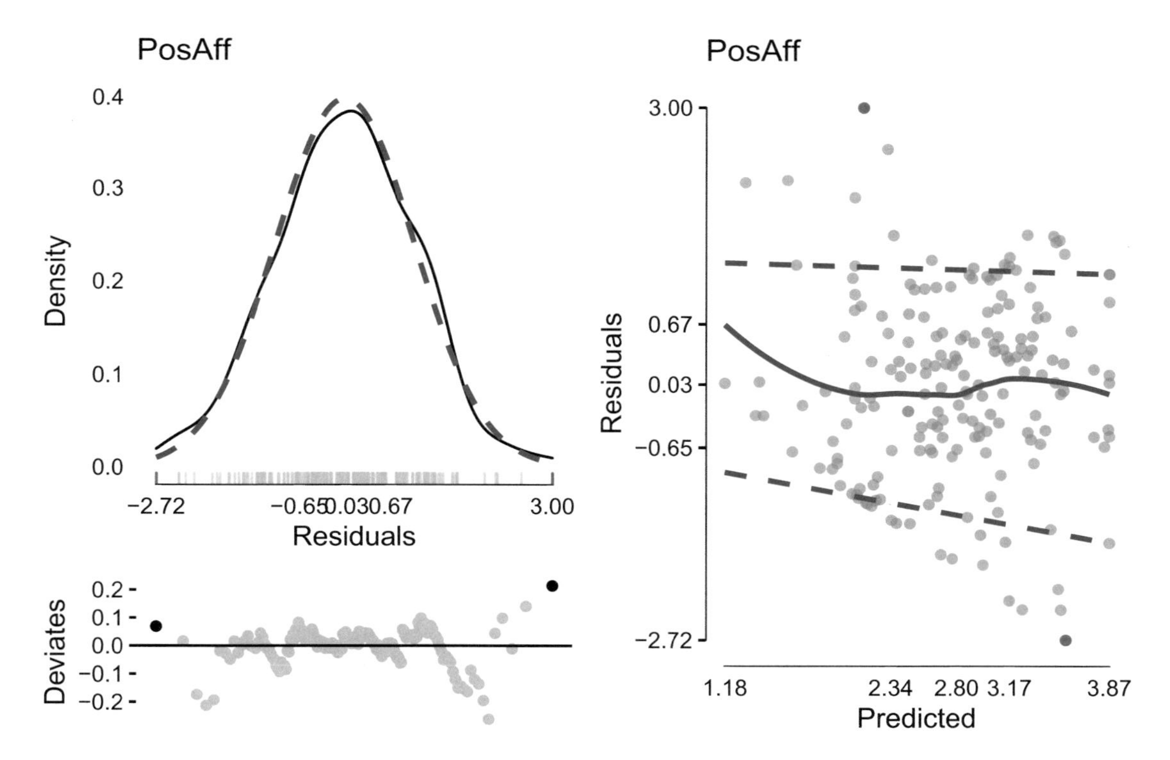

Figure 11-12. *Regression model diagnostics. Left panel shows a plot of the residuals to assess normality, and they will be highlighted black if an extreme value. The right panel shows a plot of the predicted vs. residual values to help assess the homogeneity of variance assumption*

The **right panel** of Figure 11-12 shows a scatter plot of the model-predicted values against the residuals. This can help assess the homogeneity of variance assumption. If the residual variance is homogenous, we would expect the spread of the residuals (the y-axis) to be about the same at all levels of the predicted values. Sometimes this is very easy to see visually. Sometimes it is not so easy to see. To aid in the visual interpretation, quantile regression lines are added. Quantile regression lines allow a regression model to predict the 10th and 90th percentiles. These predicts are added as dashed blue lines. Under the assumption of homogenous variance, one would expect these lines to be approximately horizontal and parallel to each other. If they are not parallel, that indicates that the variance is larger/smaller for some predicted values than for others, known as heterogeneous variance. The solid blue line in the middle is a loess smooth line, which also would ideally be about flat and about zero, to indicate there is no systematic bias in the residuals. Systematic bias can occur, for example, when for low predicted values, the residuals are always positive and for high predicted values the residuals are always negative and indicate, typically, either ceiling/floor effects in the data and/or issues that would ideally be addressed by revising the model. As with the density plot, the axes in the scatter plot show the minimum, 25th percentile; median, 75th percentile; and maximum, providing a nice quantitative summary of the residuals and predicted values.

These graphs look fairly good. They are not perfect, but real data never are and the degree of variation in the spread of residuals is not extreme. We might be reasonably satisfied that the assumption or normality and homogeneity of variance are met or at least not terribly violated.

Example

We look at another outcome measure, NegAff, which is negative affect. Again we will use STRESS and SUPPORT as predictors:

```
malt <- lm(NegAff ~ STRESS + SUPPORT, data = acesData)

## assess model diagnostics
maltd <- modelDiagnostics(malt, ev.perc = .005)
plot(maltd, ncol = 2, ask = FALSE)
```

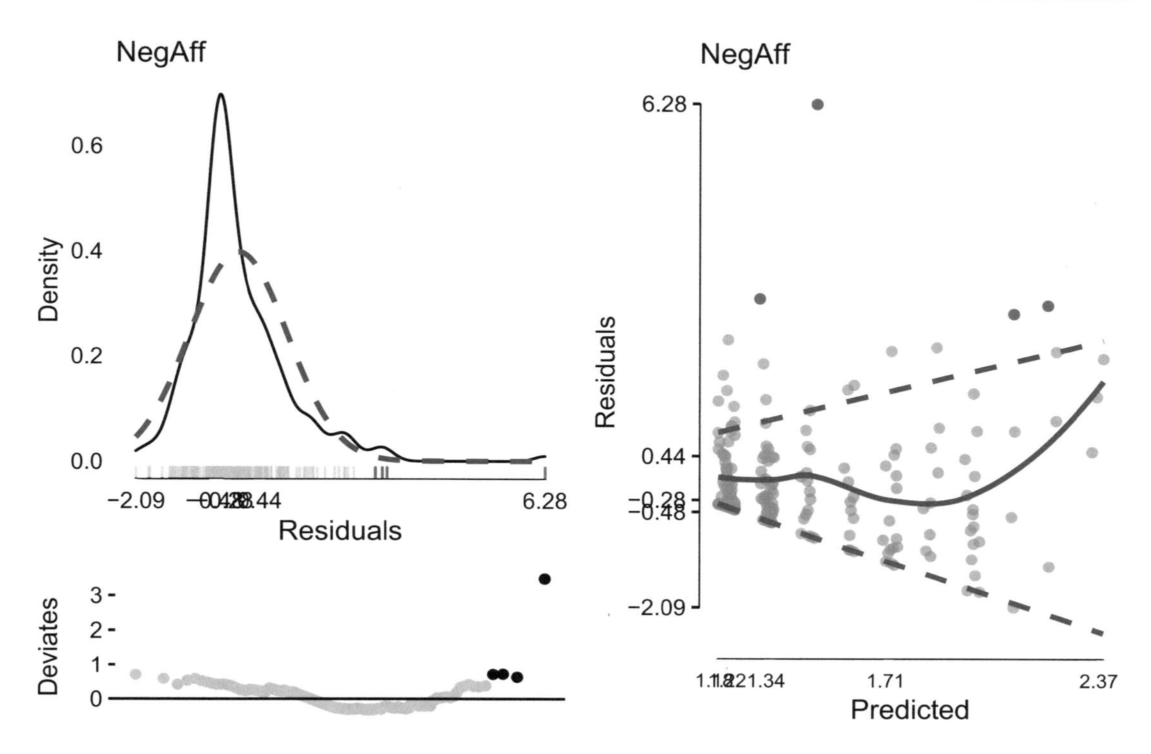

Figure 11-13. *Regression model diagnostics. Left panel shows a plot of the residuals to assess normality, and they will be highlighted black if an extreme value. The right panel shows a plot of the predicted vs. residual values to help assess the homogeneity of variance assumption*

These graphs shown in Figure 11-13 show clear problems in the model. The residuals are not at all normally distributed. There are extreme values, based on the current criteria. The homogeneity of variance plot does not appear as bad, but the extreme values on the residuals are apparent. At a minimum, we would want to remove these extreme values. However, the marked non-normality also is a concern. While the mathematical rationale is beyond the scope of this text, one way to attempt to "fix" large, positive extreme values is by using a logarithmic transformation. For the sake of example, we do this here. This "magic" of log (and other transformations) is discussed in depth in another book [22].

As is true with most magic, log transformations come at a price and are only defined if all outcomes scores are greater than zero. At zero or below, the log is not defined.

We fit a new model, malt2, using log() to get the log of negative affect and use that as our outcome. Again we plot diagnostics, shown in Figure 11-14:

```
malt2 <- lm(log(NegAff) ~ STRESS + SUPPORT, data = acesData)

## assess model diagnostics

malt2d <- modelDiagnostics(malt2, ev.perc = .005)
plot(malt2d, ncol = 2, ask = FALSE)
```

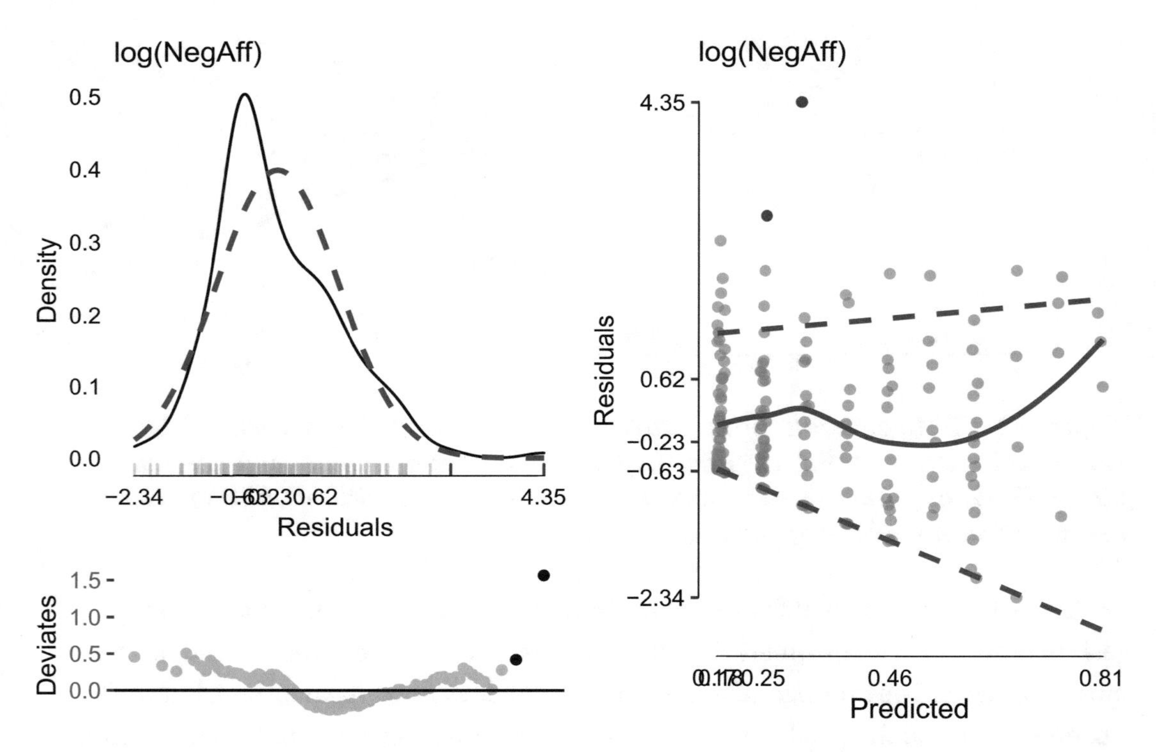

Figure 11-14. *Regression model diagnostics after log transforming the outcome*

Even after the log transformation, the extreme values remain.

What if we removed those?

Extreme values are stored and accessible within the diagnostics object output, malt2d. It shows the scores on the outcome, the index (the row in the data the extreme values come from), and the effect type where they are an extreme value, in this case, extreme on the residuals. We use the Index values to remove them from the dataset and then refit our model. We can use - in front indices to tell data table that instead of selecting those rows, we want to **exclude** those rows:

```
malt2d$extremeValues

##     log(NegAff) Index EffectType
## 1:        1.565   123  Residuals
## 2:        1.066   174  Residuals

malt3 <- lm(formula = log(NegAff) ~ STRESS,
            data = acesData[-malt2d$extremeValues$Index])

## assess model diagnostics
malt3d <- modelDiagnostics(malt3, ev.perc = .005)
plot(malt3d, ncol = 2, ask = FALSE)
```

The results are shown in Figure 11-15.

After excluding the initial extreme values, one new extreme value emerges. However, it is relatively less extreme; and overall the model diagnostics look much better now, although there is still evidence the variance of the residuals is not equal. Notice how the negative residuals are much smaller for low predicted values and get more negative as the predicted values go up. There is not much more we can do at this point to improve the model assumptions. The possible, perhaps even likely, violation of equal residual variance (homogeneity of variance) is something we would acknowledge as a limitation in any report and consider when interpreting the results from this multiple regression (i.e., the results may be slightly biased due to the assumption likely being at least somewhat violated).

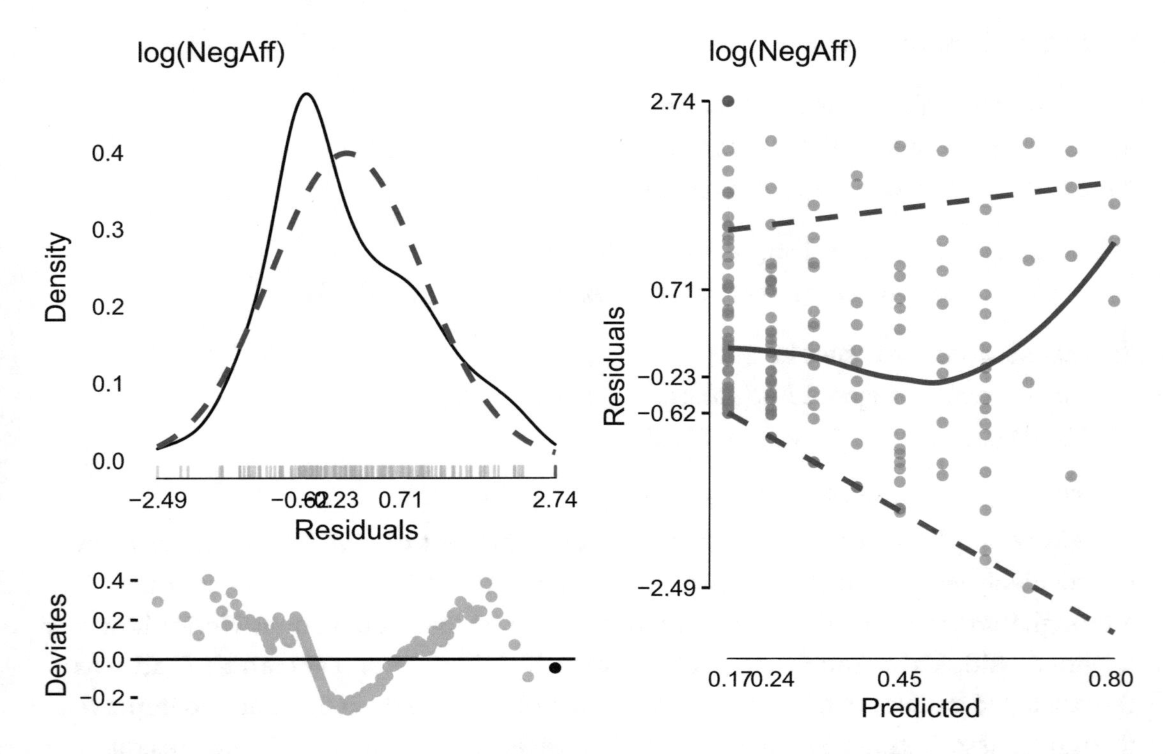

***Figure 11-15.** Regression model diagnostics after log transforming the outcome and removing extreme values*

More advanced statistics books cover alternate analyses that do not require the same assumptions, but those methods are quite a bit more complex. Indeed, logarithmic transformation and removing extrema take a deft hand. Our purpose in showing you this messier example is predominately to show that assumption checking is needed. A useful secondary purpose is to encourage you to continue learning after this book; there are generally ways of transforming data to wring out some quantity of useful information. It often simply takes more time and practice.

11.4 Categorical Predictors

In addition to continuous variables, categorical variables can be included as predictors/explanatory variables in regression. The most common approach to including a categorical predictor is to use dummy coding. The basic idea behind dummy coding is to turn a categorical predictor with *k* levels into *k* separate dummy code variables, where each dummy coded variable is coded as 1 for a particular level of the variable and

0 otherwise. The simplest case of this is a binary variable, a categorical variable with only two levels (e.g., younger adults, older adults) where you might create a new dummy coded variable where 1 = older adults and 0 = younger adults. This dummy coded variable is now a numeric predictor (it has 0s and 1s) and can be included in regression as usual. The name dummy coding comes from the idea that you create a set of substitute (i.e., dummy) variables. It also is sometimes called one-hot encoding because it creates new variables where one level of the old variable is "active" or "hot" (hot in the sense of a live wire) at a time, coded as `1`.

The only real difference between interpreting a dummy coded variable versus a regular continuous predictor is that (1) the 0 point is not arbitrary, but represents one particular group (e.g., younger adults) and (2) a one-unit change in the dummy coded variable is not arbitrary, but represents the difference between groups (because 0 to 1 **is** a one-unit change).

To see this in action, we can take a look at the first few rows of the `mtcars` dataset. There is a variable, `cyl`, which has the number of cylinders in the car: 4, 6, or 8. The following table shows the first few rows of `cyl` and what dummy coded variables would look like:

```
##                   cyl dummy_cyl4 dummy_cyl6 dummy_cyl8
## Mazda RX4           6          0          1          0
## Mazda RX4 Wag       6          0          1          0
## Datsun 710          4          1          0          0
## Hornet 4 Drive      6          0          1          0
## Hornet Sportabout 8          0          0          1
## Valiant             6          0          1          0
```

In that example, look at how, for example, `dummy_cyl4` is coded as 1 when *cyl* = 4 and otherwise is coded as zero.

Although we can always create one dummy code variable for each unique level of a categorical variable, in a regression model we only include $k - 1$ dummy code variables. The reason for this is that the dummy coded variables have multicollinearity otherwise. For example, continuing with the `cyl` example, we know that every car in that dataset has either four, six, or eight cylinders. If we include `dummy_cyl4` and `dummy_cyl6` in a model, if a car is 1 on cyl4, we know it has four cylinders. If it is 1 on cyl6, we know it has six cylinders. If it is zero on both cyl4 and cyl6, the only remaining option is for it to be an eight-cylinder car.

The general rule is that if you have a *k*-level categorical predictor, you can create *k* dummy codes and include *k* – 1 dummy codes in your regression model. In the cylinder case, there were three levels; and we can create three dummy codes, but we would only include two dummy codes in any regression model.

Our book, *Advanced R Statistical Programming and Data Models: Analysis, Machine Learning, and Visualization* [22], goes into more detail about dummy coding.

Another book is not required, however, because using dummy coding to include categorical predictors in regression is built into R. R automatically dummy codes any factor variable you enter into a regression as a predictor, so you do not have to do it manually. If the variable is not a factor, you may want to convert it to a factor first using the `factor()` function.

Take a look at a simple linear regression with a categorical predictor. We fit the model and store these results in `mcat1`. We can also graph the results as we have for previous regression models. For this example, we will use positive affect as our outcome but use education level as our predictor. EDU is a binary variable, 0/1, indicating whether people have completed a bachelor's degree (1) or not (0). First, we make it a factor variable, and then we just add it to our linear regression as usual:

```
acesData[, EDU := factor(EDU)]

mcat1 <- lm(formula = PosAff ~ EDU,
            data = acesData)

APAStyler(modelTest(mcat1))

##                   Term                       Est           Type
##  1:        (Intercept) 2.82*** [ 2.62, 3.02] Fixed Effects
##  2:               EDU1   -0.13 [-0.47, 0.20] Fixed Effects
##  3: N (Observations)                     187 Overall Model
##  4:          logLik DF                     3 Overall Model
##  5:             logLik               -284.24 Overall Model
##  6:                AIC                574.47 Overall Model
##  7:                BIC                584.16 Overall Model
##  8:                 F2                  0.00 Overall Model
##  9:                 R2                  0.00 Overall Model
## 10:             Adj R2                  0.00 Overall Model
## 11:                EDU   f2 = 0.00, p = .436  Effect Sizes
```

```
visreg(mcat1, xvar = "EDU", partial = FALSE, rug = FALSE, gg = TRUE) +
  ggtitle("EDU as a categorical predictor")
```

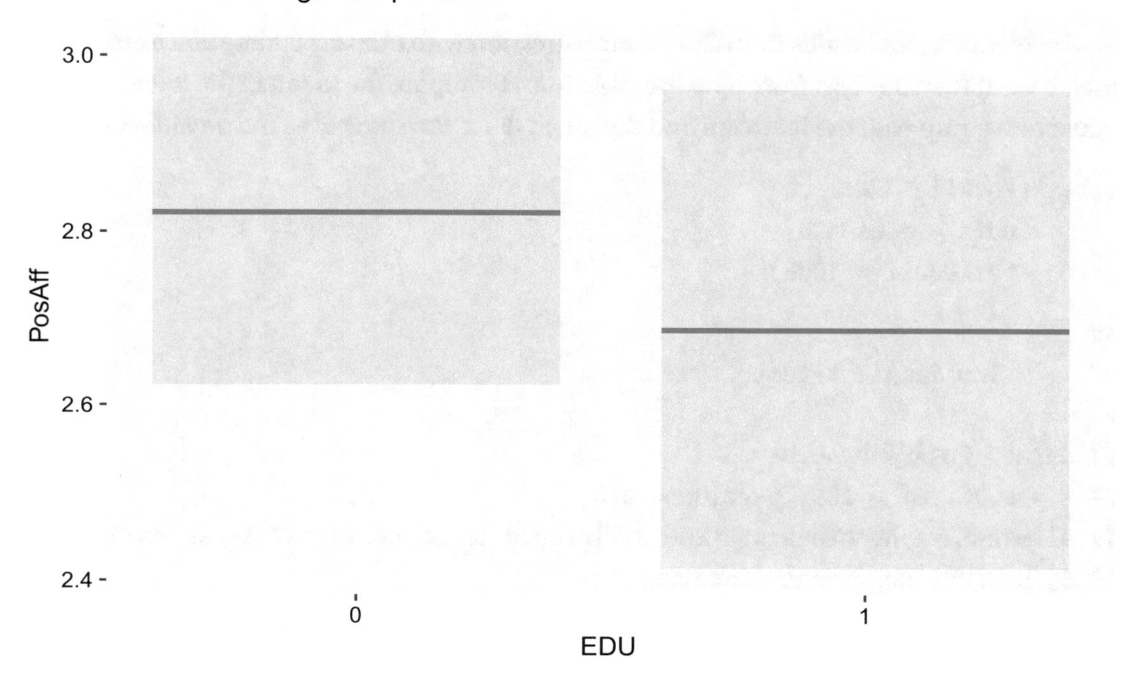

Figure 11-16. *Simple linear regression plot from visreg with a categorical predictor, education, showing the predicted positive affect and 95% confidence intervals for each education level*

Figure 11-16 shows the predicted positive affect for each level of education in the solid blue lines, and the shaded gray boxes are the 95% confidence intervals around the predicted means. In this simple example, since there are no other predictors, the intercept, the expected value of positive affect when the predictors are 0, is the mean positive affect level in people with lower education ($EDU = 0$), and the regression coefficient for `EDU` is the **difference** in positive affect on average between people with lower education and those with higher education, which we can see is not statistically significant.

It may seem strange at first to refer to the regression coefficient for `EDU1` as a mean difference. After all, we have always talked about the regression coefficients as the slope of a line. Both are actually true. The regression coefficient for `EDU1` is the slope of a line. Because 0 indicates lower education and 1 indicates higher education, a one-unit change (i.e., the slope of that line) is exactly the predicted difference between people with lower and higher education. In fact, the whole reason that dummy coding is a popular technique is because it allows us to use the way regression works, calculating slopes, to get meaningful information about categorical predictors.

Incidentally, the results from this linear regression with dummy coding will exactly match what we would get from a two-sample t-test (compare the mean in the lower education group with the intercept and the mean difference as well as the p-value for EDU1):

```
t.test(PosAff ~ EDU,
       data = acesData,
       var.equal = TRUE)

##
##      Two Sample t-test
##
## data:  PosAff by EDU
## t = 0.78, df = 185, p-value = 0.4
## alternative hypothesis: true difference in means is not equal to 0
## 95 percent confidence interval:
##  -0.2044  0.4721
## sample estimates:
## mean in group 0 mean in group 1
##           2.822           2.688
```

The reason we would want to learn how to do this in regression, however, is that unlike a t-test, regression can include multiple predictors and a mix of categorical and continuous predictors, allowing far more complex and flexible models to be fit than would be possible with a t-test.

Example

Next, consider using a categorical predictor with three levels instead of only two. At the start of this chapter, we had some code to create a categorical stress variable that was coded as stress 0, 1, or 2+ and stored as a factor variable, called StressCat.

We will use this categorical variable to predict positive affect. Note that because StressCat has three levels, even though we only add one predictor, once it is dummy coded, two dummy coded predictors get added into the model. We also will run some model diagnostics using `modelDiagnostics()`, the same as if we had continuous predictors. The results are in Figure 11-17. The residuals vs. predicted values plot likely looks a bit different from others that we have seen because with a categorical predictor, we do not really generate continuous predictions; there are only three possible predicted values, although the observed outcome data is continuous so the residuals continue to be continuous:

```
mcat2 <- lm(formula = PosAff ~ StressCat,
            data = acesData)

plot(modelDiagnostics(mcat2), ncol = 2)
```

The diagnostic results look fairly good, despite the discrete nature of predicted values (which does not violate any regression assumption – only the outcome must be continuous). Next we can again use the `modelTest()` function to get model results and effect sizes:

```
APAStyler(modelTest(mcat2))

##                    Term                         Est          Type
##  1:         (Intercept)  3.14*** [ 2.88,  3.39] Fixed Effects
##  2:          StressCat1    -0.32 [-0.74,  0.11] Fixed Effects
##  3:         StressCat2+ -0.69*** [-1.03, -0.34] Fixed Effects
##  4: N (Observations)                        187 Overall Model
##  5:           logLik DF                       4 Overall Model
## ---
##  8:                 BIC                  575.22 Overall Model
##  9:                  F2                    0.08 Overall Model
## 10:                  R2                    0.08 Overall Model
## 11:              Adj R2                    0.07 Overall Model
## 12:           StressCat      f2 = 0.08, p < .001  Effect Sizes
```

In the results, we can see that two dummy codes for `StressCat` were included. The omitted dummy code variable is `StressCat0` which means that when the other two dummy codes are zero, the intercept of the model is the predicted positive affect for someone with no stress. Thus, we now can interpret the intercept as the predicted positive affect score when all predictors are zero, which in this case means *StressCat* = 0.

The first predictor is `StressCat1` which captures the difference between people with no stress and people with a stress score of 1. The negative sign indicates that people with a stress score of 1 have lower positive affect than do those with 0 stress, but we can see it is not statistically significantly different from zero and the 95% confidence interval includes zero.

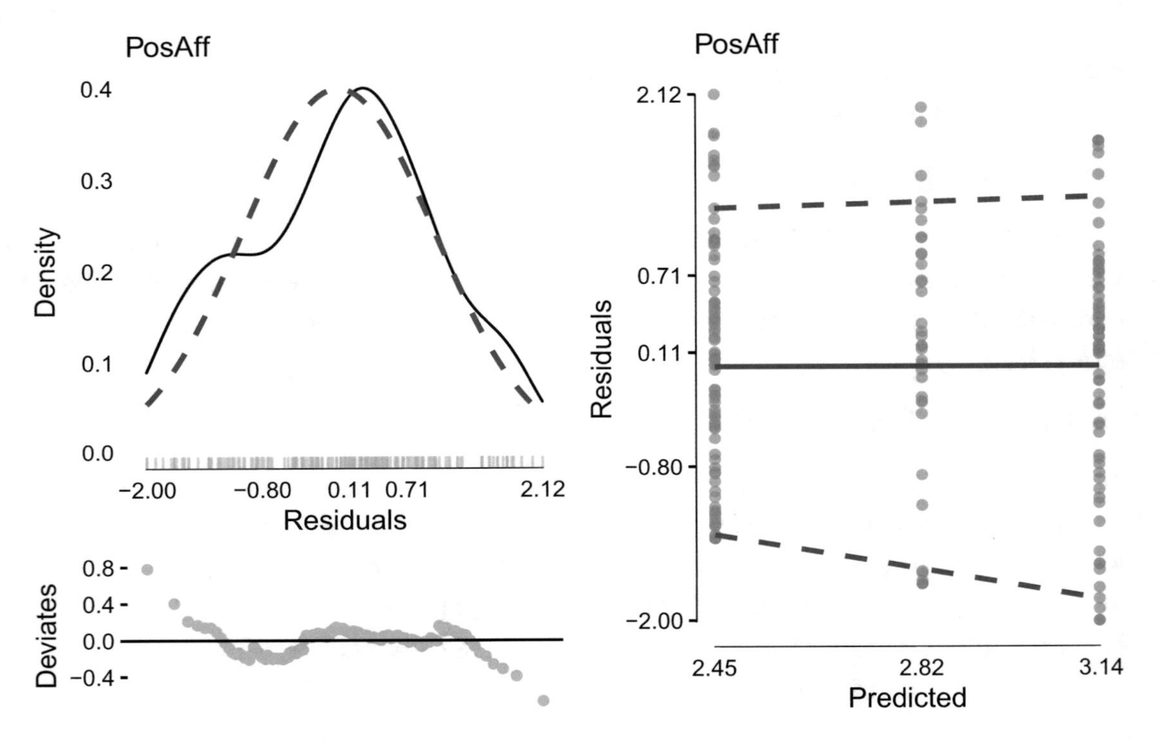

Figure 11-17. *Regression diagnostic plot when there is only a categorical predictor*

The second predictor is `StressCat2+` which captures the difference between people with no stress and people with a stress score of 2+. The negative sign indicates that people with a stress score of 2+ have lower positive affect than do those with 0 stress, and we can see it is statistically significant, $p < .001$, and the 95% confidence interval does not include zero.

From `modelTest()`, we also get an effect size estimate. Note that this is only one effect size and one p-value. The effect size and p-value shown at the end are actually for the entire `StressCat` variable. That is, it tests whether the variable as a whole (both dummy codes together) is associated with positive affect. This is a helpful way to know whether a categorical predictor as a whole is statistically significant and what its overall effect size is.

One thing you may wonder is that although both dummy codes tell us whether people with stress of 1 or 2+ differ in positive affect from people with stress of 0, we do not have a test of whether people with stress of 1 differ from those with stress of 2+. That is, we do not have all possible pairwise comparisons between levels of our categorical predictor. This is not an issue when you only have two levels, but with categorical predictors that have three or more levels, you can have many pairwise comparisons. There are generally two options. First, you can change which dummy code variable you omit, which will become the new "comparison" group and let you get other pairwise comparisons. However, this requires refitting models, and if you had say a five-level categorical predictor, you could have to do it many times. The other approach is to use the emmeans package [13] which has helpful functions for running additional tests on regression models. We will explore that option. There are two functions we need to use. First, we use the emmeans() function, which we give the regression model as the first argument and the variable we want to calculate means on as the specs argument. We save these results and print them. What is shown is the estimated positive affect mean in each group along with 95% confidence intervals:

```
mcat2.means <- emmeans(object = mcat2, specs = "StressCat")
mcat2.means
```

```
## StressCat emmean    SE  df lower.CL upper.CL
## 0           3.14 0.130 184     2.88     3.39
## 1           2.82 0.174 184     2.48     3.16
## 2+          2.45 0.119 184     2.22     2.69
##
## Confidence level used: 0.95
```

Now that we have an object with the mean positive affect for each level of stress category, we can use the pairs() function to get all possible pairwise comparisons. By default, pairs() also adjusts the p-values using Tukey's method for the multiple comparisons. In the output, we can see the mean difference between each group, labeled the "estimate" and the "p.value," which indicates whether those two groups are significantly different from each other. From the results, we can see that people with 2+

stress are significantly different from those with 0 stress (based on the p-value column in that middle row), but the other comparisons are not statistically significant:

```
pairs(mcat2.means)

## contrast estimate    SE  df t.ratio p.value
## 0 - 1        0.316 0.217 184 1.451   0.3172
## 0 - (2+)     0.685 0.177 184 3.880   0.0004
## 1 - (2+)     0.370 0.211 184 1.751   0.1893
##
## P value adjustment: tukey method for comparing a family of 3 estimates
```

Example

For our final example, we look again at StressCat but this time in a multiple regression where we also have a continuous predictor, social support, SUPPORT. We start by fitting the model in what is hopefully now a familiar way and then plotting some model diagnostics, shown in Figure 11-18:

```
mcat3 <- lm(formula = PosAff ~ StressCat + SUPPORT,
            data = acesData)

plot(modelDiagnostics(mcat3), ncol = 2)
```

The diagnostics look fairly good so we may proceed. First, we graph the results to help with understanding and interpreting them. We use the visreg() function as before. We give it our multiple regression model results stored in mcat3. The x-axis variable is SUPPORT. Generally it is best to put continuous predictors on the x-axis. We also do something new. We add StressCat to the by argument. This will create a separate regression line for each level of StressCat. The overlay=TRUE argument means we want all three lines in one plot, not separate panels:

```
visreg(mcat3, xvar = "SUPPORT", by = "StressCat",
       partial = FALSE, rug = FALSE, overlay = TRUE,
       gg = TRUE) +
  ggtitle("SUPPORT and StressCat as predictors")
```

The results are in Figure 11-19. We can see there are three lines, one for each level of stress category. However, even though we have three lines, all of them have parallel slopes. This is because the association between social support and positive affect is assumed to be the same. All that has changed is that the level - the height - of the lines is allowed to be different for different stress categories. We can see the association of stress as well by comparing the height of the three lines. For example, it becomes clear that people with 2+ stress have the lowest positive affect in general, although the slope of social support is large enough that, for example, people with support of 10 and stress of 2+ are still predicted to have higher positive affect than someone with stress of 0 but support of 0.

The coefficients for the regression model capture a few different parts of Figure 11-19. The intercept will be the expected positive affect when all predictors are zero (i.e., 0 stress and 0 support). The regression coefficients for stress category capture the difference, relative to intercept, in the height of the lines. The slope for social support captures the slope of the lines in the figure. Now we will use `modelTest()` to get the numerical model results:

```
APAStyler(modelTest(mcat3))

##                 Term                       Est          Type
## 1:       (Intercept)  2.12*** [ 1.74,  2.50] Fixed Effects
## 2:        StressCat1    -0.22 [-0.61,  0.17] Fixed Effects
```

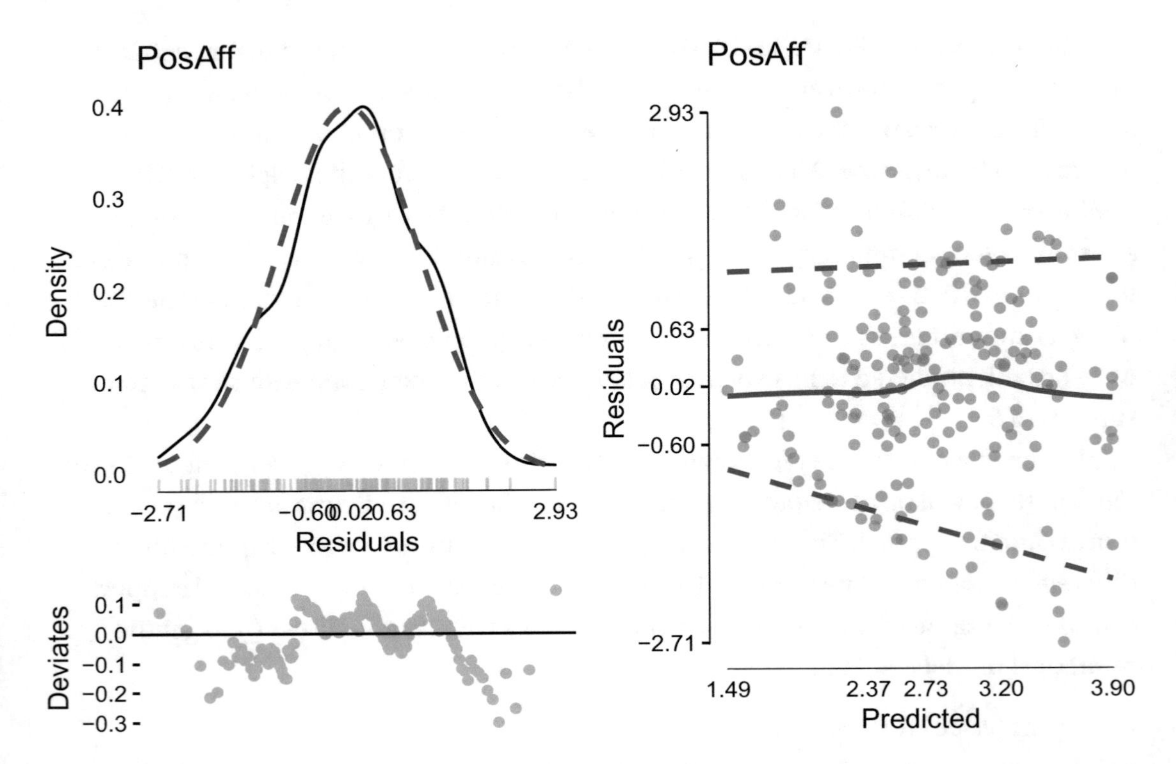

***Figure 11-18.** Regression diagnostic plot when there is a categorical predictor and a continuous predictor*

```
##  3:      StressCat2+ -0.63*** [-0.94, -0.32] Fixed Effects
##  4:          SUPPORT  0.18*** [ 0.13,  0.23] Fixed Effects
##  5: N (Observations)                     187 Overall Model
## ---
## 10:               F2                    0.35 Overall Model
## 11:               R2                    0.26 Overall Model
## 12:           Adj R2                    0.25 Overall Model
## 13:        StressCat     f2 = 0.09, p < .001  Effect Sizes
## 14:          SUPPORT     f2 = 0.25, p < .001  Effect Sizes
```

In these results, we can again see that stress category 2+ has significantly lower positive affect than stress category 0, now controlling for social support. Stress category 1 still does not differ from 0 (again, you can see this in the "Est" column that the CI contains 0). Social support has a significant, positive association with positive affect, controlling for stress category. We can also compare the effect sizes and see that social support as a variable has a medium effect size, f^2, whereas stress category overall has a small effect size.

Finally, if we wanted all pairwise comparisons between stress categories, we can recycle our earlier code. The emmeans() function calculates adjusted means holding other variables, in this case social support, at the average for the sample:

```
mcat3.means <- emmeans(object = mcat3,
                       specs = "StressCat")
mcat3.means
```

```
## StressCat emmean    SE  df lower.CL upper.CL
## 0           3.09 0.117 183     2.86     3.33
## 1           2.87 0.157 183     2.57     3.18
## 2+          2.46 0.107 183     2.25     2.67
```

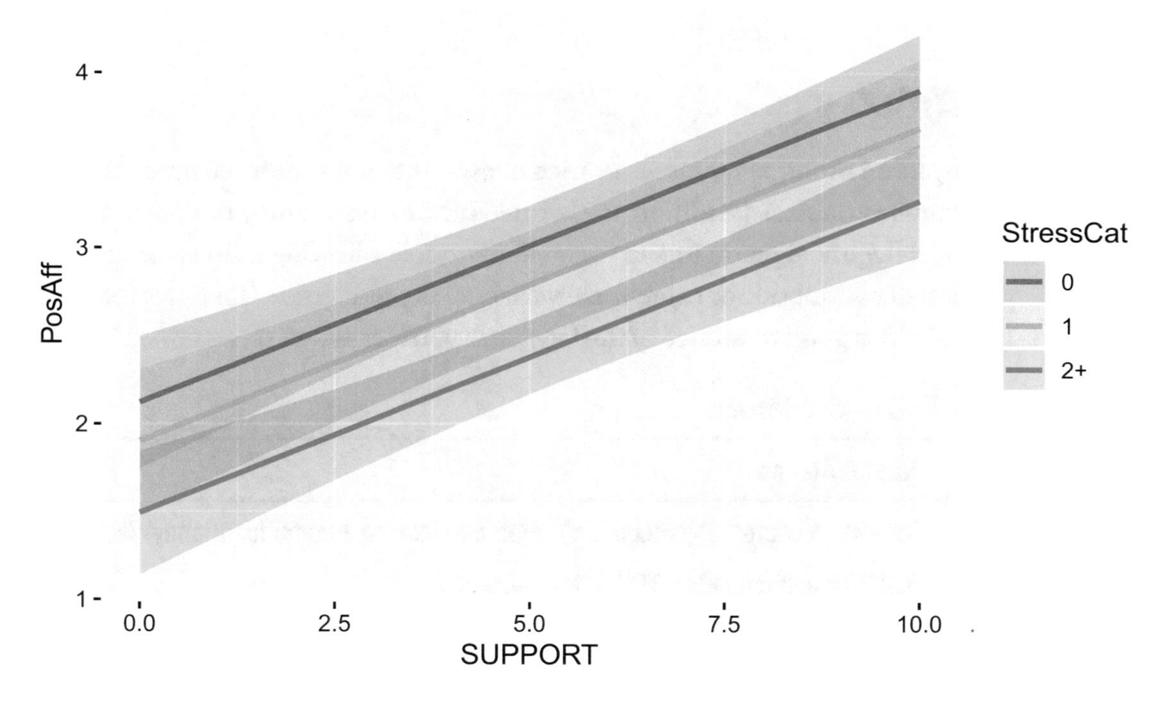

Figure 11-19. *Simple linear regression plot from visreg with a continuous predictor, SUPPORT, showing the predicted positive affect and 95% confidence intervals for each StressCat level*

```
##
## Confidence level used: 0.95

pairs(mcat3.means)

## contrast estimate    SE  df t.ratio p.value
## 0 - 1       0.219 0.196 183 1.117   0.5045
## 0 - (2+)    0.630 0.159 183 3.963   0.0003
## 1 - (2+)    0.411 0.190 183 2.164   0.0802
##

## P value adjustment: tukey method for comparing a family of 3 estimates
```

From these comparisons, we can see that comparing people with 1 vs. 2+ stress is now almost statistically significant, $p = .08$, where it was not close when we did not control for social support, so in this case, controlling for social support actually seems to have given slightly stronger results for stress, albeit still not statistically significant.

11.5 Summary

This chapter explored multiple regression which allows us to have multiple predictors of a single outcome variable. In addition, we learned how to use dummy or one-hot encoding to add categorical predictors to regression models, allowing us to examine any combination of continuous or categorical variables as predictors of an outcome. Table 11-1 should be a good reference of the key ideas you learned in this chapter.

Table 11-1. *Chapter Summary*

Idea	What It Means
`factor()`	Converts a character (categorical) column to include a numeric (dummy) factor.
`fifelse()`	Fast if/else; if (condition TRUE) do A, else do B.
Call	Repeats the linear model formula.
Residuals	Are the difference between actual and predicted values; if normal, median == 0.
Coefficients	Gives bs and p-value of each predictor.
Residual	Standard error statistical data for further calculations and missing values.

(*continued*)

Table 11-1. *(continued)*

Idea	What It Means
R-squared	Proportion of variance in y-value explained by x-value(s).
F-statistic	Includes overall model p-value for if model is overall significant.
`confint()`	Used on a model; gives CIs for *bs*.
`qt()`	Student's t-distribution quantile function; compare to `qnorm()`.
`visreg()`	Helps visualize regression models.
`annotate()`	ggplot2 function to add text (including `visreg()`).
`xlab()`	ggplot2 function to add x-axis label.
`ylab()`	ggplot2 function to add y-axis label.
`ggtitle()`	ggplot2 function to add title text.
`ev.perc =`	New argument to `modelDiagnostics()` for extreme values.
Cohen's *f2*	Approximate effect size; ¿= .02 is small, ¿= .15 is medium, and ¿= .35 is large.
`APAStyler`	Prints `modelTest()` object "neatly"; gives *f2*.
Multicollinearity	Highly correlated predictors; not usually a problem.
`log()`	Logarithmic function; used on positive, non-normal value to "make" normal.
Categorical	Predictors one-hot or dummy encoded to include in multiple regression model.
`emmeans()`	Shows mean output for given categorical predictor.
`pairs()`	Shows difference between the mean outputs for a pair of categorical predictors.

11.6 Practice for Mastery

Check your progress and grow through practice by working through some exercises. Comprehension checks ask critical thinking questions that may be best answered with a written or verbal response. Part of the art of statistics is successfully communicating results to your stakeholders or audience. Sometimes that audience is highly technical and other times very much not technical. Exercises are more direct applications of the concepts explored in the chapter.

Comprehension Checks

1. Thinking about penguins, we know Biscoe Island did not seem to change some features of Adelie penguins. Thus, we might discount islands. Furthermore, we know that bill depth did not seem to be very exciting in terms of predicting body mass. Besides flipper length, what variables might give us additional insight into body mass?
2. If any additional penguin variables are categorical, will we need to convert them to factors first? What does `str()` tell us about `penguinsData`?

Exercises

1. Suppose you decide biological sex could be a good additional predictor of penguin body mass. Looking at the summary data for `m2Penguin`, what variable do we want to swap out for `sex`? What is the median of the residuals, and what should the median of the residuals be ideally? What is the adjusted R-squared, and how can we use this to tell if `sex` is a better predictor than the variable we swap out?

```
summary(m2Penguin)
```

```
##
## Call:
## lm(formula = body_mass_g ~ flipper_length_mm + bill_depth_mm,
##     data = penguinsData)
##
## Residuals:
##     Min      1Q  Median      3Q     Max
## -1029.8  -271.5   -23.6   245.2  1276.0
##
## Coefficients:
##                      Estimate Std. Error t value Pr(>|t|)
```

```
## (Intercept)        -6541.91     540.75   -12.1   <2e-16 ***
## flipper_length_mm     51.54       1.87    27.6   <2e-16 ***
## bill_depth_mm         22.63      13.28     1.7    0.089 .
## ---
## Signif. codes:  0 '***' 0.001 '**' 0.01 '*' 0.05 '.' 0.1 ' ' 1
##
## Residual standard error: 393 on 339 degrees of freedom
##   (2 observations deleted due to missingness)
## Multiple R-squared:  0.761,  Adjusted R-squared:  0.76
## F-statistic:  540 on 2 and 339 DF,  p-value: <2e-16
```

2. Using the `lm(formula = , data = penguinsData)` function, fill in the blank after `formula` = to create a multiple regression model with a body mass outcome predicted by flipper length and sex. Assign that model to a variable cleverly named `m3Penguin`. What does `summary()` tell you about your model?

3. If your prior block of code was done correctly, the following code block should create a diagnostic graph:

   ```
   m3Penguind <- modelDiagnostics(m3Penguin, ev.perc = 0.005)
   plot(m3Penguind, ncol = 2, ask = FALSE)
   ```

   Does the graph suggest your model works well?

4. If your prior block of code was coded properly, the regression model means for the categorical variable `sex` should yield a graph similar to Figure 11-16. In that example, the mean positive affect for education levels had confidence intervals that overlapped. Does the mean body mass overlap for biological sex in penguins?

   ```
   visreg(m3Penguin, xvar = "sex", partial = FALSE, rug = FALSE,
   gg = TRUE) +
     ggtitle("Association of sex and body mass, adjusted for
     flipper length")
   ```

5. Again, if all has gone well, you should be able to see the f^2 effect sizes for flipper length and sex. Using the chart in this chapter for Cohen's f^2, what is the effect size for flipper length? What is the effect size for sex?

```
m3PenguinTest <- modelTest(m3Penguin)
print(APAStyler(m3PenguinTest), nrow = 100)
```

CHAPTER 12

Moderated Regression

Multivariate regression allows more than one variable to be predictors of interesting outcomes. This, in general, tends to yield more accurate models. However, multiple regression by itself does not allow for two predictors to interact with each other. Thinking about penguins, perhaps body mass is not only about flipper length or bill depth. Rather, there may be some unique synergy between the two that contributes to a comfortably full penguin. This chapter explores ways in which two (or more) predictors might interact with each other. Often, these methods can create more complete models.

By the end of this chapter, you should be able to

- Understand the difference between multiple and moderated regression.
- Create moderated regression models and visuals.
- Analyze the meaning of various data from moderated regression models.
- Create human-friendly graphs and communications to convey the results of moderated regression models.

12.1 R Setup

As usual, to continue practicing creating and using projects, we start a new project for this chapter.

If necessary, review the steps in Chapter 1 to create a new project. After starting RStudio, on the upper-left menu ribbon, select *File* and click *New Project*. Choose *New Directory* ➤ *New Project*, with *Directory name* `ThisChapterTitle`, and select *Create Project*. To create an `R` script file, on the top ribbon, right under the word *File*, click the small icon with a plus sign on top of a blank bit of paper, and select the *R Script* menu option. Click the floppy disk–shaped *save* icon, name this file `PracticingToLearn_XX.R` (where XX is the number of this chapter), and click *Save*.

M. Wiley and J. F. Wiley, *Beginning R 4*, https://doi.org/10.1007/978-1-4842-6053-1_12

In the lower-right pane, you should see your project's two files, and right under the *Files* tab, click the button titled *New Folder*. In the *New Folder* pop-up, type `data` and select *OK*. In the lower-right pane, click your new *data* folder. Repeat the folder creation process, making a new folder titled `ch12`.

Remember all packages used in this chapter were already installed on your local computing machine in Chapter 2. There is no need to re-install. However, this is a new project, and we are running this set of code for the first time. Therefore, you need to run the following `library()` calls:

```
library(data.table)

## data.table 1.13.0 using 6 threads (see ?getDTthreads). Latest news:
r-datatable.com

library(ggplot2)
library(visreg)
library(emmeans)
library(palmerpenguins)

library(JWileymisc)
```

We set up our dataset the same way as in Chapter 11.

```
data(aces_daily)
acesData <- as.data.table(aces_daily)[SurveyDay == "2017-03-03" &
SurveyInteger == 3]
acesData[, StressCat := factor(fifelse(STRESS >= 2,
                                       "2+",
                                       as.character(STRESS)))]
penguinsData <- as.data.table(penguins)
```

You are now free to learn more about statistics!

12.2 Moderation Theory

So far in regression, we focused on how each predictor variable is associated with the outcome variable either on its own or in the context of a larger model. This approach allowed us to examine whether two variables are associated with each other and whether this association is unique or independent of other variables. However, so far we have always assumed that the same association was true for everyone. What if the association between two variables is not the same for everyone?

For example, if you are hungry, eating food might make you feel better. If you are already full, eating more food might make you feel worse. This is an example of what is called **moderation** in statistics. The association between eating and feeling better is *moderated* by how full someone is at the beginning. For people who are not full, eating is associated with feeling better, whereas for people who are full, eating is associated with feeling worse.

As another example, antidepressant medication has been shown to be effective at reducing symptoms of depression, on average. However, that may not be true for everyone. For example, antidepressants may help people who are depressed but may provide no benefit to someone who is not depressed. Formally, we could say the beneficial effects of antidepressants on mood are moderated by presence or absence of depression.

We could cover many more examples. If you think about your own life or work, it is easy to find cases where something is enjoyed or helps one person but not another person. Thus, although the multiple regression we have learned so far is powerful already, we need a way to make it more flexible. We need to be able to account for the fact that some people may have different associations between variables than other groups of people. Moderation is the formal term for this concept, and in regression, we can *test* moderation using **interactions** between two or more variables.

An **interaction** term is the arithmetic product of two (or more) variables. In other words, we multiply two or more variables together.

To see moderation, we start with what is now a familiar multiple regression equation with two predictors, x and w, predicting an outcome, y:

$$y = b_0 + b_1 * x + b_2 * w + \varepsilon$$

We can test whether there is moderation between the two predictors, x and w, by including an additional predictor that is the arithmetic product (multiplication) of the two variables. In an equation, it looks like this:

$$y = b_0 + b_1 * x + b_2 * w + b_3 * (x * w) + \varepsilon$$

There is a third regression coefficient, b_3, for the product of the two variables, $x * w$, and b_3 is referred to as the **interaction** term.

Note it is possible to have interactions or moderation involving more than two variables. However, for this book, we focus on introducing and interpreting interactions of two variables.

You can have interactions between any two variables, whether they are both continuous, both categorical, or one categorical and one continuous. To include interactions with categorical variables, we use the same idea and create the arithmetic product of variables, but that is done on the dummy coded variables, rather than the raw categorical variable. This works because dummy coded variables are numerical so you can do things like multiplication with them.

This is a fairly brief overview of some of the ideas and concepts behind moderation and interactions. There is a short, classic book dedicated entirely to moderation and interactions in regression that is an excellent additional resource if you want to understand it better by Aiken and West (1991) [5].

For now, we focus on how to run and interpret some applied examples of moderation in R.

Moderation in R

It is very easy to include interactions in R. Inside a regression formula, we use the * operator. This single R operator (which is not quite multiplication alone) expands to give both the main effects and the interactions between these variables.

To see this expansion in action, think back to Chapter 11 and multiple regression. The (not great) *multiple* regression on flipper lengths and bill depths was achieved with the following code:

```
mPmul <- lm(formula = body_mass_g ~ flipper_length_mm + bill_depth_mm,
         data = penguinsData)
```

In particular, notice the tell-tale + between flipper length and bill depth in the preceding code.

If you look at the summary() result, you see there are values for the intercept, b_0, the flipper length's b_1 and the bill depth's b_2:

```
summary(mPmul)
```

```
##
## Call:
## lm(formula = body_mass_g ~ flipper_length_mm + bill_depth_mm,
##     data = penguinsData)
##
## Residuals:
##     Min      1Q  Median      3Q     Max
## -1029.8  -271.5   -23.6   245.2 1276.0
##
## Coefficients:
##                   Estimate Std. Error t value Pr(>|t|)
## (Intercept)       -6541.91     540.75   -12.1   <2e-16 ***
## flipper_length_mm    51.54       1.87    27.6   <2e-16 ***
## bill_depth_mm        22.63      13.28     1.7    0.089 .
## ---
## Signif. codes:  0 '***' 0.001 '**' 0.01 '*' 0.05 '.' 0.1 ' ' 1
##
## Residual standard error: 393 on 339 degrees of freedom
##   (2 observations deleted due to missingness)
## Multiple R-squared:  0.761,  Adjusted R-squared:   0.76
## F-statistic:  540 on 2 and 339 DF,  p-value: <2e-16
```

In other words, the general formula for multiple regression

$$y = b_0 + b_1 * x + b_2 * w + \varepsilon$$

becomes the specific formula

$$body\_mass\_g = -6541.91 + 51.54 * flipper\_length\_mm + 22.63 * bill\_depth\_mm + \varepsilon$$

For *moderated* regression, the function call to lm() is almost the same, except for replacing the + with a *:

```
mPmod <- lm(formula = body_mass_g ~ flipper_length_mm * bill_depth_mm,
            data = penguinsData)
```

However, the summary function returns more variables!

```
summary(mPmod)

##
## Call:
## lm(formula = body_mass_g ~ flipper_length_mm * bill_depth_mm,
##     data = penguinsData)
##
## Residuals:
##     Min      1Q Median      3Q     Max
## -938.9 -254.0  -28.2   220.7  1048.3
##
## Coefficients:
##                                  Estimate Std. Error t value Pr(>|t|)
## (Intercept)                     -36097.06    4636.27   -7.79  8.6e-14
## flipper_length_mm                 196.07      22.60    8.67  < 2e-16
## bill_depth_mm                    1771.80     273.00    6.49  3.1e-10
## flipper_length_mm:bill_depth_mm    -8.60       1.34   -6.41  4.8e-10
##
## (Intercept)                     ***
## flipper_length_mm               ***
## bill_depth_mm                   ***
## flipper_length_mm:bill_depth_mm ***
## ---
## Signif. codes:  0 '***' 0.001 '**' 0.01 '*' 0.05 '.' 0.1 ' ' 1
##
## Residual standard error: 372 on 338 degrees of freedom
##   (2 observations deleted due to missingness)
## Multiple R-squared:  0.787,  Adjusted R-squared:  0.785
## F-statistic:  416 on 3 and 338 DF,  p-value: <2e-16
```

In other words, the general formula for moderated regression

$$
\begin{aligned}
y = b_0 + b_1 * x \\
+ b_2 * w \\
+ b_3 * (x * w) + \varepsilon
\end{aligned}
$$

becomes the specific formula

$$body\_mass\_g = -36097.06 + 196.07 * flipper\_length\_mm + 1771.80 * bill\_depth\_mm - 8.60 * (flipper\_length\_mm * bill\_depth\_mm) + \varepsilon$$

This is what we mean by "expands" - R knows when you write `*`, you are doing moderated regression. Notice too that all the b_is changed from one model to the next. Despite the similarities, there is a fair bit going on mathematically in the regression.

12.3 Continuous x Categorical Moderation in R

We now turn to looking at a continuous variable moderated by a categorical variable. We do this via an example.

Example

In Chapter 11, we looked at an example where stress categories (0, 1, 2+) and social support both predicted positive affect. We will build off that example but here allow an interaction between social support and stress categories. Compared to the prior version of the model, the only change in the code is we change + to * between `StressCat` and `Support`.

Interactions between one continuous and one categorical variable are perhaps the easiest to *interpret* first, which is why we are starting with one as an example. Before we jump into the model output, we make a graph of the results. As before, we use the `visreg()` function. In fact, the `visreg()` code is identical to that used in Chapter 11, except the model has been changed to be the regression model with an interaction:

```
mint1 <- lm(formula = PosAff ~ StressCat * SUPPORT,
            data = acesData)

visreg(mint1, xvar = "SUPPORT", by = "StressCat",
       partial = FALSE, rug = FALSE, overlay = TRUE,
       gg = TRUE) +
  ggtitle("SUPPORT and StressCat as predictors")
```

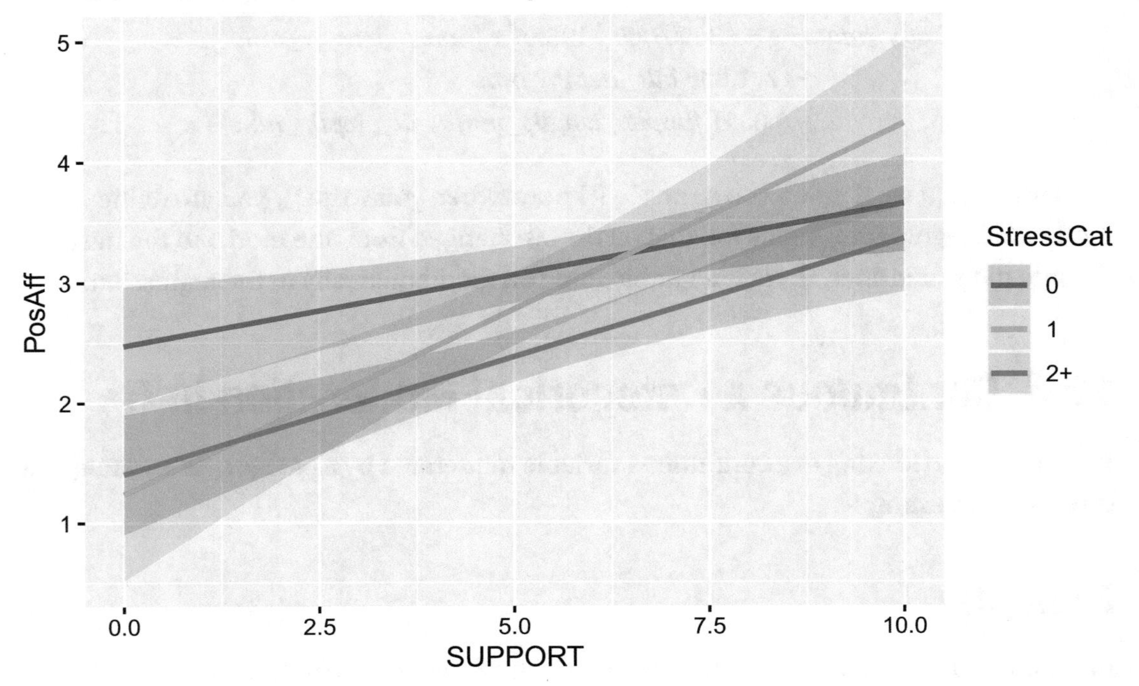

***Figure 12-1.** Moderated regression with an interaction between stress category and social support predicting positive affect. Lines are predictions in each group with 95% confidence intervals*

We see the results in Figure 12-1. It is similar to the figure from Chapter 11 except now the lines are no longer parallel. That is, the association between social support and positive affect is no longer forced to be the same for people from all stress categories. Equally, you see the differences in positive affect between different stress categories are no longer the same either. For example, people with stress of 1 are predicted to have the lowest positive affect when social support is zero but the highest positive affect when social support is 10. That the lines in Figure 12-1 are not parallel to each other is, in a picture, what is meant by *interactions* and *moderation*.

As with many things in statistics, visually there may appear to be some differences between lines, but we often want to test whether they are statistically significant (usually, $p < .05$) as a way of checking whether differences probably represent true population differences or just differences due to by chance having randomly sampled people with a stronger or weaker association in some groups. We get that test from the regression model and the tests of the coefficients for the interaction terms.

In this case, even though we only have two predictors, because stress category has three factor levels, there are two dummy codes included, and both of them interact with social support. Our model as an equation is

$$\begin{aligned} PosAff = b_0 + b_1 * StressCat1 + b_2 * StressCat2++ \\ b_3 * SUPPORT + b_4 * (StressCat1 * SUPPORT) + \\ b_5 * (StressCat2) + * SUPPORT) + \varepsilon \end{aligned}$$

As before, we can use the modelTest() function with APAStyler() to get the model results. As we have more coefficients, to make sure all the rows are printed, we use the print() function and specify the maximum number of rows we are willing to print. Take a moment to appreciate how convenient it is to use programming to switch models. Because many R functions are "intelligent" enough to detect different models, simple or no changes on the code side yield the correct calculations on the computer side:

```
mint1Test <- modelTest(mint1)
print(APAStyler(mint1Test), nrow = 100)

##                    Term                          Est           Type
##  1:         (Intercept) 2.47*** [ 1.98,  2.97] Fixed Effects
##  2:          StressCat1 -1.23** [-2.11, -0.36] Fixed Effects
##  3:         StressCat2+ -1.07** [-1.77, -0.37] Fixed Effects
##  4:             SUPPORT  0.12** [ 0.04,  0.19] Fixed Effects
##  5:  StressCat1:SUPPORT   0.19* [ 0.04,  0.34] Fixed Effects
##  6: StressCat2+:SUPPORT    0.08 [-0.04,  0.19] Fixed Effects
##  7:    N (Observations)                    187 Overall Model
##  8:          logLik DF                       7 Overall Model
##  9:              logLik                -253.26 Overall Model
## 10:                 AIC                 520.52 Overall Model
## 11:                 BIC                 543.14 Overall Model
## 12:                  F2                   0.40 Overall Model
## 13:                  R2                   0.28 Overall Model
## 14:              Adj R2                   0.26 Overall Model
## 15:           StressCat    f2 = 0.07, p = .003  Effect Sizes
## 16:             SUPPORT    f2 = 0.05, p = .003  Effect Sizes
## 17:   StressCat:SUPPORT    f2 = 0.04, p = .039  Effect Sizes
```

The interaction, product terms (b_4 and b_5 in the equation) are shown in the output as `StressCat1:SUPPORT` and `StressCat2+:Support`. Often researchers will first examine the interaction terms. If the interaction terms are not statistically significant, the researchers often drop the interaction. This is because it is much harder to interpret results when you have to report the association between variables in different groups. When only using multiple regression, you are able to say more simply something akin to "These two variables have a positive association."

When an interaction is included in a model, the regression coefficients for the variables making up that interaction often are called "simple effects" because they capture the effect (slope) at one specific value of the moderator variable. The interaction terms in our model capture the *difference* in slope of social support between different stress categories. We can do some factoring and rearranging to see what the slope is in each group.

Why do we care so much about slope? In mathematics, the slope between two values (an x value and a y value) is the rate of change. In the simpler models from linear regression back in Chapter 8, in the formula $y = b_0 + b_1 * x$, the constant value b_1 is the slope. If you change x from an input of 1 to an input of 2, you have doubled the b_1 value. This in turn increases the total size of y. Thus, by looking at b_1 – the slope – we can see the rate at which an increase in x will increase y. Of course, if b_1 is a negative value, that only tells us the rate at which an increase in x will decrease y.

So now we have a much more complex equation – there are many values all around. The solution is to use mathematical/algebraic factoring (not code factor). If we want to know how an increase in `SUPPORT` will change (increase/decrease) `PosAff`, we need to find the slope of `SUPPORT`. However, `SUPPORT` is showing up in many terms. We need to have one `SUPPORT` value by itself at the end of a bunch of numeric values. This is where mathematical or algebraic factoring comes in – the type of factoring you likely practiced in an algebra class at some point.

As a reminder of algebraic/mathematical factoring, if you had three terms that all involved a, you could factor out an a as shown:

$$2*a+3*a^2+4*a^3=\left(2+3*a+4*a^2\right)*a$$

In a similar fashion, if we wish to understand the slope (which is the rate of change) of `SUPPORT`, we can get all the terms involving `SUPPORT` together on a single line. From there, we can factor out `SUPPORT`. Take a look at the equation we saw earlier using stress categories and support as predictors for positive affect. The equation is arranged just so that the second line of it is all the terms being multiplied by SUPPORT. Then, just like the a in the preceding example, we can factor out SUPPORT. It is a bit messier than ideal; all the same, you see the tell-tale sign of factoring: $(stuff) * SUPPORT$. That "stuff" is in fact the slope of SUPPORT!

$$\begin{aligned} PosAff &= b_0 + b_1 * StressCat1 + b_2 * StressCat2+ \\ &\quad + b_3 * SUPPORT + b_4 * (StressCat1 * SUPPORT) + b_5 * (StressCat2 + * SUPPORT) + \varepsilon \\ &= b_0 + b_1 * StressCat1 + b_2 * StressCat2+ \\ &\quad + (b_3 + b_4 * StressCat1 + b_5 * StressCat2+) * SUPPORT + \varepsilon \end{aligned}$$

Yes, this process takes a bit of mathematics. Our rearranging and factoring has paid off though; factoring highlights that the slope (or rate of change) between support and positive affect is

$$(b_3 + b_4 * StressCat1 + b_5 * StressCat2+)$$

This is more complicated than any prior slope. The rate of change itself changes depending on stress! Imagine what happens to this slope once we substitute specific values for stress category. Here are all three possibilities:

- **People with 2+ stress**: $b_3 + b_4 * 0 + b_5 * 1 = b_3 + b_5 * 1 = b_3 + b_5$
- **People with 1 stress**: $b_3 + b_4 * 1 + b_5 * 0 = b_3 + b_4 * 1 = b_3 + b_4$
- **People with no stress** will be zero on both dummy code variables: $b_3 + b_4 * 0 + b_5 * 0 = b_3$

There are a few important lessons from this work. First, b_3 is the simple slope of social support when stress category is zero. That is, b_3 is the actual slope of social support. Note that the general idea here is that it would be the slope of social support for the reference group of stress category, whatever dummy code we chose to omit.

Next, we see b_4 and b_5 *are not* the slopes of social support for people with 1 or 2+ stress. Rather, b_4 and b_5 get *added* to b_3 to find the slope of social support in those other groups. b_4 and b_5 capture how *different* the slope of social support in the stress 1 group is from the stress 0 group and in the stress 2+ group from the stress 0 group. This often is a good thing. Because b_4 and b_5 capture the differences in slopes, we can test whether that difference is different from zero. If it is (a significant $p < .05$), then we can say that the two slopes are different from each other; if the difference in slopes is not significantly different from zero, we can conclude that in our sample we cannot tell if the slopes are different in reality or just in our random sample.

Capturing the interaction between two (or more) predictor values is how we moderate the slope (which is the rate of change) of `SUPPORT`. Look back at Figure 12-1. The three different lines there have three different slopes. Which three slopes? The three possibilities just discussed. Notice the slope of the line for `StressCat = 0` is the gentlest. For people who are not stressed, more social support is good, yet it does not make a huge difference. However, for people who have `StressCat = 1`, that line is quite steep. In other words, the slope is more intense, which means that a bit more social support goes a long way in helping those folks feel more positive emotions.

While we just used stress categories to help us understand the relationship between support and positive affect, interactions and moderation go both ways. Any variable can be a moderator. To show this, we rearrange the equation to highlight what the differences between stress groups are. Sadly, we must refactor. Of course, with one of us an erstwhile mathematics professor, the urge is strong to say "See, you will need to know how to factor!"

To refactor, we start with the same first line of our equation. Next, we reorder across addition to group `StressCat1` and `StressCat2+` each on their own line. Lastly, on each of their lines, we factor out the common `StressCat1` and `StressCat2+` terms to see the slope for each:

$$
\begin{aligned}
PosAff &= b_0 + b_1 * StressCat1 + b_2 * StressCat2+ \\
&\quad + b_3 * SUPPORT + b_4 * (StressCat1 * SUPPORT) + b_5 * (StressCat2+ * SUPPORT) + \varepsilon \\
&= b_0 + b_3 * SUPPORT \\
&\quad + b_1 * StressCat1 + b_4 * (StressCat1 * SUPPORT) \\
&\quad + b_2 * StressCat2+ + b_5 * (StressCat2+ * SUPPORT) + \varepsilon \\
&= b_0 + b_3 * SUPPORT_i + \\
&\quad (b_1 + b_4 * SUPPORT) * StressCat1 + \\
&\quad (b_2 + b_5 * SUPPORT) * StressCat2+ + \varepsilon
\end{aligned}
$$

What this highlights is the difference between people with no stress and 1 stress is $b_1 + b_4 * SUPPORT$ and likewise the difference between people with no stress and 2+ stress is $b_2 + b_5 * SUPPORT$.

If we set $SUPPORT = 0$, then we see b_1 can be interpreted as the difference between people with no stress and 1 stress *when social support is zero*.

Similarly, b_2 can be interpreted as the difference between people with no stress and 2+ stress *when social support is zero*.

The differences or slopes no longer apply to *all* people. With moderated regression, the slopes apply to *specific* groups of people. Thus, we must alter how we talk and write about slopes to explain which group we are describing, adding things like "when social support is zero." If we instead had an interaction between stress group and education, we would need to say "when education is zero," for example. Of course we do not have to hold the variables at zero; we can substitute different values and calculate the differences.

With all of that said, we look back at our model output and see if we can interpret the numbers. Here they are again for reference:

```
print(APAStyler(mint1Test), nrow = 100)

##                   Term                              Est          Type
##  1:        (Intercept) 2.47*** [ 1.98,  2.97] Fixed Effects
##  2:         StressCat1 -1.23** [-2.11, -0.36] Fixed Effects
##  3:        StressCat2+ -1.07** [-1.77, -0.37] Fixed Effects
##  4:            SUPPORT  0.12** [ 0.04,  0.19] Fixed Effects
##  5:  StressCat1:SUPPORT   0.19* [ 0.04,  0.34] Fixed Effects
##  6: StressCat2+:SUPPORT    0.08 [-0.04,  0.19] Fixed Effects
##  7:   N (Observations)                         187 Overall Model
##  8:         logLik DF                            7 Overall Model
##  9:             logLik                     -253.26 Overall Model
## 10:                AIC                      520.52 Overall Model
## 11:                BIC                      543.14 Overall Model
## 12:                 F2                        0.40 Overall Model
## 13:                 R2                        0.28 Overall Model
## 14:             Adj R2                        0.26 Overall Model
## 15:          StressCat    f2 = 0.07, p = .003  Effect Sizes
## 16:            SUPPORT    f2 = 0.05, p = .003  Effect Sizes
## 17:  StressCat:SUPPORT    f2 = 0.04, p = .039  Effect Sizes
```

An interpretation could look something as follows:

A multiple regression model was run on 187 people predicting positive affect from stress category, social support, and their interaction.

In people in the no stress group, each one-unit-higher social support was associated with 0.12 higher positive affect (95% CI = [0.04, 0.19]), $p < .01$.

The slope of social support and positive affect in people with a stress score of 1 was b = 0.19 [0.05, 0.34] higher than in people with no stress, $p < .05$, indicating that stress group and social support do interact.

The slope of social support and positive affect in people with a stress score of 2+ was higher, but not statistically significantly higher than in people with no stress ($p > .05$).

Overall, the stress * social support interaction had a small effect size, Cohen's $f^2 = 0.04$, but was statistically significant, $p = .039$. For people with no social support, having either stress of 1 or 2+ was associated with significantly lower positive affect (both $p < .01$).

Although by default, `APAStyler()` uses asterisks for p-values to simplify the output, if you want exact p-values for each coefficient, you can get these by using the `pcontrol()` argument. Here is an example:

```
print(APAStyler(modelTest(mint1),
  pcontrol = list(digits = 3, stars = FALSE,  includeP = TRUE,
                  includeSign = TRUE, dropLeadingZero = TRUE)),
  nrow = 100)
```

```
##                     Term                                    Est          Type
##  1:          (Intercept)  2.47p < .001 [ 1.98,  2.97] Fixed Effects
##  2:           StressCat1 -1.23p = .006 [-2.11, -0.36] Fixed Effects
##  3:          StressCat2+ -1.07p = .003 [-1.77, -0.37] Fixed Effects
##  4:              SUPPORT  0.12p = .003 [ 0.04,  0.19] Fixed Effects
##  5:   StressCat1:SUPPORT  0.19p = .012 [ 0.04,  0.34] Fixed Effects
##  6: StressCat2+:SUPPORT  0.08p = .179 [-0.04,  0.19] Fixed Effects
##  7:    N (Observations)                           187 Overall Model
##  8:           logLik DF                             7 Overall Model
##  9:              logLik                       -253.26 Overall Model
## 10:                 AIC                        520.52 Overall Model
## 11:                 BIC                        543.14 Overall Model
## 12:                  F2                          0.40 Overall Model
## 13:                  R2                          0.28 Overall Model
## 14:              Adj R2                          0.26 Overall Model
## 15:           StressCat         f2 = 0.07, p = .003  Effect Sizes
## 16:             SUPPORT         f2 = 0.05, p = .003  Effect Sizes
## 17:   StressCat:SUPPORT         f2 = 0.04, p = .039  Effect Sizes
```

This interpretation covers quite a bit, but it does leave some things lacking. For example, one may wonder what the slope of social support in the stress 1 group or stress 2+ group actually was, not just its difference. You also may note that we did not interpret the effect sizes for StressCat or SocialSupport. This is because these are the unique effects of those variables, but unique independent of the other variables in the model. Thus, for example, it would be the unique contribution of SocialSupport independent of stress group *and* the stress * social support interaction. Since the interaction includes social support, it is a bit odd to talk about the "effect size" of a variable partly independent of itself. The mathematics/statistics behind this are suited more to statistical theory perhaps than this applied discussion. In general, it is common to report the effect size for the interaction only and the simple slopes for the rest.

To get the simple slope of social support and positive affect in each stress group, we can make use of the emmeans package which has a helpful function called emtrends(). The emtrends() function calculates trends (slopes) by levels of another variable if desired. We pass three arguments: mint1 is our model as the model object, the specs is where was say we want support slopes separated by stress group, and finally the variable var which is the variable to use for simple slopes, here social support. We save the results and then use the summary() function with infer = TRUE to get p-values for each simple slope. As a sidenote, the "infer" argument comes from inferential statistics, as contrasted with descriptive statistics:

```
m1int.slopes <- emtrends(
  object = mint1,
  specs = "StressCat",
  var = "SUPPORT")

summary(m1int.slopes, infer = TRUE)

##  StressCat SUPPORT.trend     SE  df lower.CL upper.CL t.ratio p.value
##  0                 0.117 0.0389 181   0.0398    0.194 2.997   0.0031
##  1                 0.307 0.0643 181   0.1797    0.433 4.770   <.0001
##  2+                0.194 0.0425 181   0.1106    0.278 4.578   <.0001
##
## Confidence level used: 0.95
```

What the output shows is the slope of social support for stress groups 0, 1, and 2+ along with 95% confidence intervals and p-values for each slope. These results help nuance our understanding of the interaction. We now can say that social support is positively and significantly associated with positive affect for all stress groups. However, as we know from earlier, there is a significantly more positive slope in the stress group 1 vs. 0.

Likewise, we might want to know how stress groups differ from each other. Again, because we are studying the interaction, we cannot say how stress groups differ overall. Instead, it depends on the level of social support.

Unlike stress, which was categorical and thus easy to split into steps, social support SUPPORT is continuous. Because social support is continuous, we could pick many different values. Much like a histogram "bins" data into columns to more easily see what is happening, so too we must choose some levels for continuous variables. A common "default" for continuous variables are three levels: one standard deviation below the mean (low), the mean (average), and one standard deviation above the mean (high), often written $M \pm 1SD$.

We let R do the work of calculating these three values:

```
## calculate M - 1 SD, M, and M + 1 SD

m1.MeanSDlow <- mean(acesData$SUPPORT, na.rm = TRUE) -
                     sd(acesData$SUPPORT, na.rm = TRUE)

m1.MeanSD <- mean(acesData$SUPPORT, na.rm = TRUE)

m1.MeanSDhigh <- mean(acesData$SUPPORT, na.rm = TRUE) +
                     sd(acesData$SUPPORT, na.rm = TRUE)
```

Having calculated these, we use the emmeans() function to get the mean positive affect scores by stress group at each of the chosen values of social support, specified by the at argument:

```
m1int.means <- emmeans(
  object = mint1,
  specs = "StressCat",
  by = "SUPPORT",
  at = list(SUPPORT = c(m1.MeanSDlow, m1.MeanSD, m1.MeanSDhigh)))
```

Now that we have the means stored, we can get all pairwise comparisons between them by using the pairs() function. This works just as in Chapter 11 except now the results are being held at specific values of social support. By picking three good levels of social support, we can get an idea what happens as social support increases. Remember the empirical rule which states that 68% of the population lives inside $\pm 1\sigma$? By choosing these three levels of social support, we are capturing a goodly portion of the total population within and around these ranges:

```
pairs(m1int.means)
```

```
## SUPPORT = 2.79:
##  contrast estimate    SE  df t.ratio p.value
##  0 - 1        0.703 0.271 181   2.595  0.0275
##  0 - (2+)     0.852 0.223 181   3.818  0.0005
##  1 - (2+)     0.148 0.266 181   0.557  0.8429
##
## SUPPORT = 5.46:
##  contrast estimate    SE  df t.ratio p.value
##  0 - 1        0.196 0.194 181   1.008  0.5729
##  0 - (2+)     0.644 0.157 181   4.099  0.0002
##  1 - (2+)     0.448 0.188 181   2.379  0.0481
##
## SUPPORT = 8.14:
##  contrast estimate    SE  df t.ratio p.value
##  0 - 1       -0.312 0.287 181  -1.085  0.5243
##  0 - (2+)     0.436 0.217 181   2.009  0.1129
##  1 - (2+)     0.748 0.292 181   2.563  0.0299
##
## P value adjustment: tukey method for comparing a family of 3 estimates
```

From the output, we can see that at low levels of social support (one standard deviation below the mean at `SUPPORT = 2.79`), the no stress group has significantly higher positive affect than either the 1 or 2+ stress groups (the estimate on the first two rows shows the difference between 0 and either 1 or 2+).

At average levels of social support with `SUPPORT = 5.46`, the no stress group has higher positive affect than the 2+ stress group, and the 1 stress group has significantly higher positive affect than the 2+ stress group. Notice that the first row here (0 – 1) has a nonsignificant p-value and the estimate, plus or minus the standard error SE, would create a confidence interval that almost captures 0 (e.g., no difference).

At high levels of social support with SUPPORT = 8.14, the 1 stress group has higher positive affect than the 2+ stress group. No other group differences are significant for an alpha of 0.05.

Lastly, although we ignored model diagnostics, they are just as important for interaction models as any other regression model. However, moderated regressions are still multiple regression models, so all the usual assumptions and diagnostics we have learned apply the same way. Thus, in this chapter, we generally focus just on what is new: interactions, moderation, and calculating the simple effects or simple slopes.

That said, we show one example of our usual diagnostics in Figure 12-2. We would interpret these the same as usual:

```
mint1d <- modelDiagnostics(mint1)

plot(mint1d, ncol = 2, ask = FALSE)
```

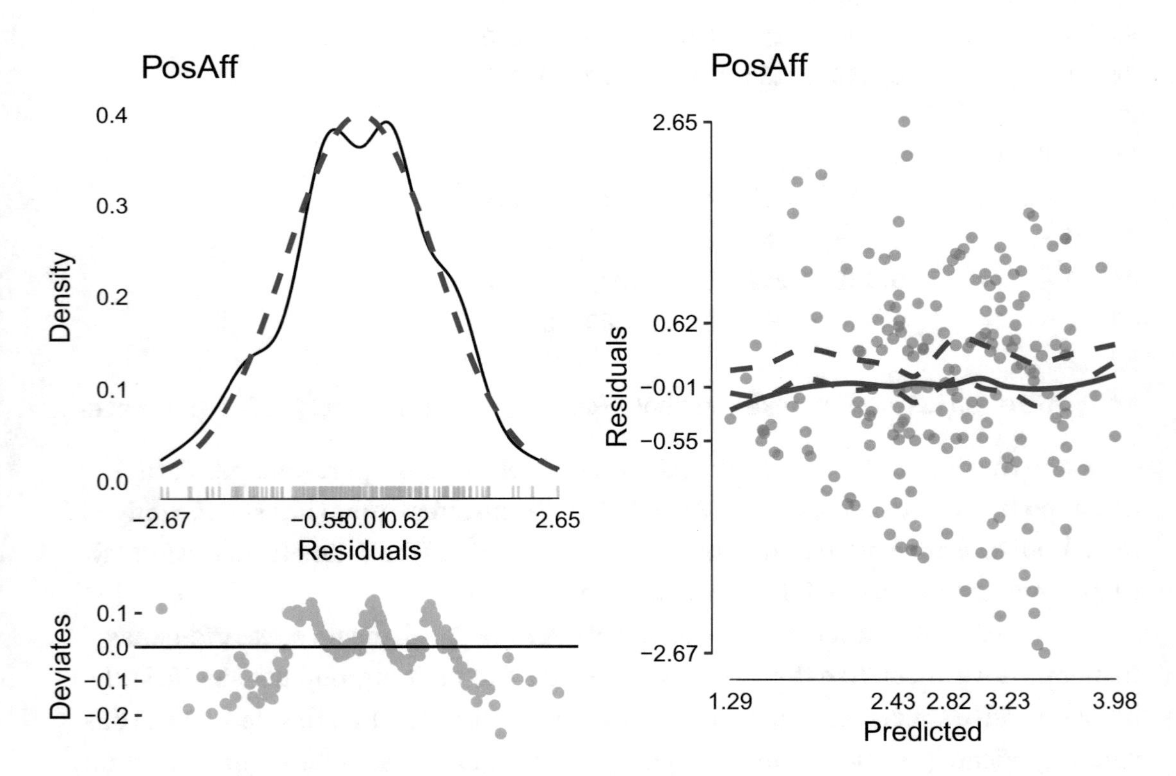

Figure 12-2. Regression diagnostic plot for a moderated regression model

One note: In the residual vs. predicted values scatter plot on the right panel of Figure 12-2, the quantile regressions failed. You can tell because clearly the dashed blue lines that normally help highlight the 10th and 90th percentiles of the residuals are not at 10th or 90th percentile. These lines usually help us visually see whether the percentiles are about parallel. That is not something to worry about; we would just ignore them in this case. The blue lines are always added just to try to help guide visual interpretation, not as a replacement for human judgement.

12.4 Continuous x Continuous Moderation in R

Two continuous predictors also can interact in regression. Their conceptual interpretation and practical implementation is very similar to the continuous x categorical interactions we looked at earlier in this chapter.

Perhaps the main difference is that because both predictors in the interaction are continuous, we cannot talk about the slope of one variable for a specific group of people. Instead, we have to talk about simple slopes at specific values of the other variable. Of course, we did see how to handle this for stress in the preceding section.

Having two continuous predictors also makes it more difficult to graph the resulting regression surface. Such a graph is technically three dimensional. That said, this is not as terrible as it might seem. This is because in real-life models, we often use multiple predictors; thus, one cannot graph the entire model in any human-visible fashion. After all, what would a seven-dimensional graph look like? So while we will show a 3D graph, it is only helpful to learn some good principles of interpretation of results. Most modern models cannot be fully graphed.

For our example, we look at stress, age, and their interaction predicting positive affect. Actually fitting the moderated regression model follows the same format as all our other regression models. We call this one `mint2`:

```
mint2 <- lm(formula = PosAff ~ STRESS * Age,
            data = acesData)
```

Before getting into more code, equations, interpretation, and output, we take a look at what this regression surface looks like. You may recall from Chapter 11 we had a couple examples of a 3D regression surface. It was perfectly flat, like a piece of paper. When we add interactions, we are saying that the regression surface is not flat. It is warped. The predicted positive affect values across levels of stress and age from the model we just fit, `mint2`, are shown in Figure 12-3. You can see the surface is not flat. If you were to look at the slope of age and positive affect, it would be different depending on what value of stress is used. Likewise, if you were to look at the slope of stress and positive affect, it would differ depending on age. That is what is meant by an interaction. When two continuous variables interact, there are no natural "breaks" for one variable so we cannot just plot three lines and capture everything. To capture everything, we would need to plot a regression surface, like the one shown in Figure 12-3.

The rest of this section will try to unpack how we can understand, interpret, test, and present results from a moderated regression model, how we can take Figure 12-3 and try to put it into words and numbers, with tests for what differences are reliable and which may just be noise. This is an important skill to develop because most real-life models are not completely graphable. Thus, we must build our mental understanding to the level that the numeric outputs "make sense."

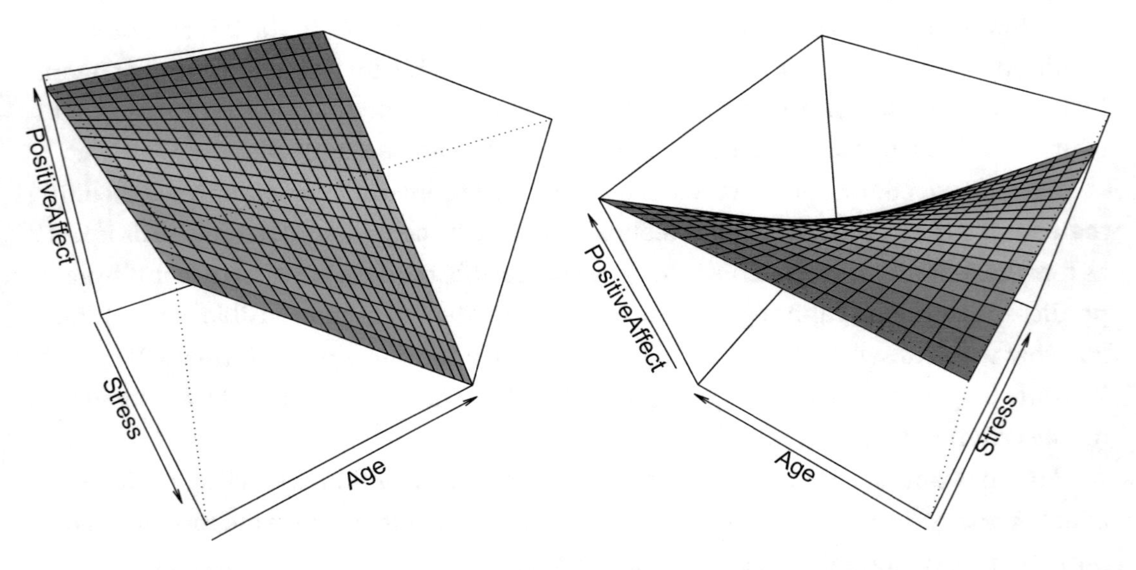

Figure 12-3. *3D graph of a moderated multiple regression surface*

As an equation, `mint2` would look like this:

$$PosAff = b_0 + b_1 * STRESS + b_2 * Age + b_3 * (STRESS * Age) + \varepsilon$$

The regression coefficient for the interaction term, b_3, is how different the simple slope of stress or age is for a one-unit shift in the other variable. If we rearrange and factor out stress or age, we can get the following equivalent equations:

$$
\begin{aligned}
PosAff &= b_0 + b_1 * Stress + b_2 * Age \\
&\quad + b_3 * (Stress * Age) + \varepsilon \\
&= b_0 + b_1 * Stress \\
&\quad + (b_2 + b_3 * Stress) * Age + \varepsilon \\
&= b_0 + b_2 * Age \\
&\quad + (b_1 + b_3 * Age) * Stress + \varepsilon
\end{aligned}
$$

What these highlight is that with the interaction in the model, the simple slope of stress is $b_1 + b_3 * Age$. b_1 is the simple slope of stress, *when age is zero*.

Similarly, the simple slope of age is $b_2 + b_3 * Stress$. b_2 is the simple slope of age, *when stress is zero*.

Although stress ranges from zero to ten, so zero is not unreasonable, age ranges from 18 to 26 in this sample, so finding the simple slope of stress for a zero-year-old is pretty unreasonable extrapolation from the sample data.

We can get model information and effect sizes just as all our other regression models:

```
mint2Test <- modelTest(mint2)
APAStyler(mint2Test)
```

```
##                   Term                      Est          Type
##  1:        (Intercept) 2.82** [ 0.86, 4.77] Fixed Effects
##  2:             STRESS   0.56 [-0.05, 1.16] Fixed Effects
##  3:                Age   0.01 [-0.08, 0.10] Fixed Effects
##  4:         STRESS:Age -0.03* [-0.06, 0.00] Fixed Effects
##  5: N (Observations)                    187 Overall Model
## ---
## 11:                 R2                 0.13 Overall Model
## 12:             Adj R2                 0.12 Overall Model
## 13:             STRESS  f2 = 0.02, p = .073  Effect Sizes
## 14:                Age  f2 = 0.00, p = .801  Effect Sizes
## 15:         STRESS:Age  f2 = 0.03, p = .024  Effect Sizes
```

In the output, we can see that b_3, the interaction coefficient, is -.03 and that it is statistically significant, $p < .05$ (based on the *). The negative sign indicates that as stress or age increases, the simple slope of the other decreases.

Neither simple slope is statistically significant, that is, the simple slope of stress when age is zero (i.e., b_1, 0.56) is not different than zero nor is the simple slope of age when stress is zero (i.e., b_2, 0.01).

We see the interaction has a small effect size, $f^2 = 0.03$. As for categorical x continuous interactions, we do not interpret the effect sizes for the other variables as they are essentially simple effect sizes when the other variable is zero. For example, the interpretation of the stress effect size alone is the effect size for the effect of stress when age is zero.

We can improve the *interpretation* of the simple effect sizes through **centering** variables. Think about taking a picture of a tree. The tree is firmly planted in the ground, yet, by moving your camera around, you can ensure the tree is **centered**. Doing this does not change one thing about the tree. However, it helps your friends understand the tree better than if you took an off-centered photo.

In our case, although stress can be zero for our participants, age cannot:

```
range(acesData$Age,
      na.rm = TRUE)

## [1] 18 26

summary(acesData$Age)
##    Min. 1st Qu.  Median    Mean 3rd Qu.    Max.    NA's
##    18.0    20.0    21.0    21.6    23.0    26.0       2
```

We notice the mean is 21.6, and thus we might decide to center or adjust our age variable so that 0 indicates a 22-year-old instead of a zero-year-old. Just like centering a large tree in a photo does not necessarily mean the tree is perfectly symmetric, so too we can decide where and how to center data. While the idea might sound complex, it is quite simple to code. We create a new column, named *Age Centered at 22* or `Agec22` for short. This is a column operation, and we want every age for each participant to be 22 smaller than it was originally:

```
acesData[, Agec22 := Age - 22]
```

Done!

As a sidenote, much like the `log()` **transformation** we did in the prior chapter, so too centering is a type of transformation of data to help us either use or understand our data and results better. If, as you approach the end of this book, you are already looking for additional topics to learn next, transformation would be a good thing to add to your list.

Transformations must occur *before* the model is fit. So we refit our regression model with this newly centered variable and run `modelTest()` again. Remember the tree is still the tree! Nothing has actually changed, except we now get simple effects of stress including the simple slope and simple effect size when Agec22 is zero, which is for a 22-year-old (which makes sense), instead of for a zero-year-old (which does not make sense):

```
mint2b <- lm(formula = PosAff ~ STRESS * Agec22,
             data = acesData)

mint2bTest <- modelTest(mint2b)
APAStyler(mint2bTest)

##                   Term                          Est           Type
##  1:        (Intercept)  3.07*** [ 2.87,  3.27] Fixed Effects
##  2:             STRESS -0.15*** [-0.21, -0.09] Fixed Effects
##  3:             Agec22     0.01 [-0.08,  0.10] Fixed Effects
##  4:      STRESS:Agec22   -0.03* [-0.06,  0.00] Fixed Effects
##  5: N (Observations)                       187 Overall Model
## ---
## 11:                 R2                      0.13 Overall Model
## 12:             Adj R2                      0.12 Overall Model
## 13:             STRESS     f2 = 0.12, p < .001  Effect Sizes
## 14:             Agec22     f2 = 0.00, p = .801  Effect Sizes
## 15:      STRESS:Agec22     f2 = 0.03, p = .024  Effect Sizes
```

Now, when we say nothing has changed, what we are saying is that nothing *substantive* has changed. After all, if you took an off-centered photo of a tree, you might miss some branches. Centered, you might see a building behind the tree you did not before. Thus, you will notice that the intercept has changed, and there will be some other numbers that look different. However, the interaction between stress and age – regardless of age being centered or not – *will not change*, just like the interaction between a tree and the ground would not change, no matter how you framed your photo.

As with other interactions, a common way to facilitate interpretation is to calculate simple slopes and to graph or plot them. Both variables are continuous so we do not have any naturally defined "groups." However, we can use the approach we used before to define "low," "average," and "high" as one standard deviation below the mean, the mean, and one standard deviation above the mean.

Despite our effort to center the model, if we are calculating simple slopes via R based on standard deviations, there is no real benefit to centering our variables, so we will just work with our original model, mint2.

First, we will calculate the simple slopes of age at different levels of stress. This actually identifies a case where $M \pm 1SD$ does not always work so well. STRESS varies from 0 to 10, but one standard deviation below the mean is negative (as we can see from the variable m2.MeanSDlow in the global environment). Negative stress values are impossible, so instead we manually set to 0, the lowest possible:

```
m2.MeanSDlow <- mean(acesData$STRESS, na.rm = TRUE) -
                  sd(acesData$STRESS, na.rm = TRUE)

m2.MeanSD <- mean(acesData$STRESS, na.rm = TRUE)

m2.MeanSDhigh <- mean(acesData$STRESS, na.rm = TRUE) +
                   sd(acesData$STRESS, na.rm = TRUE)

m2.MeanSDlow <- 0
```

The emtrends() code is the same format as we used in the categorical x continuous interaction earlier this chapter:

```
m2int.slopes <- emtrends(
  object = mint2,
  specs = "STRESS",
  var = "Age",
  at = list(STRESS = c(m2.MeanSDlow,
                       m2.MeanSD,
                       m2.MeanSDhigh)))

summary(m2int.slopes, infer = TRUE)
##  STRESS Age.trend     SE  df lower.CL upper.CL t.ratio p.value
##    0.00    0.0114 0.0453 183   -0.078   0.1008   0.252  0.8014
##    2.12   -0.0565 0.0343 183   -0.124   0.0111  -1.649  0.1009
##    4.55   -0.1342 0.0486 183   -0.230  -0.0384  -2.762  0.0063
##
## Confidence level used: 0.95
```

Looking at the results, we can see that age is not associated with positive affect for people with lower levels of stress (the p-value is bigger than alpha for that first row).

Only when stress is a standard deviation above the mean do we see that people who are older have lower positive affect. Specifically, for people with stress of 4.6 (one standard deviation above the mean), each year older they are, they are predicted to have -0.136 positive affect, $p = .010$.

Next, we can do the same thing calculating the simple slopes of stress at different ages. However, this time we round our ages to the nearest whole year. This simply helps us make interpretations better:

```
m2.MeanSDAgelow <- round(mean(acesData$Age, na.rm = TRUE) -
  sd(acesData$Age, na.rm = TRUE), 0)

m2.MeanSDAge <- round(mean(acesData$Age, na.rm = TRUE), 0)

m2.MeanSDAgehigh <- round(mean(acesData$Age, na.rm = TRUE) +
  sd(acesData$Age, na.rm = TRUE), 0)
```

Using our age cut-offs, we let `emtrends()` perform the calculation, and we inspect the `summary()`:

```
m2intAge.slopes <- emtrends(
  object = mint2,
  specs = "Age",
  var = "STRESS",
  at = list(Age = c(m2.MeanSDAgelow,
                    m2.MeanSDAge,
                    m2.MeanSDAgehigh))
)

summary(m2intAge.slopes,
        infer = TRUE)

##  Age STRESS.trend     SE  df lower.CL upper.CL t.ratio p.value
##   19      -0.0521 0.0504 183   -0.151   0.0473 -1.034  0.3024
##   22      -0.1481 0.0317 183   -0.211  -0.0856 -4.676  <.0001
##   24      -0.2121 0.0443 183   -0.300  -0.1247 -4.786  <.0001
##
## Confidence level used: 0.95
```

From these results, we see for all except the youngest participants, higher stress is associated with significantly lower positive affect ($p < .001$). The direction is the same for 19-year-olds (one standard deviation below the mean of Age), but the simple slope for 19-year-olds is not statistically significant (p-value is larger than alpha of 0.05).

We can use the `visreg()` function to plot simple slopes from our model. We need one new argument, `breaks`, which specifies the break points to use because our moderator variable is continuous. When we had a categorical moderator, `visreg()` would just use the distinct levels of that categorical variable, but now it's best if we choose and specify the break points we want. The results are in Figure 12-4, which shows graphically the simple slope tests we conducted earlier with `emtrends()`. Often, doing both numerical simple slope tests and graph is helpful. The graph helps people understand the pattern, and the numerical tests give specific values and p-values to the visual presentation:

```
visreg(mint2, xvar = "STRESS", by = "Age",
       breaks = c(m2.MeanSDAgelow, m2.MeanSDAge, m2.MeanSDAgehigh),
       partial = FALSE, rug = FALSE, overlay = TRUE,
       gg = TRUE) +
  ggtitle("STRESS and Age as predictors")
```

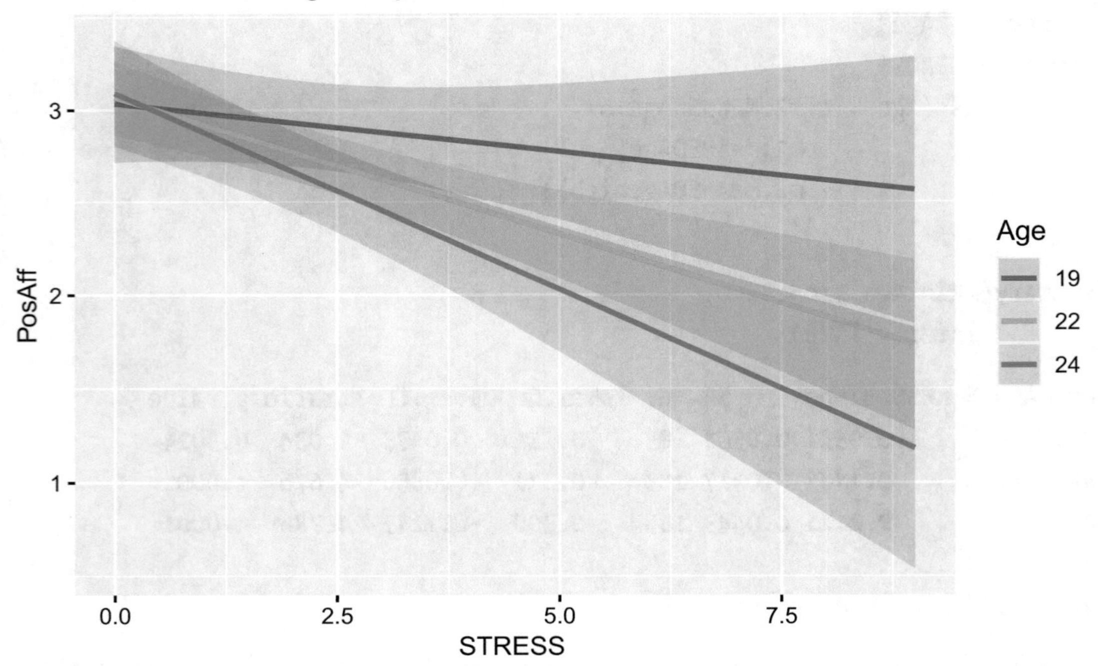

***Figure 12-4.** Moderated regression with an interaction between stress and age predicting positive affect. Lines are predictions at M – 1 SD, Mean, and M + 1 SD of age with 95% confidence intervals*

In linear models, the differences in slopes will be linear, so the mean will be roughly halfway between the low and high (Mean – 1 SD and Mean + 1 SD) values. Sometimes to make the graph appear visually cleaner, people will leave out the mean simple slope line and just show the extremes. We also can use the `annotate()` function in the `ggplot2` package to add some text labels. Here we add the simple slopes and p-values calculated from `emtrends()` so that the numerical summaries are in the figure, and we use the `ylab()` function to relabel the y-axis to be a bit cleaner. The results are in Figure 12-5:

```
visreg(mint2, xvar = "STRESS", by = "Age",
       breaks = c(19,  24),
       partial = FALSE, rug = FALSE, overlay = TRUE,
       gg = TRUE) +
  ggtitle("Stress x Age interaction on Positive Affect") +
  annotate("text", x = 5, y = 3, label = "Age M - 1 SD (19): b = -.05,
  p = .302") +
  annotate("text", x = 5, y = 1.6, label = "Age M + 1 SD (24): b = -.21,
  p < .001") +
  ylab("Positive Affect")
```

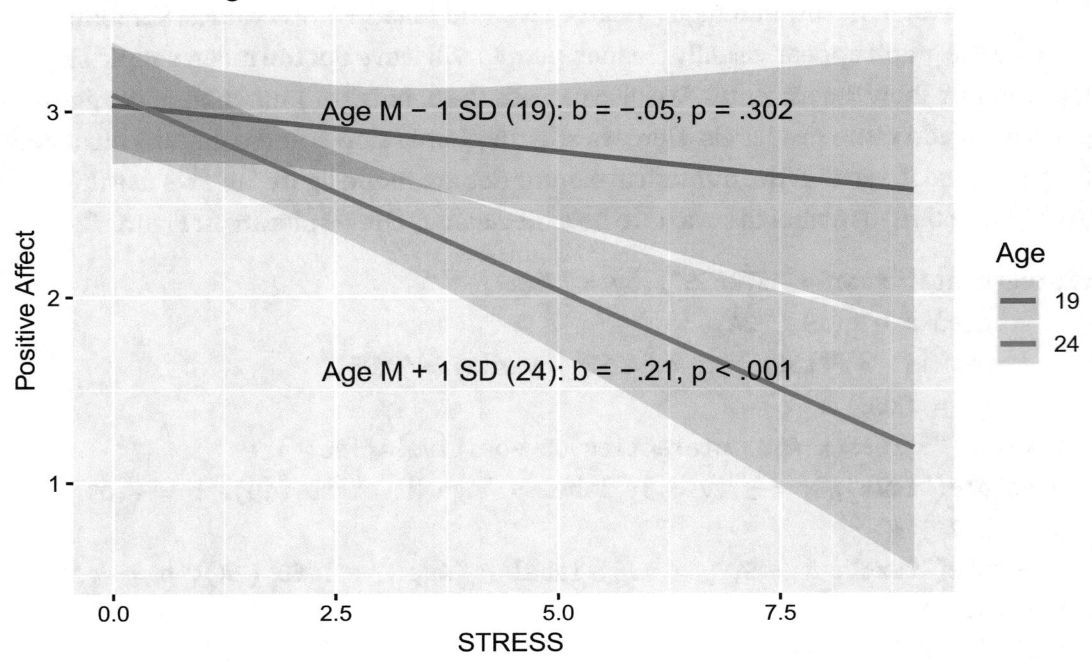

Figure 12-5. *Moderated regression with an interaction between stress and age predicting positive affect. Lines are predictions at M – 1 SD and M + 1 SD of age with 95% confidence intervals*

As we noted before, there is no directionality in an interaction. One variable is not the predictor and the other the moderator. You could choose to plot either variable on the x-axis.

Thus, it makes sense to also break the simple slope of age down by stress as shown in Figure 12-6:

```
visreg(mint2, xvar = "Age", by = "STRESS",
       breaks = c(m2.MeanSDlow,
                  m2.MeanSD,
                  m2.MeanSDhigh),
       partial = FALSE, rug = FALSE, overlay = TRUE,
       gg = TRUE) +
  ggtitle("STRESS and Age as predictors")
```

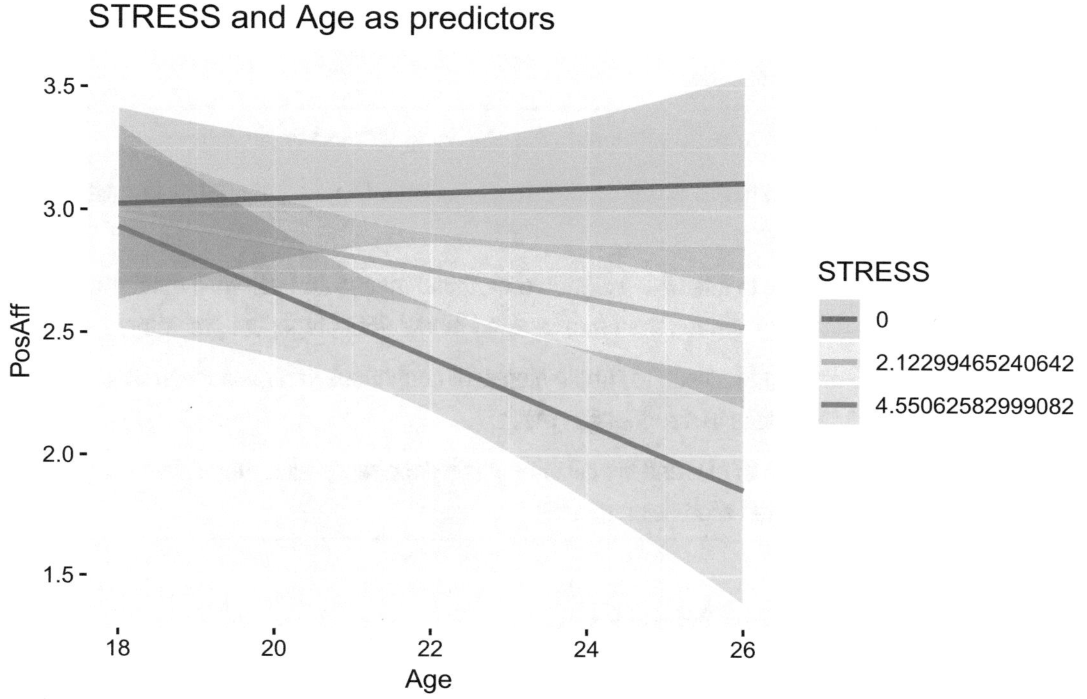

STRESS and Age as predictors

3.5
3.0
2.5
2.0
1.5
PosAff
18 20 22 24 26
Age
STRESS
0
2.12299465240642
4.55062582999082

Figure 12-6. *Moderated regression with an interaction between stress and age predicting positive affect. Lines are predictions at M – 1 SD, Mean, and M + 1 SD of stress with 95% confidence intervals*

12.5 Summary

This chapter explored what statistical moderation is and how it can be tested in regression using interaction terms. You should be able to understand how to test moderation and when there is a significant interaction how to interpret and present it using simple slopes or effects and graphs. Table 12-1 should be a good reference of the key ideas you learned in this chapter.

Table 12-1. *Chapter Summary*

Idea	What It Means
`lm()`	Function to fit regression model, including for moderated regression.
`modelTest()`.	Function to nicely summarize regression model with confidence intervals and effect sizes.
`emmeans()`	Function to calculate estimated marginal means and pairwise differences in means by groups from a regression model, for categorical predictors.
`emtrends()`	Function to calculate simple slopes of continuous variables when there are interactions in a regression model.
`visreg()`	Function to plot and visualize results from a regression model, including moderated regression models.

12.6 Practice for Mastery

Check your progress and grow through practice by working through some exercises. Comprehension checks ask critical thinking questions that may be best answered with a written or verbal response. Part of the art of statistics is successfully communicating results to your stakeholders or audience. Sometimes that audience is highly technical and other times very much not technical. Exercises are more direct applications of the concepts explored in the chapter.

Comprehension Checks

1. The slope of a line is the rate-of-change relationship between the x-axis input and the y-axis output. The steeper the slope, the faster the rate of change. Look at Figure 12-6. Is there a level of stress for which age is a good thing? Which line has the steepest slope? For the slope with the steepest line, for a two-year *increase* in age, about how much does positive affect *decrease*?

2. When do we use `emmeans()` and `pairs()` vs. `emtrends()` and `summary()`? Do categorical vs. continuous predictors play a role?

Exercises

1. Return to our example at the beginning of the chapter that showed the difference between multiple and moderated regression with penguin body mass predicted by flipper length and bill depth. Is the interaction significant?

```
mPmod <- lm(formula = body_mass_g ~ flipper_length_mm *
bill_depth_mm,
            data = penguinsData)

summary(mPmod)

##
## Call:
## lm(formula = body_mass_g ~ flipper_length_mm * bill_depth_mm,
##     data = penguinsData)
##
## Residuals:
##    Min     1Q  Median    3Q     Max
## -938.9 -254.0   -28.2 220.7  1048.3
##
## Coefficients:
##                                  Estimate Std. Error t value
Pr(>|t|)
## (Intercept)                     -36097.06    4636.27   -7.79
8.6e-14
## flipper_length_mm                  196.07      22.60    8.67
< 2e-16
## bill_depth_mm                     1771.80     273.00    6.49
3.1e-10
## flipper_length_mm:bill_depth_mm     -8.60       1.34   -6.41
4.8e-10
##
## (Intercept)                     ***
## flipper_length_mm               ***
```

```
## bill_depth_mm                      ***
## flipper_length_mm:bill_depth_mm   ***
## ---
## Signif. codes:  0 '***' 0.001 '**' 0.01 '*' 0.05 '.' 0.1 ' ' 1
##
## Residual standard error: 372 on 338 degrees of freedom
##   (2 observations deleted due to missingness)
## Multiple R-squared:  0.787,  Adjusted R-squared:  0.785
## F-statistic:  416 on 3 and 338 DF,  p-value: <2e-16
```

2. Using `modelTest()` and `APAStyler()`, what is the interaction coefficient? Is it statistically significant?

3. Can flipper length or bill depth ever be 0? Might this model be a candidate for centering?

4. Suppose we choose to not center these data. Complete the following lines of code correctly to get the $M \pm 1SD$ and the M:

```
#exercise 3
mPmod.MeanSDlowBill <- mean(penguinsData$ , na.rm = TRUE) -
                       sd(penguinsData$ , na.rm = TRUE)

mPmod.MeanSDBill <- mean(penguinsData$ , na.rm = TRUE)


mPmod.MeanSDhighBill <- mean(penguinsData$ , na.rm = TRUE) +
                        sd(penguinsData$ , na.rm = TRUE)
```

5. Done correctly, the preceding problem ought to allow the following code to generate the shown output:

```
mPmod.slopes <- emtrends(
  object = mPmod,
  specs = "bill_depth_mm",
  var = "flipper_length_mm",
  at = list(bill_depth_mm = c(mPmod.MeanSDlowBill,
                              mPmod.MeanSDBill,
                              mPmod.MeanSDhighBill)))

summary(mPmod.slopes, infer = TRUE)
```

```
##  bill_depth_mm flipper_length_mm.trend   SE  df lower.CL
upper.CL
##           15.2                     65.6 2.81 338     60.1
71.2
##           17.2                     48.6 1.82 338     45.0
52.2
##           19.1                     31.7 3.57 338     24.6
38.7
##  t.ratio p.value
##  23.311  <.0001
##  26.710  <.0001
##   8.877  <.0001
##
## Confidence level used: 0.95
```

As bill depth gets longer, what do we notice about the flipper length trends?

6. Look cautiously at Figure 12-5 as well as the code that created that graph to see how the annotations were created. Notice the annotations required manual determination of the (x, y) coordinates to points that made the annotation position well on that graph. Additionally notice that the breaks were modified in that example to remove the middle age value (e.g., contrasted with Figure 12-4).

   Modify the following code to remove the middle bill length break. Additionally, using `annotate()`, add annotations. Be sure to include the correct slopes and p-values! Compare and contrast Figure 12-7 to your image. Which more clearly shows the significant yet small difference in bill depth interaction?

```
visreg(
  mPmod,
  xvar = "flipper_length_mm",
  by = "bill_depth_mm",
  breaks = c(mPmod.MeanSDlowBill,
```

```
                 mPmod.MeanSDBill,
                 mPmod.MeanSDhighBill),
    partial = FALSE,
    rug = FALSE,
    overlay = TRUE,
    gg = TRUE
  ) +
    ggtitle("flipper_length_mm and bill_depth_mm as predictors")
```

7. Compare and contrast these exercises with the examples in this chapter. In particular, notice the variable choices in the various blocks of code. See what gets changed and what fits where, and keep in mind that most good programmers start by copying code that works, making tiny edits, and then running the edited code to see what happens.

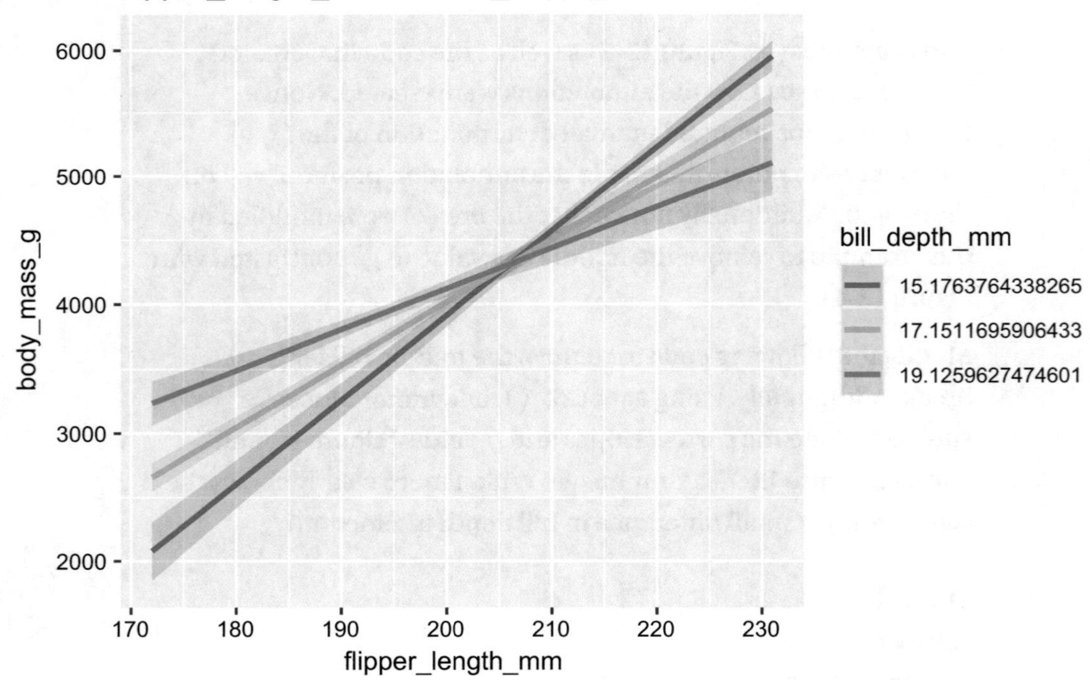

Figure 12-7. *Moderated regression with an interaction between flipper length and bill depth predicting body mass. Lines are predictions at M – 1 SD, Mean, and M + 1 SD of bill depth with 95% confidence intervals*

CHAPTER 13

Analysis of Variance

The penguins you have been studying live on three different islands. We initially explored the Adelie penguins of Biscoe Island – and compared them to all Adelie penguins. However, that method might be a trifle flawed. What we really want to do is measure the average weight of each island's Adelie penguins – and see if any three of the islands are not like the rest. Such a test – measuring the change or *variance* between groups – is called analysis of variance (ANOVA for short). This chapter will help you use ANOVA to detect differences between categorical groups – or, as we say in `R`, `factor()`s.

By the end of this chapter, you should be able to

- Evaluate data structures and research questions that lend themselves to ANOVA.
- Create visualizations to understand ANOVA results.
- Evaluate the difference (and when to use) between one-way and factorial ANOVA.

13.1 R Setup

As usual, to continue practicing creating and using projects, we start a new project for this chapter.

If necessary, review the steps in Chapter 1 to create a new project. After starting RStudio, on the upper-left menu ribbon, select *File* and click *New Project*. Choose *New Directory* ➤ *New Project*, with *Directory name* `ThisChapterTitle`, and select *Create Project*. To create an `R` script file, on the top ribbon, right under the word *File*, click the small icon with a plus sign on top of a blank bit of paper, and select the *R Script* menu option. Click the floppy disk–shaped *save* icon, name this file `PracticingToLearn XX.R` (where XX is the number of this chapter), and click *Save*.

M. Wiley and J. F. Wiley, *Beginning R 4*, https://doi.org/10.1007/978-1-4842-6053-1_13

In the lower-right pane, you should see your project's two files, and right under the *Files* tab, click the button titled *New Folder*. In the *New Folder* pop-up, type data and select *OK*. In the lower-right pane, click your new *data* folder. Repeat the folder creation process, making a new folder titled ch13.

Remember all packages used in this chapter were already installed on your local computing machine in Chapter 2. There is no need to re-install. However, this is a new project, and we are running this set of code for the first time. Therefore, you need to run the following library() calls:

```
library(data.table)

## data.table 1.13.0 using 6 threads (see ?getDTthreads). Latest news:
r-datatable.com

library(ggplot2)
library(visreg)
library(emmeans)
library(palmerpenguins)
library(ez)

## Registered S3 methods overwritten by 'lme4':
##   method                          from
##   cooks.distance.influence.merMod car
##   influence.merMod                car
##   dfbeta.influence.merMod         car
##   dfbetas.influence.merMod        car

library(JWileymisc)
```

We set up our dataset the same way as in Chapter 11.

```
acesData <- as.data.table(aces_daily)[SurveyDay == "2017-03-03" &
SurveyInteger == 3]
acesData[, StressCat := factor(fifelse(STRESS >= 2,
                                        "2+",
                                        as.character(STRESS)))]

penguinsData <- as.data.table(penguins)
```

You are now free to learn more about statistics!

13.2 ANOVA Background

Analysis **o**f **va**riance (ANOVA) is a set of statistical models that work by partitioning the variation in a continuous outcome and estimating how much of that variation can be attributable to different groups. ANOVAs are a subset of the generalized linear model, so can be seen as special cases of regression. Specifically, ANOVAs allow a continuous variable as an outcome and categorical variables as predictors. ANOVAs do not work with continuous predictors. ANOVAs have been a particularly popular technique for analyzing experimental data. ANOVAs often work well for experimental data as the main predictor of interest is the group or condition randomized to, which will be categorical.

Historically, ANOVAs had many benefits as the focus on continuous outcomes and categorical predictors allowed many computational shortcuts that make ANOVAs much easier to calculate by hand or calculator compared to other methods like regression. Even now that computational capacity means there is no need for the simpler calculations of ANOVAs over regression, some procedures and goals are more natural in ANOVAs so they continue to be used.

There are many questions that can be answered by an ANOVA. Here are some examples:

- Does crop growth differ depending whether crops are watered in the morning, afternoon, or overnight?
- Do people eat fewer calories when nutritional information is prominently displayed vs. when it is buried or not available?
- Does loneliness in children vary by attachment style (secure, anxious-preoccupied, fearful-avoidant, dismissive-avoidant)?
- Do people spend more time active if randomized to exercise at a gym or play team sports?
- Do Adelie penguins seem to get better food on some islands than others?

ANOVA is like a t-test in that it compares means between groups. Whereas t-tests are limited to comparing only two groups, ANOVAs can test the difference between two or more group means. There are different kinds of ANOVAs, but generally, ANOVA tests the null hypothesis that the means (denoted with the Greek letter μ [mu]) for all groups are identical. That is

$$H_0 : \mu_1 = \mu_2 = \mu_3 = \ldots = \mu_k$$

A statistically significant ANOVA indicates not all means are identical. In other words, a statistically significant ANOVA indicates that at least one group mean differs from the rest of the group means. Note that a significant ANOVA also could indicate that more than one group mean may be different; all we know for sure is at least one is different.

If we only had two groups, then a t-test and ANOVA would give the same result. Thus, one question may be: Could we just make do with t-tests? If we wanted to compare multiple group means, could we compare pairs of groups using t-tests? Although it is possible to break even many group comparisons down by comparing only two groups at a time with t-tests, this approach has some drawbacks. First, running many tests inflates the Type I error rate. Second, you cannot take into account the effect of more than one variable at a time. Third, when the assumptions are met, for more than two groups, ANOVA provides more power than pairwise t-tests. More power means less Type II errors (i.e., failing to reject a null hypothesis that is, in fact, false).

One reason for greater power in ANOVA than t-test for data with more than two groups is that t-tests include an estimate of the pooled variance across groups. In ANOVAs, the pooled variance is estimated across all groups, and the degrees of freedom for this draw on the overall sample size. Larger sample results in higher power to detect statistically significant differences. By using all the groups simultaneously, the ANOVA is able to effectively have a larger sample size than are the series of pairwise t-tests. If there are only two groups, it does not matter which you choose.

If you are reading about ANOVAs or see them used, you will often see terms such as a “one-way” ANOVA. When labeling ANOVAs, the number of independent variables (i.e., factors) denotes the number of “ways” for the ANOVA:

- **One-way**: One independent variable or factor (i.e., one predictor)
- **Two-way**: Two independent variables or factors (i.e., two predictors)
- **Three-way**: Three independent variables or factors
- **Higher-way**: This process keeps going, although most ANOVAs in practice are one-way or two-way; still, sometimes you see higher-“way” ANOVAs.

By default, in ANOVAs with more than one independent variable, all possible interactions between predictor variables are included. If you need a refresher on interactions, see Chapter 12.

As ANOVAs are a special case of the generalized linear model, they have the essentially the same assumptions as do linear regressions. Key assumptions of ANOVAs are the following:

- **Independence**: The observations are independent of each other (although note that this can be relaxed in repeated measures ANOVA or mixed effects ANOVA, just as it is relaxed in mixed effects regression; topics that are often covered in more advanced books).
- **Continuous**: The outcome variable is an interval or ratio type (i.e., continuous).
- **Normality**: The parameter sampling distributions are normally distributed, typically assessed by checking the residual distributions.
- **Homogeneity of variance**: The variance in the outcome must be approximately equal (homogenous) for each group. If the variances are unequal (heterogeneous), this assumption is violated.

As the name suggests, ANOVAs work by analyzing (partitioning) variance. For example, suppose we have an ANOVA with one independent variable, group, and there are three different levels of group. An ANOVA would break the total variability in the outcome down as shown in Figure 13-1.

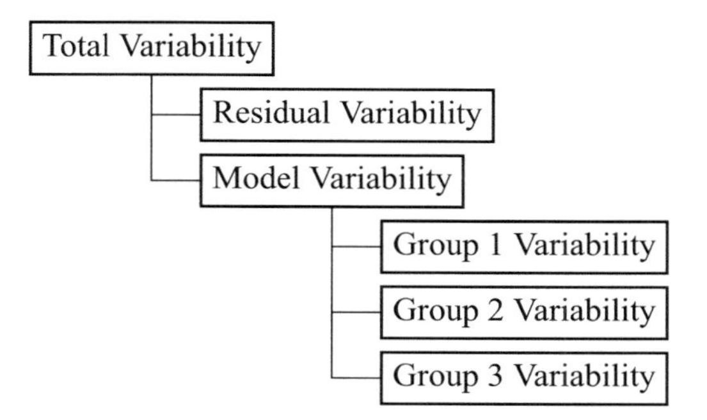

Figure 13-1. *Diagram of variance partitioning for an ANOVA with one independent variable that has three different groups (levels)*

Formal Mathematics

We can formalize this general idea of variability partitioning. Suppose that we have a continuous outcome variable, *y*, measured on *k* different groups and there are n_k people in each group. We indicate the outcome for a particular person, *i*, in a particular *j* as y_{ij}. We indicate the overall mean of the outcome for all observations as $\bar{y}$. Finally, we indicate the mean of *y* in each group as $\bar{y}_j$. With those definitions, ANOVA formally defines three broad sources of variability. Specifically, it works with what are called the sums of squares (SS), which are the sum of the squared deviations. The following equations show these three sources:

$$SS_{Total} = \sum_{j=1}^{k} \sum_{i=1}^{n_k} \left(y_{ij} - \bar{y}\right)^2$$

$$SS_{Model} = \sum_{j=1}^{k} n_k \left(\bar{y}_j - \bar{y}\right)^2$$

$$SS_{Residual} = \sum_{j=1}^{k} \sum_{i=1}^{n_k} \left(y_{ij} - \bar{y}_j\right)^2$$

- SS_{Total} is the total sum of squared deviations of all the observations from the grand mean for all observations. This is basically the total variability in the outcome, *y*.
- SS_{Model} is the sum of square deviations of each group's mean from the overall grand mean, weighted by the number of observations in that group. It captures how much of the total variability in the outcome, SS_{Total}, can be explained by differences between groups.
- $SS_{Residual}$ is the sum of squared deviations of the observations from their own individual group means. This captures how much variability there is "left over" or residual in observations, even after taking into account to which group an observation belongs. This is often thought of as the unexplained variability, sometimes also called the error. It is akin to the residual variance in regression.

A visual diagram of these three sums of squares is shown in Figure 13-2. Panel A shows SS_{Total}, Panel B shows SS_{Model}, and Panel C shows $SS_{Residual}$. The sums of squares are what would happen if you took each vertical line, squared it, and then summed all of them.

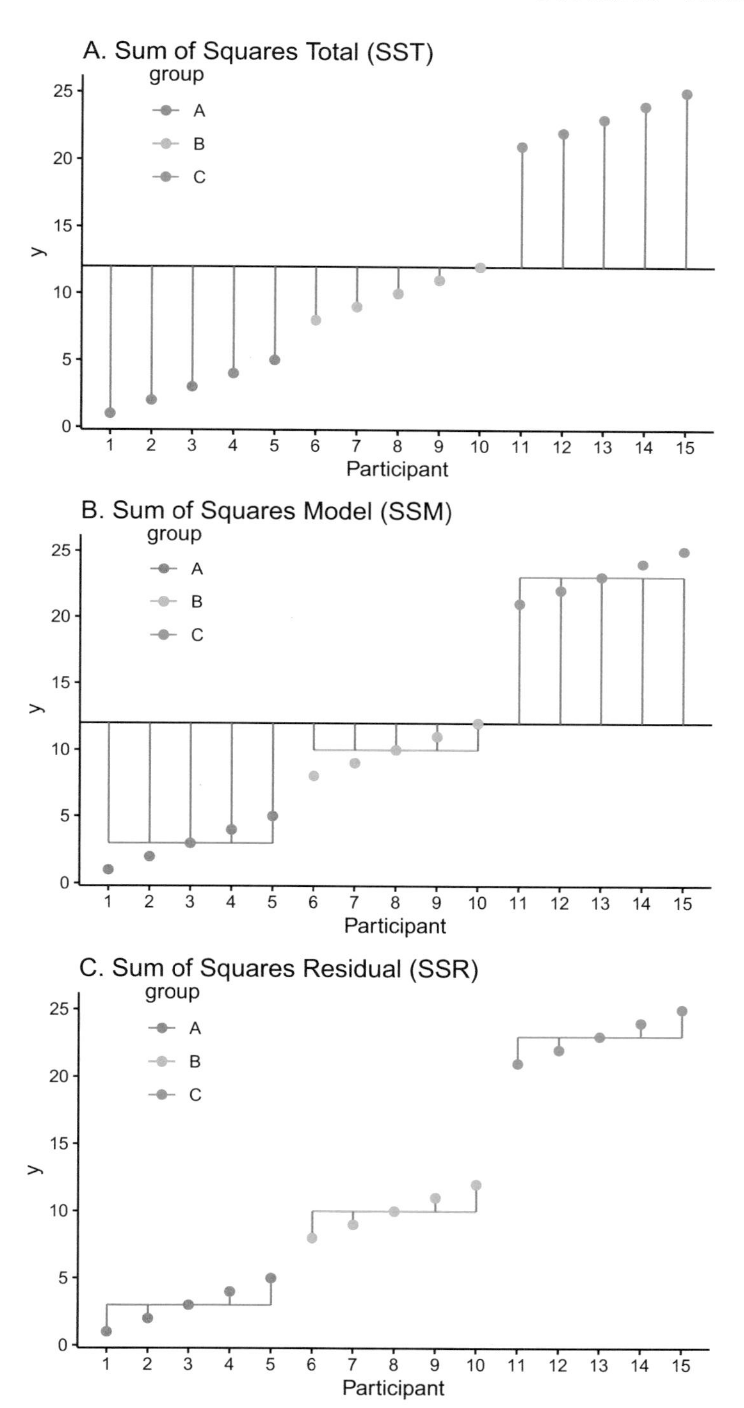

***Figure 13-2.** Visual explanation of the different kinds of sums of squares in an ANOVA*

To actually conduct an ANOVA, we need just a few more pieces of information. Rather than use the raw sums of squares, the ANOVA uses mean squares, which are the sums of squares divided by the degrees of freedom for each term. When working with a regular ANOVA, there are two different degrees of freedom (df):

$$DF_{Model} = k - 1$$
$$DF_{Residual} - N - k$$

For example, if we use the example shown in Figure 13-2, there are 15 observations total, so $N = 15$, and there are three groups, so $k = 3$. That would mean $DF_{Model} = 3 - 1 = 2$ and $DF_{Residual} = 15 - 3 = 12$.

The mean squares for the model and residual are defined as follows:

$$MS_{Model} = \frac{SS_{Model}}{DF_{Model}}$$
$$MS_{Residual} = \frac{SS_{Residual}}{DF_{Residual}}$$

To determine whether the group means are significantly different, we compare the amount of variability explained by the model to the residual variability. This ratio is called the F-ratio:

$$F = \frac{MS_{Model}}{MS_{Residual}} = \frac{\text{MSBetween Groups}}{\text{MSWithin Groups}}$$

We say the F-ratio is the ratio of the treatment or group effect (MS_{Model}) to the differences we would expect due to random chance ($MS_{Residual}$). When the treatment (group) has no effect, then the differences between treatments (numerator) are entirely due to chance: $F = \frac{0}{MS_{Residual}}$.

If the difference is due to chance, then the numerator will be much smaller than the denominator, and the F-ratio will go to zero and we would conclude that there is no treatment effect.

When the treatment does have an effect, causing differences between the samples, the between-treatment differences (numerator) should be larger than chance (denominator): $F = \frac{\text{treatment(group)effect}}{\text{differencesduetochance}}$.

A large F-ratio indicates that the differences between treatments are greater than chance. The treatment does have a statistically significant effect.

The F-ratio follows the F distribution, another statistical distribution a bit like the normal distribution or the t-distribution that we have seen previously.

If the F-ratio is larger than the critical value for our desired statistical significance, it indicates that one or more of the treatment (group) means are statistically significantly different from each other. In the end, we use the F-ratio to get a p-value and use that for hypothesis testing, same as we would use a p-value from a t-test, correlation, or regression model.

What drives a large F-ratio? F can be large if either MS_{Model} is large, meaning there are large differences in the group means, or if $MS_{Residual}$ is small, indicating very little variability within groups. This would mean if you were designing an experiment and you wanted to ensure significant results, you could either try to make the groups more different, such as by having a very potent treatment, or make people within each group as similar as possible, such as by getting a very similar set of people to be in the study.

The F-ratio tells us only that the direct or indirect manipulation was successful. That is, that group means were different. It does not tell us specifically which group means differ from which. We need additional tests to find out where the group differences are. Often we find out which groups are different by carrying out pairwise comparisons of the group means after the ANOVA has been run. These comparisons can be made using the pooled residual variance, $MS_{Residual}$ from the ANOVA, so they are still more powerful than conducting t-tests.

One note here is if there are many levels of the groups; for example, suppose there were four different groups (named A, B, C, and D). Then there are six unique pairwise comparisons. How so? The R function `combn()` provides all (unordered) **combinations** provided. The elements of the first argument (in our case, a variable named groups that has the letters A–D) are combined in sets of size `m` (in this case, m = 2 because we are discussing pairs):

```
groups <- LETTERS[1:4]
combn(groups,
      m = 2)

##      [,1] [,2] [,3] [,4] [,5] [,6]
## [1,] "A"  "A"  "A"  "B"  "B"  "C"
## [2,] "B"  "C"  "D"  "C"  "D"  "D"
```

Running six hypothesis tests increases the Type I error rate. As that is not ideal, people often try to do some adjustments to control the overall error rate. Perhaps the simplest and most conservative option is the Bonferroni method. A Bonferroni correction is easy to apply. You take your desired alpha value, say $\alpha = 0.05$, and divide it by the number of tests performed to get the new alpha value. For example, $Bonferroni_{\alpha} = .05/6 = .0083$. This is a very easy method as it does not require any complex calculations, but it is known to be overly conservative in that it controls the Type I error rate lower than the actual desired overall .05 error rate. Still, it is popular because it is both safe and easy.

Null hypothesis significance testing is all or none. That is, results are either statistically "significant" if $p < .05$ or are not significant if $p > .05$. Real life is not so all or none. Furthermore, statistical significance does not necessarily indicate practical significance (i.e., real-world importance). Significance testing rules out chance. But how big is the effect? Questions like "how big" or "how much" can be conveyed by effect sizes. A common effect size for ANOVA is the proportion of variance explained, which, although it is called R^2 in regression, often is referred to as η^2 (eta squared) in ANOVA, or ω^2 (omega squared) which is basically the same as regression's adjusted R^2. With all that background, we jump into conducting and interpreting some ANOVAs in R.

13.3 One-Way ANOVA

As you may have come to expect, in R, these calculations become rather easy. We start with one-way ANOVA, that is, an ANOVA with only one independent or predictor variable. This predictor variable is a categorical variable and will need to be a `factor()` in R.

Example

In this example, we look at whether stress category (0, 1, 2+) explains variation in positive affect. We will make use of the ez package in R to run ANOVAs. In particular, the `ezANOVA()` function can run independent measures, repeated measures, and mixed model ANOVA and provides assumption tests. In this book, we are only learning about independent measures ANOVAs, but know that if you need more complex ANOVAs, the `ezANOVA()` function will still work. The `ezANOVA()` function requires several arguments to work:

- `data`, the dataset to analyze. It should be a data frame, and there should not be missing data, so we remove missing data first.

- dv, the variable name of the outcome variable.
- wid, a variable that has the IDs for each person, preferably stored as a factor for ezANOVA().
- between, the independent, predictor variable(s). Note that in more complex ANOVAs with repeated measures, you might use the within argument for variables too. This also needs to be a factor.
- type, the type of sums of squares to calculate. A common choice is type 3, although there are other types and not everyone agrees on what is best, particularly when there are unbalanced group sizes (i.e., not all groups have an identical number of people).

In addition, although these are not needed, we use the detailed argument to get more detailed output and the return_aov argument to make it easier to conduct some post hoc pairwise comparisons.

We create a dataset without missing data on our variables by using negation ! with is.na(). Additionally, we convert the UserID to a factor():

```
aovdata <- acesData[!is.na(PosAff) & !is.na(StressCat)]
aovdata[, UserID := factor(UserID)]
```

It is not required to reduce our dataset to only the columns used for the ANOVA. However, it may be helpful to see just the columns that R will use for the ANOVA. In particular, notice which columns are factors (e.g., categorical data) and which are numeric:

```
str(aovdata[, .(UserID, StressCat, PosAff)])

## Classes 'data.table' and 'data.frame':      187 obs. of  3 variables:
##  $ UserID   : Factor w/ 187 levels "1","2","3","4",..: 1 2 3 4 5 6 7 8 9 10 ...
##  $ StressCat: Factor w/ 3 levels "0","1","2+": 3 2 3 3 1 1 1 3 1 2 ...
##  $ PosAff   : num  1.55 3.17 3.51 1.7 2.22 ...
##  - attr(*, ".internal.selfref")=<externalptr>
```

With our data properly "cleaned" and in the correct format, we fit the ANOVA using the ezANOVA() function as described. We store the results in anova1. Finally we can print() to view the output from the ANOVA:

```
anova1 <- ezANOVA(
  data = aovdata,
  dv = PosAff,
  wid = UserID,
  between = StressCat,
  type = 3,
  detailed = TRUE,
  return_aov = TRUE)

## Warning: Data is unbalanced (unequal N per group). Make sure you
specified a well-considered value for the type argument to ezANOVA().
## Coefficient covariances computed by hccm()

print(anova1)

## $ANOVA
##        Effect DFn DFd     SSn   SSd        F         p p<.05     ges
## 1 (Intercept)   1 184 1325.43 212.2 1149.429 4.643e-81     * 0.86201
## 2   StressCat   2 184   17.46 212.2    7.569 6.934e-04     * 0.07602
##
## $'Levene's Test for Homogeneity of Variance'
##   DFn DFd    SSn   SSd      F      p p<.05
## 1   2 184 0.3051 76.43 0.3673 0.6931
##
## $aov
## Call:
##    aov(formula = formula(aov_formula), data = data)
##
## Terms:
##                 StressCat Residuals
## Sum of Squares      17.46    212.17
## Deg. of Freedom         2       184
##
```

```
## Residual standard error: 1.074
## Estimated effects may be unbalanced
```

The main output we want to look at is under the heading, ANOVA. Although the intercept is included, we are mostly interested in the second row which is for our group variable, `StressCat`. We can see the numerator degrees of freedom, labeled "DFn," is 2 (3 groups - 1). The denominator degrees of freedom (DFd) are the residual degrees of freedom ($N - k = 187 - 3 = 184$). We have the sums of squares (SS) for stress and the residual sum of squares. Because stress category is our only predictor, that is the model sum of squares. We have the F-ratio, here about 7.6, and then using the F ratio and the degrees of freedom, it has looked up the p-value based on the assumed F distribution. The p-value for stress category is very small (.00069), indicating that at least one stress category's average positive affect differs. Notice that we do not have any information about the average positive affect in any group or which groups differ at this point. We will get that information later.

In other words, all we know right now is at least one of these groups is not like the others.

The last column, "ges," is a generalized effect size measure, but for simple, independent measures ANOVAs like this one, it will be eta^2, the same as R^2 in regression, that is the proportion of total variation in the outcome explained by that predictor. We can see that stress category can explain about 7.6% of the variance in positive affect.

After the ANOVA heading is Levene's test for homogeneity of variance. This is a test for whether the variance (the square of standard deviation) of positive affect is the same in all of our groups, which in this example are the stress categories `StressCat`.

Because we want the variances to be equal (this is an assumption of ANOVA), we want the p-value *not* to be significant. A significant p-value for Levene's test would indicate that the variances differed between groups and thus the homogeneity of variance assumption was *violated*. This sort of test is not possible in regression because with continuous predictors, you cannot define groups to calculate the variance within.

We also can use the `modelDiagnostics()` function you used for regression. We do not call it on the `anova1` object directly; rather we call it on the `aov` object stored within `anova1`. Other than that, we can use the diagnostic plots as usual to help identify extreme values and normality. The results are in Figure 13-3. In this case, the results look rather good. The residuals are about normally distributed with no clear outliers or extreme values. Homogeneity of variance also appears to be met; we know this was also supported by Levene's test, which we saw in the earlier text output:

```
plot(modelDiagnostics(anova1$aov), ncol = 2)
```

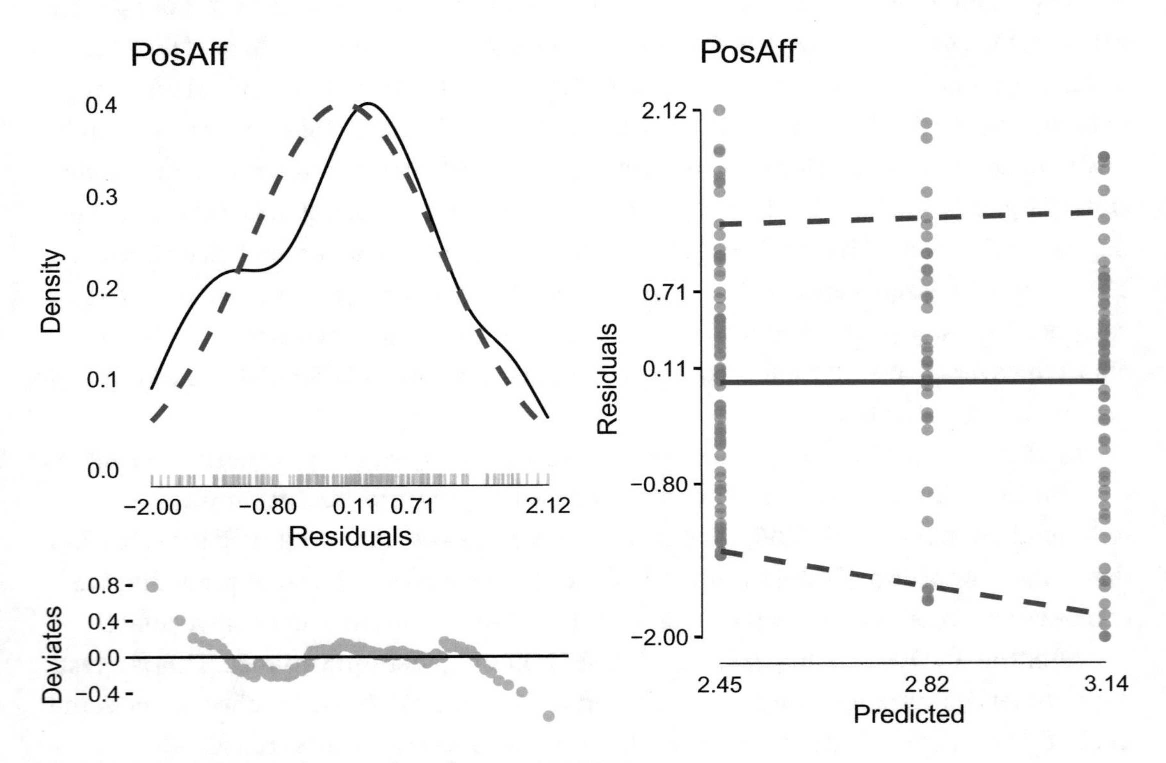

Figure 13-3. *Diagnostic plots for an ANOVA*

At this point, what we know is the mean positive affect is not the same in all stress categories, F(2, 184) = 7.6, $p < .001$. However, we do not know what the means are or which groups differ. To get a finer-grained understanding of our results, we turn again to the emmeans package. We store the means in an object, `anova1.means`.

```
anova1.means <- emmeans(object = anova1$aov,
                        specs = "StressCat")
```

We can use the `summary()` function to see the means and test whether each is different from zero (although that is not a very interesting hypothesis in this case). In other words, the p-values in this case are not of great interest. The means themselves, on the other hand, are:

```
summary(anova1.means, infer = TRUE)
```

```
##  StressCat emmean    SE  df lower.CL upper.CL t.ratio p.value
##  0           3.14 0.130 184     2.88     3.39 24.090  <.0001
##  1           2.82 0.174 184     2.48     3.16 16.197  <.0001
##  2+          2.45 0.119 184     2.22     2.69 20.549  <.0001
##
## Confidence level used: 0.95
```

Viewing the means can help us to understand what the impact of stress is, beyond just the F-ratio telling us that stress category is associated with positive affect.

To understand better, we can present the means graphically. We use the `plot()` function on the means object, which uses the `ggplot2` package behind the scenes. Because it is a `ggplot2` graph, we can use familiar code to modify it a bit and label it. We use one new line of code, `coord_flip()`, to flip the coordinates so `PosAff` is on the y-axis and `StressCat` is on the x-axis. The final results are in Figure 13-4. It becomes clearer which two groups are not like each other – the two with non-overlapping confidence bands (namely, groups 0 and 2+):

```
plot(anova1.means) +
  xlab("Positive Affect") +
  ylab("Stress Category") +
  coord_flip() +

  ggtitle("Estimated Means and 95% Confidence Intervals from One-Way ANOVA")
```

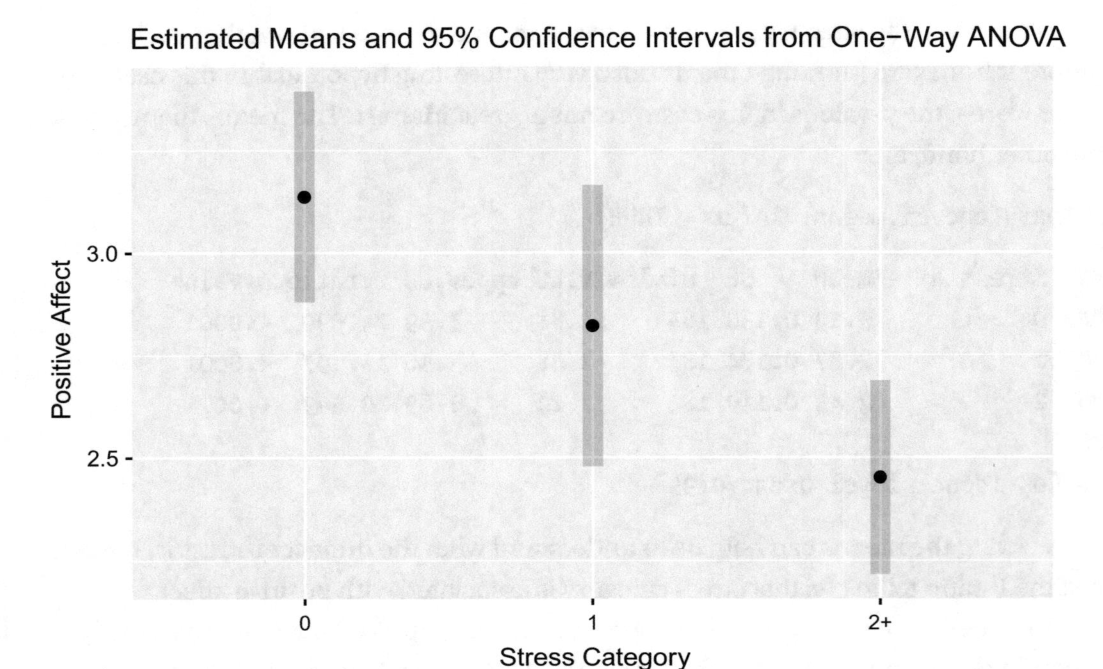

Figure 13-4. *ANOVA means plot*

Finally, to confirm our visual observation, we get all pairwise comparisons between stress groups using the `pairs()` function. This is the same approach we used for categorical predictors in regression in Chapter 11:

```
pairs(anova1.means)

##   contrast estimate    SE  df t.ratio p.value
##   0 - 1       0.316 0.217 184 1.451   0.3172
##   0 - (2+)    0.685 0.177 184 3.880   0.0004
##   1 - (2+)    0.370 0.211 184 1.751   0.1893
##
## P value adjustment: tukey method for comparing a family of 3 estimates
```

The `pairs()` function does some more complex calculations so that we get p-values (here the p-values are again of interest) adjusted for multiple comparisons using Tukey's method. Tukey's method is better, albeit more complex, than the Bonferroni correction mentioned earlier. The results show people with 0 vs. 2+ stress have significantly different average positive affect values ($p < .001$). No other pairwise comparisons were statistically significant based on p-value.

Now we have seen a complete one-way ANOVA example. We have seen how to check assumptions, estimate the ANOVA, conduct follow-up tests, and get effect sizes.

Example

Now that you have seen all the mechanics of this process, you are ready to try another analysis. You turn your attention to penguins, in particular, the Adelie penguins which inhabit all three islands. You confirm this by using `.N` in the column operation and a suitable by statement:

```
penguinsData[order(species, island),
             .N,
             by = .(species, island)]
```

```
##      species    island   N
## 1:    Adelie    Biscoe  44
## 2:    Adelie     Dream  56
## 3:    Adelie Torgersen  52
## 4: Chinstrap     Dream  68
## 5:    Gentoo    Biscoe 124
```

Your question remains: "Is one island better suited for Adelie penguins to find the most food?"

You quickly save just the Adelie penguins to a dataset for ANOVA, taking the time to make sure there are no missing values in your two variables of interest, islands and body mass:

```
aovPenguins <- penguinsData[ species == "Adelie" &
                             !is.na(body_mass_g) &
                             !is.na(island)]
```

The penguin data set lacks an ID variable. Keeping in mind `.N` counts the total number of values and `1:10` would list the numbers 1–10, you create a new column named `ID` that is numbered from 1 to the total number of Adelie penguins. You set this column to a factor variable (as required for `ezANOVA()`) and take a quick look at the variables you intend to use:

```
aovPenguins[ , ID := 1:.N]
```

```
aovPenguins[, ID := factor(ID)]

str(aovPenguins[, .(ID, island,body_mass_g)])

## Classes 'data.table' and 'data.frame':      151 obs. Of  3 variables:
##  $ ID         : Factor w/ 151 levels "1","2","3","4",..: 1 2 3 4 5 6 7 8
9 10 ...
##  $ island     : Factor w/ 3 levels "Biscoe","Dream",..: 3 3 3 3 3 3 3 3
3 3 ...
##  $ body_mass_g: int  3750 3800 3250 3450 3650 3625 4675 3475 4250 3300 ...
##  - attr(*, ".internal.selfref")=<externalptr>
```

You are now ready to use the ezANOVA() function and assign that output to anovaP1. Be sure to note that your factored, categorical predictor is island, while your numeric outcome is body_mass_g:

```
anovaP1 <- ezANOVA(
  data = aovPenguins,
  dv = body_mass_g,
  wid = ID,
  between = island,
  type = 3,
  detailed = TRUE,
  return_aov = TRUE)

## Warning: Data is unbalanced (unequal N per group). Make sure you
specified a well-considered value for the type argument to ezANOVA().

## Coefficient covariances computed by hccm()

print(anovaP1)

## $ANOVA
##        Effect DFn DFd       SSn      SSd         F            p p<.05
## 1 (Intercept)   1 148 2.049e+09 31528779 9.616e+03 1.533e-136     *
## 2      island   2 148 1.365e+04 31528779 3.205e-02 9.685e-01
##         ges
## 1 0.9848427
## 2 0.0004329
```

```
##
## $'Levene's Test for Homogeneity of Variance'
##   DFn DFd   SSn      SSd      F     p p<.05
## 1   2 148 16967 11425914 0.1099 0.896
##
## $aov
## Call:
##    aov(formula = formula(aov_formula), data = data)
##
## Terms:
##                   island Residuals
## Sum of Squares     13655  31528779
## Deg. of Freedom        2       148
##
## Residual standard error: 461.6
## Estimated effects may be unbalanced
```

In the `print()` output, you see that `island` does not have a significant p-value (0.9685). This rather ends your hope of there being an ideal island for Adelie penguins to enjoy an endless feast (or else, all islands are equally full of food). Levene's test at least shows a nonsignificant p-value of 0.896; the variances are most likely equal (which is good).

Running model diagnostics is perhaps overkill. However, just to be sure, you inspect `emmeans()` and view the `summary()` results. While the p-values are not of interest in this view, it is worth noting that the emmeans are all rather close together:

```
anovaP1.means <- emmeans(object = anovaP1$aov,
                         specs = "island")

summary(anovaP1.means, infer = TRUE)
```

```
##  island    emmean   SE  df lower.CL upper.CL t.ratio p.value
##  Biscoe      3710 69.6 148     3572     3847 53.310  <.0001
##  Dream       3688 61.7 148     3567     3810 59.800  <.0001
##  Torgersen   3706 64.6 148     3579     3834 57.350  <.0001
##
## Confidence level used: 0.95
```

A quick visual inspection of Figure 13-5 shows that all these confidence intervals overlap. There is simply no difference in the penguins on each island in terms of body mass:

```
plot(anovaP1.means) +
  xlab("body mass in grams") +
  ylab("islands") +
  coord_flip() +
  ggtitle("Estimated Means and 95% Confidence Intervals from One-Way ANOVA")
```

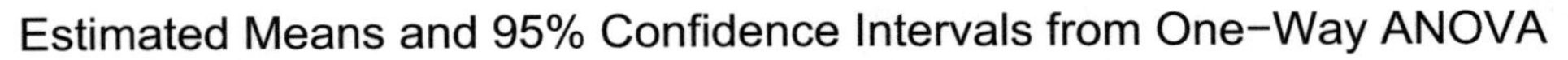

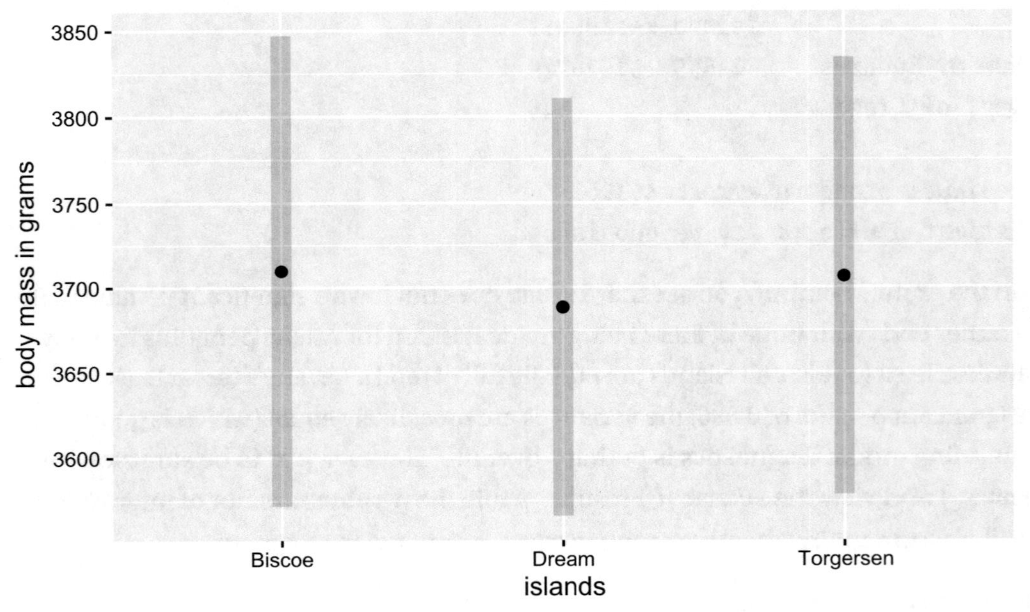

Figure 13-5. *ANOVA means plot*

A brief look at pairs() confirms it. There are no significant p-values between the pairs:

```
pairs(anovaP1.means)

## contrast            estimate   SE  df t.ratio p.value
## Biscoe - Dream         21.27 93.0 148  0.229  0.9716
```

```
##   Biscoe - Torgersen     3.29 95.0 148  0.035  0.9993
##   Dream - Torgersen    -17.98 89.3 148 -0.201  0.9779
##
## P value adjustment: tukey method for comparing a family of 3 estimates
```

Notice that, once you become familiar with these types of analysis, it does not take all that long to check a hypothesis in R. In the next section, we consider examples with two independent variables.

13.4 Factorial ANOVA

A factorial ANOVA is an ANOVA with more than one independent variable where all possible combinations of all independent variables are included. It is a catch-all term in that "more than one" fits into this category.

Example

For this next example, we consider both stress category (i.e., `StressCat`) and biological sex (i.e., `Female`), which is coded so females are 1 and males are 0. `ezANOVA()` prefers if all predictors are stored as factors, so we will convert `Female` to a factor first:

```
aovdata[, Female := factor(Female)]
```

To see the crossed factorial structure, we could use our standard `data.table` techniques. In particular, we might count up the number of study participants in each pair of categories via the `.N` column operation and a suitable `by =` statement:

```
aovdata[order(StressCat, Female),
        .N,
        by = .(StressCat, Female)]

##    StressCat Female  N
## 1:         0      0 26
## 2:         0      1 42
## 3:         1      0 13
## 4:         1      1 25
## 5:        2+      0 34
## 6:        2+      1 47
```

However, cross-classifying factors is a frequent enough goal that base R has a special function named xtabs(). This function gives a frequency cross-tabulation showing the count (frequency) of people in all cells using a familiar formula structure. Either way, notice the cells or the rows for all combinations are the same values (e.g., StressCat of 1 and Female of 1 is 25):

```
xtabs(formula = ~ StressCat + Female,
      data = aovdata)
```

```
##            Female
## StressCat   0  1
##          0 26 42
##          1 13 25
##         2+ 34 47
```

The results show that although some cells have fewer people than others, the smallest cell has 13 people (males with stress category of 1), and the largest cell has 47 people (females with stress category of 2+).

To fit a two-way (because we have two independent variables), factorial ANOVA, we again use the ezANOVA() function, and the code is almost identical to our previous example, except we have added Female as another variable to the between argument. Again we can print the results to see the output:

```
anova2 <- ezANOVA(
  data = aovdata,
  dv = PosAff,
  wid = UserID,
  between = .(StressCat, Female),
  type = 3, detailed = TRUE,
  return_aov = TRUE)
```

```
## Warning: Data is unbalanced (unequal N per group). Make sure you
specified a well-considered value for the type argument to ezANOVA().
```

```
## Coefficient covariances computed by hccm()
```

```
print(anova2)
```

```
## $ANOVA
##             Effect DFn DFd       SSn   SSd         F         p p<.05
## 1      (Intercept)   1 181 1205.8395 205.5 1062.0816 1.186e-77     *
## 2        StressCat   2 181   15.8228 205.5    6.9682 1.215e-03     *
## 3           Female   1 181    0.6124 205.5    0.5394 4.636e-01
## 4 StressCat:Female   2 181    6.6638 205.5    2.9347 5.568e-02
##        ges
## 1 0.854394
## 2 0.071492
## 3 0.002971
## 4 0.031409
##
## $'Levene's Test for Homogeneity of Variance'
##   DFn DFd   SSn   SSd      F      p p<.05
## 1   5 181 1.378 73.52 0.6787 0.6401
##
## $aov
## Call:
##    aov(formula = formula(aov_formula), data = data)
##
## Terms:
##                 StressCat Female StressCat:Female Residuals
## Sum of Squares      17.46   0.01             6.66    205.50
## Deg.of Freedom          2      1                2       181
##
## Residual standard error: 1.066
## Estimated effects may be unbalanced
```

In the main ANOVA table result, we now see results for `StressCat`, `Female`, and their interaction, `StressCat:Female`. The interaction term is close, but not quite statistically significant, $p = .056$.

However, unlike multiple regression with interactions we saw in Chapter 12, in factorial ANOVAs, even if the interaction is not statistically significant, we do not drop it. Also, unlike multiple regression, the results for `StressCat` and `Female` are not simple effects at only specific values of the moderator. They are still main effects, capturing the overall main effect of that variable, so we can still interpret them. The results show that independent of biological sex, positive affect means (averages) differ significantly across stress categories. We can also see that there is no main effect of biological sex. The effect size measures, "ges," can be roughly interpreted as the proportion of total variance explained by the main effect of stress category, the main effect of biological sex, and their interaction. They are not simple effects as in regression.

Results from Levene's test for homogeneity of variance are not statistically significant, indicating that our model does not violate the assumption of homogeneity. This is good.

We again use `modelDiagnostics()` to view graphs to examine the other assumptions. The normality assumption and homogeneity of variance assumption continue to be met, and no new extreme values appear to be present, as shown in Figure 13-6:

```
plot(modelDiagnostics(anova2$aov), ncol = 2)
```

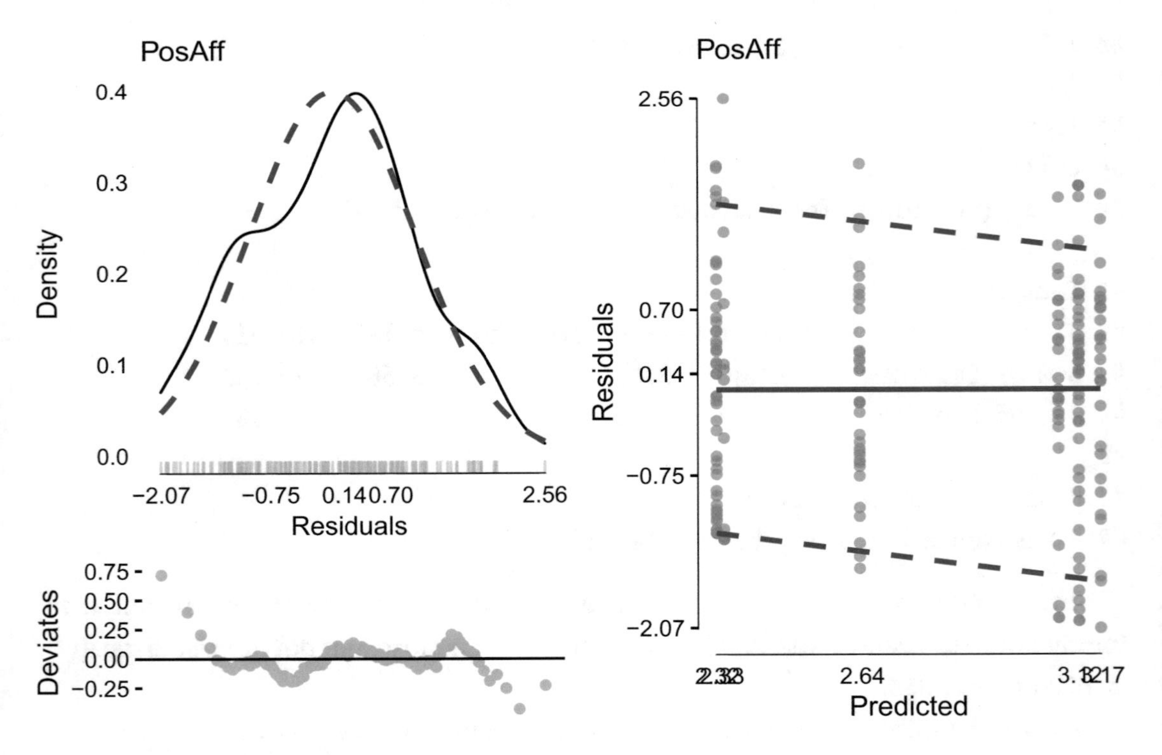

Figure 13-6. *Diagnostic plots for an ANOVA*

With diagnostics done, we can move on to getting estimated means and comparing different groups to see where differences are. However, with a factorial ANOVA, we have quite a few possible choices. We will briefly outline the options and go through a quick example of each option:

- **A**: Calculate the means in each individual "cell," the crossing of stress category and biological sex. Compare differences between stress categories separately for females and males.
- **B**: Calculate the means in each individual "cell," the crossing of stress category and biological sex. Compare differences between females and males separately for each stress category.
- **C**: Calculate the means for each stress category, averaging across biological sex. Compare pairwise differences in stress categories.
- **D**: Calculate the means for each biological sex category, averaging across stress category. Compare pairwise differences in biological sex categories.

Options A and B both take the interaction into account and then focus pairwise tests on different variables. Options C and D both ignore the interaction and focus on the main effect of each independent variable, which is not unreasonable considering that the interaction is not statistically significant.

We now look at the results for each option. To help make it clear which option we are on, we will put all the code together in one chunk per option.

Option A: Calculate the means in each individual "cell," the crossing of stress category and biological sex. Compare differences between stress categories separately for females and males using the `by = "Female"` argument to `emmeans()`:

```
anova2.meansA <- emmeans(object = anova2$aov,
                         specs = "StressCat",
                         by = "Female")

summary(anova2.meansA, infer = TRUE)

## Female = 0:
##  StressCat emmean    SE  df lower.CL upper.CL t.ratio p.value
##  0           3.17 0.209 181     2.75     3.58 15.153  <.0001
##  1           2.33 0.296 181     1.75     2.92  7.903  <.0001
```

```
## 2+          2.63 0.183 181    2.27    3.00 14.420 <.0001
##
## Female = 1:
## StressCat emmean    SE  df lower.CL upper.CL t.ratio p.value
## 0           3.12 0.164 181    2.79    3.44 18.969 <.0001
## 1           3.07 0.213 181    2.65    3.50 14.426 <.0001
## 2+          2.32 0.155 181    2.01    2.63 14.922 <.0001
##
## Confidence level used: 0.95
```

The graph of the means in Figure 13-7 has a separate panel for females and males, with stress category on the x-axis and positive affect on the y-axis. We can see a slightly different pattern of average positive affect by stress category, with males having the lowest positive affect for stress category 1 and females having the lowest positive affect for stress category 2+:

```
plot(anova2.meansA) +
  xlab("Positive Affect") +
  ylab("Stress Category") +
  coord_flip() +
  ggtitle("Estimated Means and 95% Confidence Intervals from Factorial
  ANOVA")
```

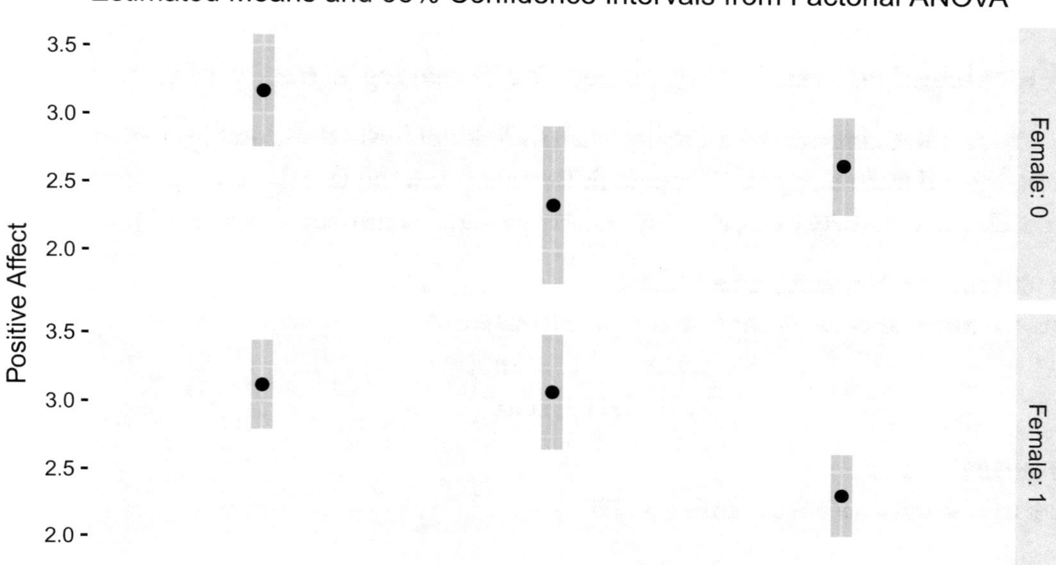

Figure 13-7. *Factorial ANOVA means plot – Option A*

In the pairwise comparisons, none of the stress category pairwise comparisons quite reach statistical significance for males (`Female = 0`). For females (`Female = 1`), stress category 2+ has significantly lower positive affect than both stress categories 0 and 1 (both $p \leq .01$).

```
pairs(anova2.meansA)

## Female = 0:
##  contrast estimate    SE  df t.ratio p.value
##  0 - 1      0.8311 0.362 181  2.296  0.0589
##  0 - (2+)   0.5314 0.278 181  1.914  0.1375
##  1 - (2+)  -0.2997 0.347 181 -0.863  0.6646
##
## Female = 1:
##  contrast estimate    SE  df t.ratio p.value
##  0 - 1      0.0445 0.269 181  0.165  0.9850
##  0 - (2+)   0.7996 0.226 181  3.534  0.0015
```

```
##  1 - (2+)    0.7551 0.264 181  2.863  0.0130
##
## P value adjustment: tukey method for comparing a family of 3 estimates
```

Option B: Calculate the means in each individual "cell," the crossing of stress category and biological sex. Compare differences between females and males separately for each stress category using the by = "StressCat" argument to emmeans():

```
## calculate the estimated means
anova2.meansB <- emmeans(object = anova2$aov,
                         specs = "Female",
                         by = "StressCat")

## print the means
summary(anova2.meansB, infer = TRUE)

## StressCat = 0:
##  Female emmean    SE  df lower.CL upper.CL t.ratio p.value
##  0        3.17 0.209 181     2.75     3.58 15.153  <.0001
##  1        3.12 0.164 181     2.79     3.44 18.969  <.0001
##
## StressCat = 1:
##  Female emmean    SE  df lower.CL upper.CL t.ratio p.value
##  0        2.33 0.296 181     1.75     2.92  7.903  <.0001
##  1        3.07 0.213 181     2.65     3.50 14.426  <.0001
##
## StressCat = 2+:
##  Female emmean    SE  df lower.CL upper.CL t.ratio p.value
##  0        2.63 0.183 181     2.27     3.00 14.420  <.0001
##  1        2.32 0.155 181     2.01     2.63 14.922  <.0001
##
## Confidence level used: 0.95
```

The graph of the means in Figure 13-8 has a separate panel for each stress category, with biological sex on the x-axis and positive affect on the y-axis. We can see a slightly different pattern of average positive affect by stress category. Males and females seem to only differ when stress category is 1. Note these are actually the same means and confidence intervals as in **Option A**, only rearranged and presented in a different layout:

```
plot(anova2.meansB) +
  xlab("Positive Affect") +
  ylab("Female") +
  coord_flip() +
  ggtitle("Estimated Means and 95% Confidence Intervals from Factorial
  ANOVA")
```

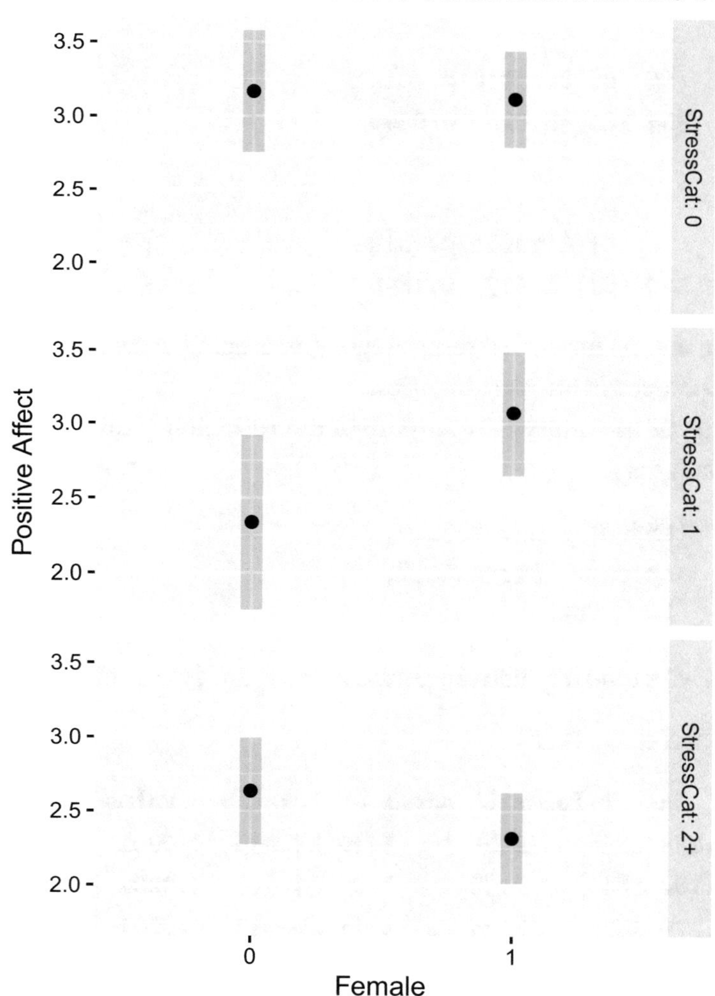

Figure 13-8. Factorial ANOVA means plot – Option B

In the pairwise comparisons, we confirm males and females are only significantly different on positive affect when stress category is 1 ($p = .040$), with the pattern that when stress category is 1, males have lower positive affect than do females:

```
pairs(anova2.meansB)
## StressCat = 0:
##  contrast estimate    SE  df t.ratio p.value
##  0 - 1      0.0477 0.266 181  0.180  0.8577
##
## StressCat = 1:
##  contrast estimate    SE  df t.ratio p.value
##  0 - 1     -0.7389 0.364 181 -2.028  0.0440
##
## StressCat = 2+:
##  contrast estimate    SE  df t.ratio p.value
##  0 - 1      0.3159 0.240 181  1.317  0.1896
```

Option C: Calculate the means for each stress category, averaging across biological sex. Compare pairwise differences in stress categories.

This option averages means and pairwise comparisons across biological sex. All that is left is the effect of stress category:

```
anova2.meansC <- emmeans(object = anova2$aov,
                         specs = "StressCat",
                         by = NULL)

## NOTE: Results may be misleading due to involvement in interactions

summary(anova2.meansC, infer = TRUE)

##  StressCat emmean    SE  df lower.CL upper.CL t.ratio p.value
##  0           3.14 0.133 181     2.88     3.40 23.639  <.0001
##  1           2.70 0.182 181     2.35     3.06 14.848  <.0001
##  2+          2.48 0.120 181     2.24     2.71 20.652  <.0001
##
## Results are averaged over the levels of: Female
## Confidence level used: 0.95
```

None of these means will be exactly the same as those in Option A or B because the individual cell means have been collapsed and these are "marginal" means in that they are on the "margin" of a table averaging across.

The plot of means in Figure 13-9 now only shows means by stress category and does not separate by biological sex. We can see that overall, positive affect goes down as stress category increases:

```
plot(anova2.meansC) +
  xlab("Positive Affect") +
  ylab("Stress Category") +
  coord_flip() +
  ggtitle("Estimated Means and 95% Confidence Intervals, averaged across
  sex")
```

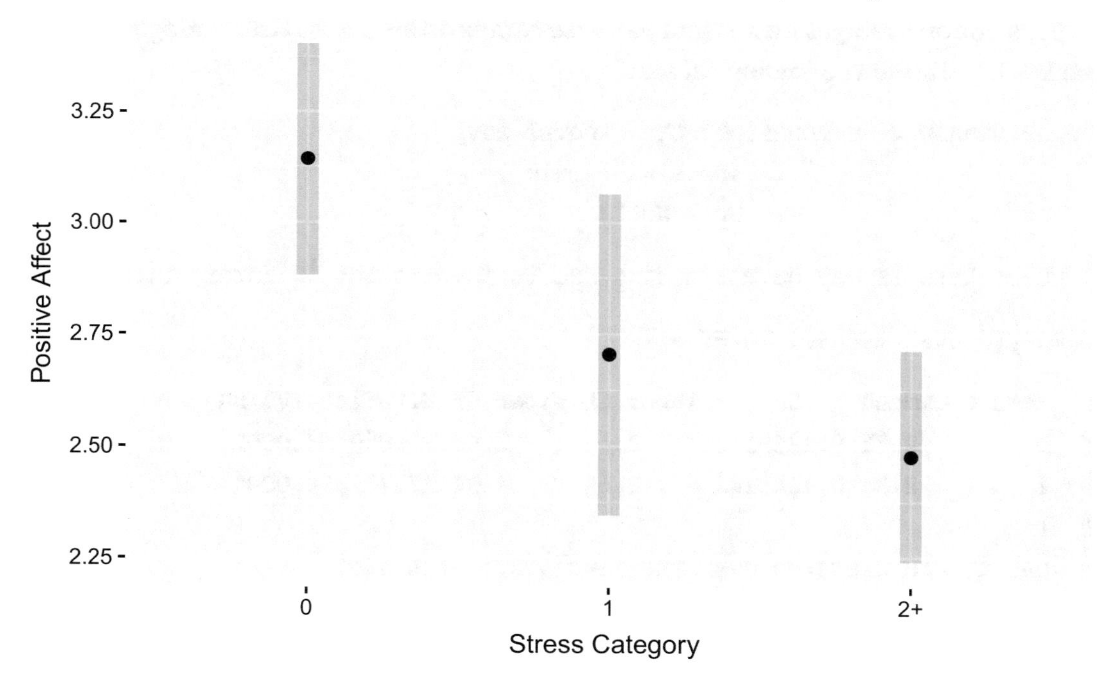

Figure 13-9. *Factorial ANOVA means plot – Option C*

The pairwise comparisons also show that overall, only stress categories 0 and 2+ are statistically significantly different from each other, with the other pairwise comparisons not reaching statistical significance:

```
pairs(anova2.meansC)

## contrast estimate    SE  df t.ratio p.value
## 0 - 1       0.438 0.226 181 1.941   0.1301
## 0 - (2+)    0.665 0.179 181 3.717   0.0008
## 1 - (2+)    0.228 0.218 181 1.044   0.5502
##
## Results are averaged over the levels of: Female
## P value adjustment: tukey method for comparing a family of 3 estimates
```

Option D: Calculate the means for each biological sex category, averaging across stress category. Compare pairwise differences in biological sex categories.

This option averages means and pairwise comparisons across stress category. All that is left is the effect of biological sex:

```
anova2.meansD <- emmeans(object = anova2$aov,
                         specs = "Female",
                         by = NULL)

## NOTE: Results may be misleading due to involvement in interactions

summary(anova2.meansD, infer = TRUE)

## Female emmean    SE  df lower.CL upper.CL t.ratio p.value
## 0        2.71 0.135 181     2.45     2.98 20.069  <.0001
## 1        2.84 0.104 181     2.63     3.04 27.387  <.0001
##
## Results are averaged over the levels of: StressCat
## Confidence level used: 0.95
```

None of these means will be exactly the same as those in Option A or B because the individual cell means have been collapsed and these are "marginal" means in that they are on the "margin" of a table averaging across. The plot of means in Figure 13-10 now only shows means by biological sex and does not separate by stress category. We can see that overall, positive affect is higher in females than in males:

```
plot(anova2.meansD) +
  xlab("Positive Affect") +
  ylab("Female") +
  coord_flip() +
  ggtitle("Estimated Means and 95% Confidence Intervals, averaged across
  stress category")
```

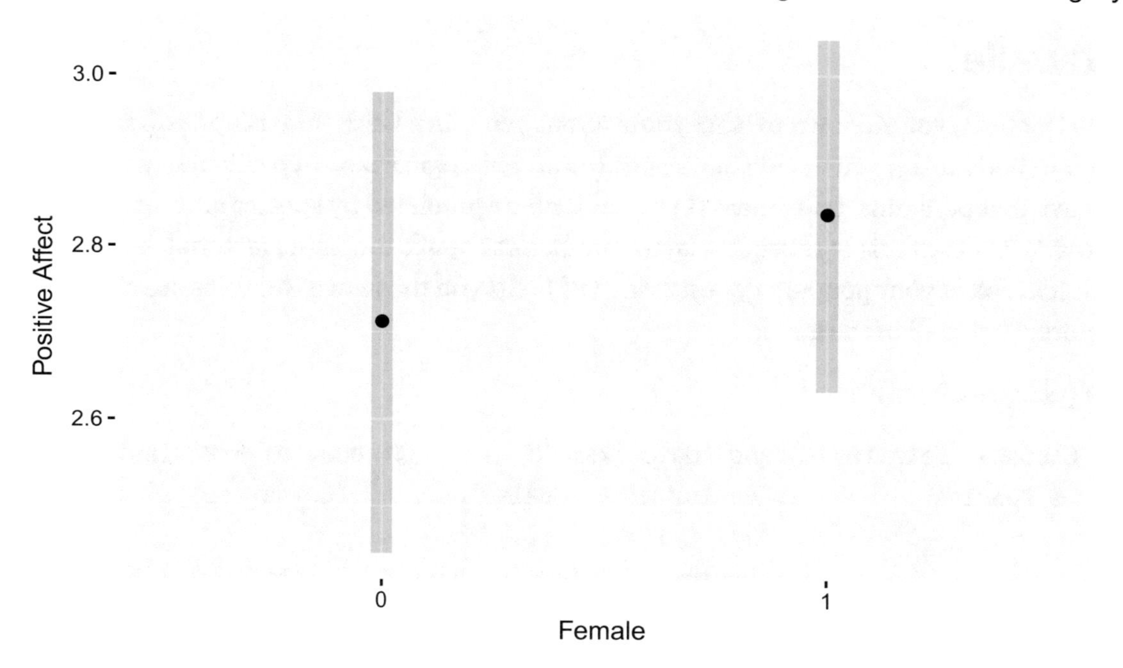

Figure 13-10. *Factorial ANOVA means plot – Option CD*

The pairwise comparison between females and males reveals that although the means are not exactly identical, they are not statistically significantly different:

```
pairs(anova2.meansD)
```

```
##  contrast estimate   SE  df t.ratio p.value
##  0 - 1      -0.125 0.17 181 -0.735  0.4636
##
## Results are averaged over the levels of: StressCat
```

That is everything for a factorial ANOVA. In practice, people would not calculate and report all options (A –D) that we showed for the estimated means and pairwise comparisons. However, we included all those options for completeness and to show what is possible. Typically, people will choose one or perhaps two options that are the most straightforward to interpret and are appropriate for the data.

For example, if the interaction was not even close to statistically significant, it may make sense to focus on Option C or D. If the interaction is significant, average across one variable may not make sense, so it is typically best to choose between Options A and B, which are actually the same means, just arranged differently. It is up to the person analyzing data to decide what is the most informative way to present the results.

Example

One last time, you sit down to learn more about penguins. Fairly sure islands do not explain body mass, you decide to try species and biological sex. In particular, you believe that penguins' body mass (in grams) can be predicted by species and sex. You know ANOVAs of this type require factor predictors (hence the name factorial ANOVA). A quick look at your penguin data set via `str()` tells you those two variables are already factors. That is all to the good:

```
str(penguinsData)

## Classes 'data.table' and 'data.frame':      344 obs. of 8 variables:
##  $ species          : Factor w/ 3 levels "Adelie","Chinstrap",.. : 1 1 1
                          1 1 1 1 1 1 1 ...
##  $ island           : Factor w/ 3 levels "Biscoe","Dream",..: 3 3 3 3 3
                          3 3 3 3 3 ...
##  $ bill_length_mm   : num  39.1 39.5 40.3 NA 36.7 39.3 38.9 39.2 34.1 42 ...
##  $ bill_depth_mm    : num  18.7 17.4 18 NA 19.3 20.6 17.8 19.6 18.1 20.2 ...
##  $ flipper_length_mm: int  181 186 195 NA 193 190 181 195 193 190 ...
##  $ body_mass_g      : int  3750 3800 3250 NA 3450 3650 3625 4675 3475
                          4250 ...
##  $ sex              : Factor w/ 2 levels "female","male": 2 1 1 NA 1 2 1
                          2 NA NA ...
##  $ year             : int  2007 2007 2007 2007 2007 2007 2007 2007 2007
                          2007 ...
##  - attr(*, ".internal.selfref")=<externalptr>
```

You also know missing values need to be excluded. Since this time you want all penguin species, the data set from earlier examples will not work. Using row selection operations and negation (via !) of `is.na()`, you are able to remove any penguins who are missing one or more of body mass, species, or sex:

```
aovPenguinsFull <- penguinsData[!is.na(body_mass_g) &
                                  !is.na(species) &
                                    !is.na(sex)]
```

You also remember that the `ezANOVA()` function requires an ID variable (that must also be a factor). Recycling your prior solution to this, you add an ID column to your data set:

```
aovPenguinsFull[ , ID := 1:.N]
aovPenguinsFull[, ID := factor(ID)]
```

Using the `xtabs()` function, you notice there seem to be plenty of penguins of each type in the frequency cross-tabulation table:

```
xtabs(formula = ~ species + sex,
      data = aovPenguinsFull)

##              sex
## species     female male
##   Adelie        73   73
##   Chinstrap     34   34
##   Gentoo        58   61
```

From there, it is a simple re-purposing of the code from the preceding example. A few variable swaps and you have a two-way, factorial ANOVA:

```
anovaP2 <- ezANOVA(
  data = aovPenguinsFull, dv = body_mass_g,
  wid = ID,
  between = .(species, sex),
  type = 3, detailed = TRUE,
  return_aov = TRUE)

## Warning: Data is unbalanced (unequal N per group). Make sure you
specified a well-considered value for the type argument to ezANOVA().
```

```
## Coefficient covariances computed by hccm()

print(anovaP2)

## $ANOVA
##        Effect DFn DFd       SSn      SSd         F          p p<.05
## 1 (Intercept)   1 327 5.233e+09 31302628 54661.828  0.000e+00     *
## 2     species   2 327 1.430e+08 31302628   746.924 1.185e-122     *
## 3         sex   1 327 2.985e+07 31302628   311.838  1.761e-49     *
## 4 species:sex   2 327 1.677e+06 31302628     8.757  1.973e-04     *
##       ges
## 1 0.99405
## 2 0.82041
## 3 0.48813
## 4 0.05084
##
## $'Levene's Test for Homogeneity of Variance'
##   DFn DFd    SSn      SSd     F      p p<.05
## 1   5 327 240535 11310528 1.391 0.2272
##
## $aov
## Call:
##    aov(formula = formula(aov_formula), data = data)
##
## Terms:
##                   species       sex species:sex Residuals
## Sum of Squares  145190219  37090262     1676557  31302628
## Deg.of Freedom          2         1           2       327
##
## Residual standard error: 309.4
## Estimated effects may be unbalanced
```

Unlike the prior example, the interaction term, species:sex, is also significant for $\alpha = 0.05$. You of course still keep that term; this time, you know that the difference between species depends in part on sex.

You remember that for Levene's test, ideally the p-value is not significant. This is true in this case, and you are pleased.

A quick check of Figure 13-11 shows there are no outliers in the lower left, the deviates are mostly on the line, and the right side is roughly parallel. All seems in order:

```
plot(modelDiagnostics(anovaP2$aov), ncol = 2)
```

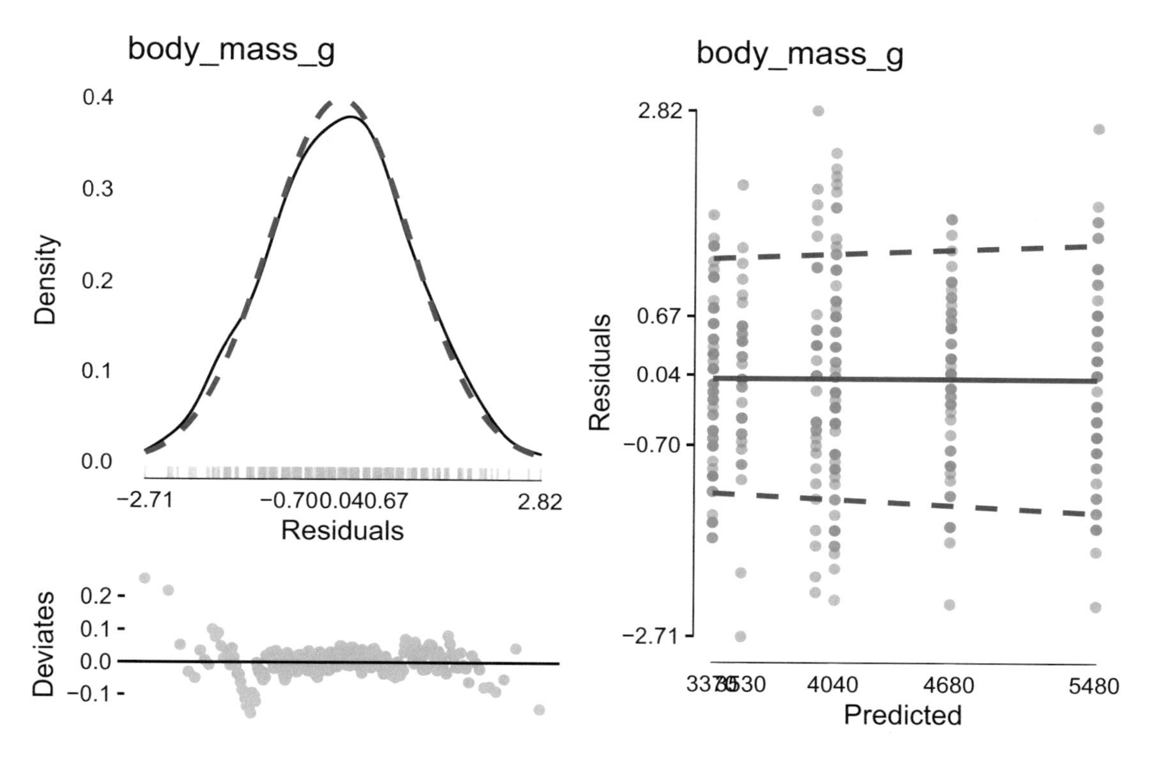

Figure 13-11. *Diagnostic plots for an ANOVA*

Because the interaction is significant, it makes sense to choose one of the option A or B choices discussed in the preceding example. You decide to use the "A" model, where you calculate the mean in each individual "cell," the crossing of species and biological sex. You will compare the differences between species categories (e.g., Adelie, Chinstrap, and Gentoo) separately for females and males.

```
anovaP2.meansA  <- emmeans(object = anovaP2$aov,
                           specs = "species",
                           by = "sex")
summary(anovaP2.meansA, infer = TRUE)
```

```
## sex = female:
##  species   emmean   SE  df lower.CL upper.CL  t.ratio p.value
##  Adelie      3369 36.2 327     3298     3440   93.030 <.0001
##  Chinstrap   3527 53.1 327     3423     3632   66.470 <.0001
##  Gentoo      4680 40.6 327     4600     4760  115.190 <.0001
##
## sex = male:
##  species   emmean   SE  df lower.CL upper.CL t.ratio p.value
##  Adelie      4043 36.2 327     3972     4115 111.660 <.0001
##  Chinstrap   3939 53.1 327     3835     4043  74.230 <.0001
##  Gentoo      5485 39.6 327     5407     5563 138.460 <.0001
##
## Confidence level used: 0.95
```

As always, while the summary is numerically nice (and you can see the p-values are all significant), a graph often shows best what is happening. In Figure 13-12, you see the visual of the interaction between species and sex. Looking at the graph, Gentoo males have much more mass than the other species and even than Gentoo females. It is the combination of species and sex that makes Gentoo males so large:

```
plot(anovaP2.meansA) +
  xlab("Body Mass") +
  ylab("species") + coord_flip() +
  ggtitle("Estimated Means and 95% Confidence Intervals from Factorial ANOVA")
```

In the pairwise comparison, except for Adelie-Chinstrap males, all are significant differences. On the graph in Figure 13-12, you can see that both Adelie and Chinstrap males are both quite close to the 4000 line:

```
pairs(anovaP2.meansA)
```

```
## sex = female:
##  contrast            estimate   SE  df t.ratio p.value
##  Adelie - Chinstrap      -158 64.2 327  -2.465 0.0377
##  Adelie - Gentoo        -1311 54.4 327 -24.088 <.0001
##  Chinstrap - Gentoo     -1152 66.8 327 -17.246 <.0001
##
```

```
## sex = male:
##  contrast             estimate   SE  df t.ratio p.value
##  Adelie - Chinstrap        104 64.2 327   1.627  0.2357
##  Adelie - Gentoo         -1441 53.7 327 -26.855 <.0001
##  Chinstrap - Gentoo      -1546 66.2 327 -23.345 <.0001
##
## P value adjustment: tukey method for comparing a family of 3 estimates
```

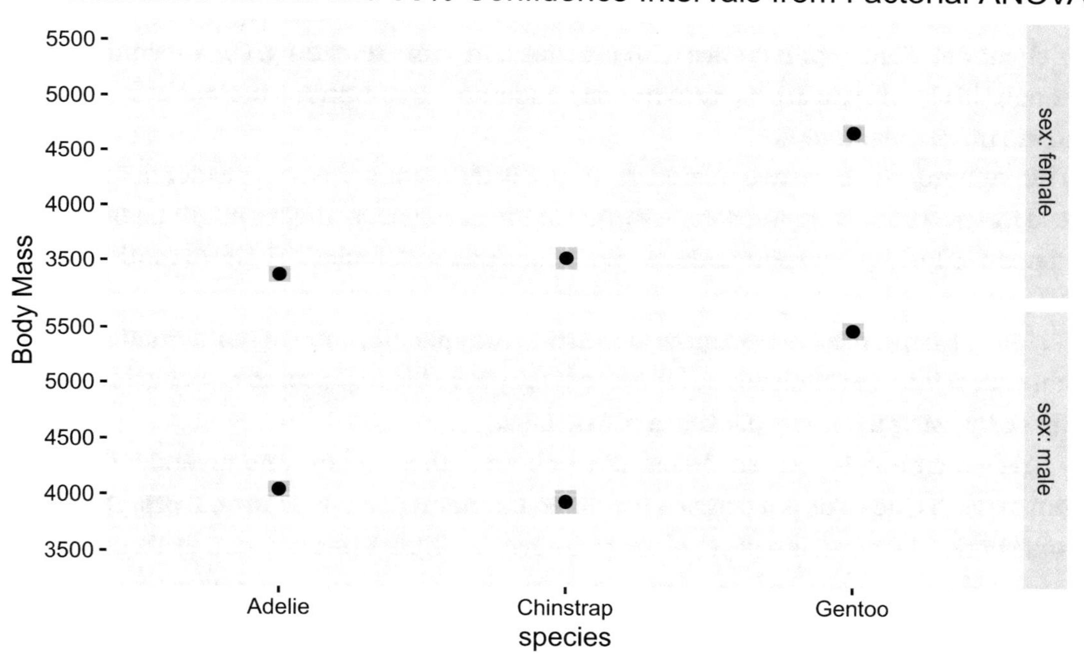

***Figure 13-12.** Factorial ANOVA means plot – Option A*

You have found (at last) a good way to predict penguin mass.

13.5 Summary

With this chapter's exploration of one-way and two-way (factorial) analysis of variance, you have reached the end of this book. As with most things, the more you practice, the more fluent you become in using these various analytic techniques. More importantly, you now think like a data scientist. You are quite familiar with RStudio and R – two things to be sure to add to your resume or CV.

In particular, you have noticed that R is quite good at following a common framework. The same `formula =` argument is a frequent feature of functions that compute statistical models. The tougher skill to earn is understanding when to use which model. That in part comes with practice and some rereading. On a second pass through this book, you are likely to notice details and assumptions that make more sense and are more meaningful.

From here, you have two related yet possibly different paths to consider taking as you move forward with studying data. A formal study of statistical theory might be of value if you need to be highly expert in the assumptions and mathematics behind models. That road will eventually take you through calculus. Another road is to continue your journey in R. Many R experts have backgrounds outside the specific arena of mathematics or computer science or statistics. All the same, they are able to do some rather astounding science by using the many packages of R on CRAN.

The good news is you can choose one or both of these roads (and meander between them at will). One of us is a classically trained mathematician with formal programming coursework. The other is a psychologist who has multiple R packages on CRAN. One of the very nice features of the R community is its welcoming nature toward new, lifelong learners and its diversity.

As you practice these last few skills on your journey from novice to professional, Table 13-1 should be a good reference of the key ideas you learned in this chapter.

Table 13-1. *Chapter Summary*

Idea	What It Means
`ezANOVA()`	Function to fit one-way and factorial ANOVAs.
`emmeans()`	Get estimated cell or marginal means based off of an ANOVA.
`xtab()`	Shows a summary count between two or more categories.

13.6 Practice for Mastery

Check your progress and grow through practice by working through some exercises. Comprehension checks ask critical thinking questions that may be best answered with a written or verbal response. Part of the art of statistics is successfully communicating results to your stakeholders or audience. Sometimes that audience is highly technical and other times very much not technical. Exercises are more direct applications of the concepts explored in the chapter.

Comprehension Checks

1. In ANOVA, if an interaction is not significant, does one remove it?
2. Can a predictor variable be ratio-level data for ANOVA?

Exercises

1. Would it be possible to complete the final example, only swap out `species` for `island`? Experiment with the code, and see what happens.
2. Rework the final example about penguin species and biological sex. However, replace body mass with flipper length. What do you see? Which of methods A, B, C, and D (thinking of the penultimate example) makes the most promising route to explore?
3. Rework the final example about penguin species and biological sex. However, replace body mass with bill depth. What do you see? Which of methods A, B, C, and D (thinking of the penultimate example) makes the most promising route to explore?

Bibliography

[1] Motor trend car road tests. *Motor Trend*, 1974.

[2] *Comprehensive R Archive Network CRAN*, Accessed February 15, 2020.

[3] *Texas Dept. of State Health Services*, Accessed February 16, 2020.

[4] *American Community Survey (ACS) Summary File*, Accessed February 29, 2020.

[5] Leona S. Aiken and Stephen G. West. *Multiple regression: Testing and interpreting interactions*. Sage, 1991.

[6] Heds 1 at English Wikipedia/Public domain. *The Normal Distribution*, Accessed July 16, 2020.

[7] Patrick Breheny and Woodrow Burchett. Visualization of regression models using visreg. *The R Journal*, 9(2):56–71, 2017.

[8] Jacob Cohen. *Statistical Power Analysis for the Behavioral Sciences*. Lawrence Erlbaum Associates, USA, 1988.

[9] Matt Dowle and Arun Srinivasan. *data.table: Extension of 'data.frame'*, 2019. R package version 1.12.8.

[10] Richard Iannone. *DiagrammeR: Graph/Network Visualization*, 2020. R package version 1.0.6.1.

[11] Gorman KB, Williams TD, and Fraser WR. Ecological sexual dimorphism and environmental variability within a community of antarctic penguins (genus pygoscelis). *PLoS ONE*, 9(3)(e90081):-13, 2014.

[12] Michael A. Lawrence. *ez: Easy Analysis and Visualization of Factorial Experiments*, 2016. R package version 4.4-0.

[13] Russell Lenth. *emmeans: Estimated Marginal Means, aka Least-Squares Means*, 2020. R package version 1.4.6.

[14] Jeroen Ooms. *writexl: Export Data Frames to Excel 'xlsx' Format*, 2019. R package version 1.2.

[15] R Core Team. *R: A Language and Environment for Statistical Computing*. R Foundation for Statistical Computing, Vienna, Austria, 2019.

M. Wiley and J. F. Wiley, *Beginning R 4*, https://doi.org/10.1007/978-1-4842-6053-1

[16] R Core Team. *R: A Language and Environment for Statistical Computing.* R Foundation for Statistical Computing, Vienna, Austria, 2020.

[17] Hadley Wickham. *ggplot2: Elegant Graphics for Data Analysis.* Springer-Verlag New York, 2016.

[18] Hadley Wickham and Jennifer Bryan. *readxl: Read Excel Files*, 2019. R package version 1.3.1.

[19] Hadley Wickham and Evan Miller. *haven: Import and Export 'SPSS,' 'Stata' and 'SAS' Files*, 2019. R package version 2.2.0.

[20] Joshua F. Wiley. *extraoperators: Extra Binary Relational and Logical Operators*, 2019. R package version 0.1.1.

[21] Joshua F. Wiley. *JWileymisc: Miscellaneous Utilities and Functions*, 2020. R package version 1.1.1.

[22] M. Wiley and J. F. Wiley. *Advanced R Statistical Programming and Data Models: Analysis, Machine Learning, and Visualization.* Apress, 2019.

[23] M. Wiley and J. F. Wiley. *Advanced R 4 Data Programming and the Cloud: Using PostreSQL, AWS, and Shiny.* Apress, 2020.

[24] Yang Yap, Danica C. Slavish, Daniel J. Taylor, Bei Bei, and Joshua F. Wiley. Bi-directional relations between stress and self-reported and actigraphy-assessed sleep: a daily intensive longitudinal study. *Sleep*, 43(3):zsz250, 2020.

Index

A

B

C

M. Wiley and J. F. Wiley, *Beginning R 4*, https://doi.org/10.1007/978-1-4842-6053-1

N

O

P

Q

R

S

T

U

V

W

X

Y

Z

Printed in the United States
By Bookmasters